An Expanding World
Volume 17

Agriculture, Resource Exploitation, and Environmental Change

AN EXPANDING WORLD

The European Impact on World History, 1450–1800

General Editor: A.J.R. Russell-Wood

EXPANSION, INTERACTION, ENCOUNTERS
1 **The Global Opportunity** *Felipe Fernández-Armesto*
2 **The European Opportunity** *Felipe Fernández-Armesto*
3 **The Globe Encircled and the World Revealed** *Ursula Lamb*
4 **Europeans in Africa and Asia** *Anthony Disney*
5 **The Colonial Americas** *Amy Turner Bushnell*

TECHNOLOGY AND SCIENCE
6 **Scientific Aspects of European Expansion** *William Storey*
7 **Technology and European Overseas Enterprise** *Michael Adas*

TRADE AND COMMODITIES
8 **Merchant Networks in the Early Modern World** *Sanjay Subrahmanyam*
9 **The Atlantic Staple Trade** (Parts I & II) *Susan Socolow*
10 **European Commercial Expansion in Early Modern Asia** *Om Prakash*
11 **Spices in the Indian Ocean World** *M.N. Pearson*
12 **Textiles: Production, Trade and Demand** *Maureen Mazzaoui*
13 **Interoceanic Trade in European Expansion** *Pieter Emmer and Femme Gaastra*
14 **Metals and Monies in a Global Economy** *Dennis O. Flynn and Arturo Giráldez*
15 **Slave Trades** *Patrick Manning*

EXPLOITATION
16 **From Indentured Servitude to Slavery** *Colin Palmer*
17 **Agriculture, Resource Exploitation, and Environmental Change** *Helen Wheatley*
18 **Plantation Societies** *Judy Bieber*
19 **Mines of Silver and Gold in the Americas** *Peter Bakewell*

GOVERNMENT AND EMPIRE
20 **Theories of Empire** *David Armitage*
21 **Government and Governance of Empires** *A.J.R. Russell-Wood*
22 **Imperial Administrators** *Mark Burkholder*
23 **Local Government in European Empires** *A.J.R. Russell-Wood*
24 **Warfare and Empires** *Douglas M. Peers*

SOCIETY AND CULTURE
25 **Settlement Patterns in Early Modern Colonization** *Joyce Lorimer*
26 **Biological Consequences of the European Expansion** *Kenneth Kiple and Stephen V. Beck*
27 **European and non-European Societies** (Parts I & II) *Robert Forster*
28 **Christianity and Missions** *J.S. Cummins*
29 **Families in the Expansion of Europe** *Maria Beatriz Nizza da Silva*
30 **Changes in Africa, America and Asia** *Murdo MacLeod and Evelyn Rawski*

THE WORLD AND EUROPE
31 **Europe and Europe's Perception of the World** (Parts I & II) *Anthony Pagden*

Please note titles may change prior to publication

An Expanding World
The European Impact on World History 1450–1800

Volume 17

Agriculture, Resource Exploitation, and Environmental Change

edited by
Helen Wheatley

VARIORUM
1997

Published by VARIORUM
Ashgate Publishing Limited
Gower House, Croft Road
Aldershot, Hampshire GU11 3HR
Great Britain

Ashgate Publishing Company
Old Post Road
Brookfield, Vermont 05036
USA

ISBN 0-86078-514-9

British Library CIP data

Agriculture, Resource Exploitation, and Environmental Change. (An Expanding World: The European Impact on World History, 1450–1800: Vol. 17). 1. Agriculture–History. 2. Natural resources–Management–History. I. Wheatley, Helen.
333.7' 09

US Library of Congress CIP data

Agriculture, Resource Exploitation, and Environmental Change / edited by Helen Wheatley. p.cm. – (An Expanding World: The European Impact on World History, 1450–1800: Vol. 17).
Includes bibliographical references.
1. Environmental risk assessment–History. 2. Agriculture–Environmental aspects–History. 3. Landscape assessment–Environmental aspects–History. 4. Land use–Environmental aspects–History. 5. Natural resources–Environmental aspects–History. I. Wheatley, Helen. II. Series.
GE145. A37 1997 96-30063
333.7' 14–dc21 CIP

This book is printed on acid free paper.

Printed and bound by Athenaeum Press, Ltd., Gateshead, Tyne & Wear.

AN EXPANDING WORLD 17

Contents

Acknowledgements

The chapters in this volume are taken from the sources listed below, for which the editor and publishers wish to thank their authors, original publishers or other copyright holders for permission to use their material as follows:

Chapter 1: Jean Andrews, ‘Diffusion of MesoAmerican Food Complex to Southeastern Europe’, *Geographical Review* LXXXIII, no. 2 (1993), pp. 194–204. Copyright © 1993 by The American Geographical Society. Reprinted by permission of The American Geographical Society, New York.

Chapter 2: J.E. Spencer, ‘The Rise of Maize as a Major Crop Plant in the Philippines’, *Journal of Historical Geography*, I, no. 1 (1975), pp. 1–16. Copyright © 1975 by Academic Press Limited. Reprinted by permission of the publisher Academic Press Limited, London.

Chapter 3: Daniel W. Gade, ‘Landscape, System and Identity in the Post-Conquest Andes’, *Annals of the Association of American Geographers* LXXXII, no. 3 (1992), pp. 460–477. Copyright © 1992 by Basil Blackwell, Inc.

Chapter 4: Dan Flores, ‘Bison Ecology and Bison Diplomacy: The Southern Plains from 1800–1850’, *The Journal of American History* LXXVIII, no. 2 (1991), pp. 465–485. Copyright © 1991 by the Organization of American Historians.

Chapter 5: Elinor G.K. Melville, ‘Environmental Change and Social Change in the Valle del Mezquital, Mexico, 1521–1600’, *Comparative Studies in Society and History* XXXII, no. 1 (1990), pp. 24–53. Copyright © 1990 by The Society for the Comparative Study of Society and History and Cambridge University Press.

Chapter 6: Jacques Barrau, ‘Indigenous and Colonial Land-Use Systems in Indo-Oceanian Savannas: the Case of New Caledonia’, in ed. David R. Harris, *Human Ecology in Savanna Environments* (London, 1980), pp. 253–265. Copyright © 1980 by Academic Press Limited. Reprinted by permission of the publisher Academic Press Limited, London.

Chapter 7: Leonard Guelke and Robert Shell, ‘Landscape of Conquest: Frontier Water Alienation and Khoikhoi Strategies of Survival, 1652–1780’, *Journal of Southern African Studies* XVIII, no. 4 (1992), pp. 803–824. Copyright © 1992 by Carfax Publishing Company, P.O. Box 25, Abingdon, Oxfordshire, OX14 3UE.

Chapter 8: Shawn W. Miller, 'Fuelwood in Colonial Brazil: The Economic Consequences of Fuel Depletion for the Bahian Recôncavo, 1549–1820', *Forest and Conservation History* XXXVIII (1994), pp. 181–192. Copyright © 1994 by the Forest History Society, Inc, Durham, North Carolina.

Chapter 9: David O. Percy, 'Ax or Plow? Significant Colonial Landscape Alteration Rates in the Maryland and Virginia Tidewater', *Agricultural History* LXVI, no. 2 (1992), pp. 66–74. Copyright © 1992 by the University of California Press. Reprinted by permission.

Chapter 10: James J. Parsons, 'Human Influences on the Pine and Laurel Forests of the Canary Islands', *Geographical Review* LXXI, no. 3 (1981), pp. 253–271. Copyright © 1981 by The American Geographical Society. Reprinted by permission of The American Geographical Society, New York.

Chapter 11: Judith A. Carney, 'From Hands to Tutors: African Expertise in the South Carolina Rice Economy', *Agricultural History* LXVII, no. 3 (1993), pp. 1–30. Copyright © 1993 by the University of California Press. Reprinted by permission.

Chapter 12: Robert MacCameron, 'Environmental Change in Colonial New Mexico', *Environmental History Review* XVIII, no. 2 (1994), pp. 17–39. Copyright © 1994 by the American Society for Environmental History, Newark, New Jersey.

Chapter 13: Chesley W. Sanger, '"Saw Several Finners But No Whales": The Greenland Right Whale (Bowhead) – An Assessment of the Biological Basis of the Northern Whale Fishery During the Seventeenth, Eighteenth and Nineteenth Centuries', *International Journal of Maritime History* III, no. 1 (1991), pp. 127–154. Copyright © 1991 by the International Journal of Maritime History, St Johns, Newfoundland.

Chapter 14: Arthur J. Ray, 'Some Conservation Schemes of the Hudson's Bay Company, 1821–50: An Examination of the Problems of Resource Management in the Fur Trade', *Journal of Historical Geography* I, no. 1 (1975), pp. 49–68. Copyright © 1975 by Academic Press. Reprinted by permission of the publisher Academic Press Limited, London.

Chapter 15: Peter Boomgaard, 'Forest Management and Exploitation in Colonial Java, 1677–1897', *Forest and Conservation History* XXXVI, no. 1 (1992), pp. 4–14. Copyright © 1992 by the Forest History Society Inc., Durham, North Carolina.

Chapter 16: Richard Grove, 'Conserving Eden: The (European) East India Companies and their Environmental Policies on St. Helena, Mauritius and in Western India, 1600–1854', *Comparative Studies in Society and History* XXXV, no. 2 (1993), pp. 318–351. Copyright © 1993 by The Society for the Comparative Study of Society and History and Cambridge University Press.

Every effort has been made to trace all the copyright holders, but if any have been inadvertently overlooked the publishers will be pleased to make the necessary arrangement at the first opportunity.

General Editor's Preface

A.J.R. Russell-Wood

An Expanding World: The European Impact on World History, 1450–1800 is designed to meet two objectives: first, each volume covers a specific aspect of the European initiative and reaction across time and space; second, the series represents a superb overview and compendium of knowledge and is an invaluable reference source on the European presence beyond Europe in the early modern period, interaction with non-Europeans, and experiences of peoples of other continents, religions, and races in relation to Europe and Europeans. The series reflects revisionist interpretations and new approaches to what has been called 'the expansion of Europe' and whose historiography traditionally bore the hallmarks of a narrowly Eurocentric perspective, focus on the achievements of individual nations, and characterization of the European presence as one of dominance, conquest, and control. Fragmentation characterized much of this literature: fragmentation by national groups, by geography, and by chronology.

The volumes of *An Expanding World* seek to transcend nationalist histories and to examine on the global stage rather than in discrete regions important selected facets of the European presence overseas. One result has been to bring to the fore the multicontinental, multi-oceanic and multinational dimension of the European activities. A further outcome is compensatory in the emphasis placed on the cross-cultural context of European activities and on how collaboration and cooperation between peoples transcended real or perceived boundaries of religion, nationality, race, and language and were no less important aspects of the European experience in Africa, Asia, the Americas, and Australia than the highly publicized confrontational, bellicose, and exploitative dimensions. Recent scholarship has not only led to greater understanding of peoples, cultures, and institutions of Africa, Asia, the Americas, and Australasia with whom Europeans interacted and the complexity of such interactions and transactions, but also of relations between Europeans of different nationalities and religious persuasions.

The initial five volumes reflect the changing historiography and set the stage for volumes encompassing the broad themes of technology and science, trade and commerce, exploitation as reflected in agriculture and the extractive industries and through systems of forced and coerced labour, government of empire, and society and culture in European colonies and settlements overseas. Final volumes examine the image of Europe and Europeans as 'the other' and the impact of the wider world on European *mentalités* and mores.

An international team of editors was selected to reflect a diversity of educational backgrounds, nationalities, and scholars at different stages of their professional careers. Few would claim to be 'world historians', but each is a recognized authority in his or her field and has the demonstrated capacity to ask the significant questions and provide a conceptual framework for the selection of articles which combine analysis with interpretation. Editors were exhorted to place their specific subjects within a global context and over the *longue durée*. I have been delighted by the enthusiasm with which they took up this intellectual challenge, their courage in venturing beyond their immediate research fields to look over the fences into the gardens of their academic neighbours, and the collegiality which has led to a generous informal exchange of information. Editors were posed the daunting task of surveying a rich historical literature and selecting those essays which they regarded as significant contributions to an understanding of the specific field or representative of the historiography. They were asked to give priority to articles in scholarly journals; essays from conference volumes and *Festschriften* were acceptable; excluded (with some few exceptions) were excerpts from recent monographs or paperback volumes. After much discussion and agonizing, the decision was taken to incorporate essays only in English, French, and Spanish. This has led to the exclusion of the extensive scholarly literature in Danish, Dutch, German and Portuguese. The ramifications of these decisions and how these have had an impact on the representative quality of selections of articles have varied, depending on the theme, and have been addressed by editors in their introductions.

The introduction to each volume enables readers to assess the importance of the topic *per se* and place this in the broader context of European activities overseas. It acquaints readers with broad trends in the historiography and alerts them to controversies and conflicting interpretations. Editors clarify the conceptual framework for each volume and explain the rationale for the selection of articles and how they relate to each other. Introductions permit volume editors to assess the impact on their treatments of discrete topics of constraints of language, format, and chronology, assess the completeness of the journal literature, and address *lacunae*. A further charge to editors was to describe and evaluate the importance of change over time, explain differences attributable to differing geographical, cultural, institutional, and economic circumstances and suggest the potential for cross-cultural, comparative, and interdisciplinary approaches. The addition of notes and bibliographies enhances the scholarly value of the introductions and suggests avenues for further enquiry.

I should like to express my thanks to the volume editors for their willing participation, enthusiasm, sage counsel, invaluable suggestions, and good judgment. Evidence of the timeliness and importance of the series was illustrated by the decision, based on extensive consultation with the scholarly community, to expand a series, which had originally been projected not to exceed eight volumes, to more than thirty volumes. It was John Smedley's initiative which gave rise to discussions as to the viability and need for such a series and he has overseen the publishing, publicity, and marketing of *An Expanding World*. As

General Editor, my task was greatly facilitated by the assistance of Dr Mark Steele who was initially responsible for the 'operations' component of the series as it got under way; latterly this assistance has been provided by staff at Variorum.

The Department of History,
The Johns Hopkins University

Introduction

Helen Wheatley

'Forest history, rightly understood, is everywhere on this planet one of exploitation and destruction'. So wrote Warren Dean in his posthumously published history of the forests of Brazil, *With Broadax and Firebrand*. Europeans were not unique in their love for domesticated landscapes and loathing of wilderness. Nor did they stand alone in their propensity to transform the natural world by exploiting its riches. Yet Dean points to the unique significance of a moment engraved in the memory of every Brazilian schoolchild: the Portuguese sacrifice of a tree in 1500 to fashion 'a rude cross, for them the symbol of the salvation of mankind'. Fixed upon this cross was the Portuguese claim to its new territory.[1] With the arrival of the Europeans, the forests of the Americas stood on the brink of their greatest transformation since the last Ice Age.

The ecological impact of European expansion in the New World was immediate and profound. Islands, including the Azores and Canaries as well as those of the Caribbean, were rapidly transformed by European settlement and resource extraction. Forests must be counted as both a major attraction and impediment for European traders and settlers. The Portuguese poet Camões proclaimed that the island of Madeira 'was like a gem and the gem was its trees'.[2] The British Royal Navy laid claim to the tallest white pines of New England for its ship masts.[3] Exotic woods and die woods attracted European traders, while brisk local trade in wood products for fuel, construction and shipping sprang up throughout the new European settlements. Meanwhile, settlers everywhere cleared the forests eagerly to make way for agriculture. The combination of exploitation and clearance brought deforestation on a scale long familiar to Europeans, but new to many of the societies with whom they came into contact. The ecologies of entire regions were transformed where demand was especially high or where European settlement was especially pronounced. The dramatic transformation of American landscapes into what Alfred Crosby has called 'neo-Europes' provides a striking example of how deep and lasting those environmental changes could be.

Environmental historians have tended to neglect the era of European expansion before 1800, preferring to set their sights on the even more dramatic changes that accompanied industrialization, expansion and imperialism in the

[1] Warren Dean, *With Broadax and Firebrand: The Destruction of the Brazilian Atlantic Forest* (Berkeley, 1995), pp. 5, 41.

[2] John Perlin, *A Forest Journey* (Cambridge, MA, 1989), p. 250.

[3] William Cronon, *Changes in the Land* (New York, 1983), p. 110.

nineteenth century.[4] One reason for this neglect has been the relative lack of information available to historians about earlier ecological systems, and even in this volume some nineteenth-century histories must serve as examples of processes that can be traced back to earlier periods. Another reason may be that, aside from the dramatic linkage of the Americas to the rest of the world, Europeans did not change the patterns of exploitation in ways that could be regarded as unique. Before the nineteenth century, the ecological impact of European trade and settlement followed patterns familiar throughout much of the world. Land clearance for agriculture, consumption of wood for fuel and building materials, and the extraction of particular resources for trade could all have dramatic effects on local ecosystems.

Like European settlers, historians have often drawn a sharp line between 'wilderness' and 'civilization'. Frederick Jackson Turner's notion of the Atlantic frontier as the 'hither edge of a free land' served, especially for historians in North America, as an appropriate guide for situating the study of the environmental effects of European expansion. The observations of literate and often admiring European chroniclers provided the basis for contrasting wilderness to the effects of trade and settlement. Turner himself quoted an 1837 publication, Peck's *New Guide to the West*, which described the process of ecological change as a sequence of settlement and migrations that 'like waves of the ocean, have rolled one after the other'. The first settler was a pioneer who subsisted off the natural landscape and transformed it with a rude form of agriculture; the next cleared the land and developed an infrastructure that exhibited 'the picture and forms of plain frugal, civilized life'. Finally, 'the men of capital and enterprise' arrived to transform the rural village into the thoroughly humanized landscape of the metropolis.[5]

Yet it is no longer sufficient to describe the arc of European movement outward along a frontier of wilderness, for we can no longer be sure where wilderness ended and cultivation began. Landscapes believed to be natural by contemporary European observers were often the products of human modification. Europeans did not readily recognize the effects of swidden agriculture, nor did they realize that the landscape might contain hunting gardens, patches of flora maintained for their ability to attract prey. The spread and protection of favoured species within a 'wild' landscape could escape the notice of a sixteenth-century European – or a modern historian – bound to sharp cultural distinctions between the cultivated and uncultivated.

Historians have become more adept at comprehending cultivated landscapes that did not fit the European mold. They now recognize, for example, that fire

[4] See for example Richards and Tucker, *Global Deforestation* (Durham, 1983).

[5] Frederick Jackson Turner in ed. Harold P. Simonson, *The Significance of the Frontier in American History,* (New York, 1963), pp. 28, 41–3.

has been no less a tool of cultivation than the digging stick or plough. Indeed, over the long span of human history, fire may be the most important tool of all.[6] Historians understand that the slash-and-burn techniques of swidden agriculture, as well as the reliance of farmers on trees for fuel and fodder, erased the barrier between field and forest. The recognition of long-term modification of environments formerly assumed to be wild has made the challenge of assessing the ecological impact of European expansion even more daunting.[7] Nor can we assert that conquest and settlement stamped an indelible European mark upon the land, because we see the incorporation of indigenous techniques not only throughout the realm of European expansion, but transported back to Europe itself.

Despite the ambiguities and uncertainties embedded in the evaluation of landscapes before their encounter, European expansion and the incorporation of new areas into the global market economy resulted in substantial and lasting transformations of ecosystems throughout the world. These transformations were marked by the decline and extinction of indigenous species and by changes in the behavior of still more survivors of the encounter. Ecological change went hand in hand with changes in systems of land and resource use.

Historians have approached the problem of agriculture, resource exploitation and cross-cultural encounter in three broad ways. They have analyzed the process of biological exchange which resulted from European expansion; they have evaluated the impact of land and resource exploitation on the environment and on indigenous societies; and they have examined Europeans' own responses to expansion and change, especially in terms of European attitudes toward nature and the emergence of conservationism as a response to severe resource depletion and environmental degradation.

Biological exchange

The most striking impact of European expansion was the sheer volume and variety of exchange of biota that resulted from contact between ecologies that had formerly been isolated from one another. The phenomenon of biological exchange was not unique to European expansion. Europe itself had already experienced the effects, both subtle and dramatic, of the introduction of exotic species. World

---

6 The firefighter turned historian Stephen J. Pyne has worked the hardest to advance this view among his colleagues. See his *World Fire: The Culture of Fire on Earth* (New York, 1995).

7 Dean feels safe in asserting that South America provides the most accessible example of the encounter between Europeans and wilderness: '...Of all the tropical continents South America was the last to be invaded by humans, and human dominance of its forests was much less intense and of much shorter duration than that exercised in Asia, Africa and Australia. Hence the Europeans confronted in their New World a nature more pristine that that which they encountered elsewhere in the tropics....South America, then, is the forest historian's freshest battleground, where all the fallen still lie sprawled and unburied and where the victors still wander about, looting and burning the train', *With Broadax and Firebrand*, p. 5.

historians have made much of some especially striking examples, such as the accidental introduction of the plague from Asia or the deliberate spread of sugar cane into Asia and Africa from its original home in New Guinea.[8] Advances in navigation brought the level of interaction to a new scale, as the peripatetic Europeans ended the biological isolation of islands and continents.

The Americas, harbouring important centres of biological diversity and agricultural complexes, have garnered the lion's share of scholarly attention. Historians now refer to the momentous biological meeting between Old World and New as the Columbian Exchange. They study the spread of American plant species to other parts of the world. They also study the transfer of what Alfred Crosby has called the European 'portmanteau of biota', including livestock, to the Americas and elsewhere as part of the process of creating 'neo-Europes' abroad.[9]

Both the spread of American species abroad and the importation of Old World species to the Americas have been studied as elements of cultural and technological diffusion. Carl O. Sauer and the diffusionist school of historical geography did much to refine this theoretical approach in the 1940s and 1950s.[10] With able contemporary practitioners such as Crosby, diffusionism has continued to dominate the environmental field of world history.

The diffusion of plant life was a key element of global exchange, and it also provides an excellent means of tracing cultural interactions. Biology and technology are fused in agriculture. The first task of the diffusionists was to trace

[8] On the spread of disease, start with William H. McNeill, *Plagues and Peoples* (New York, 1976). There is now an enormous literature on the role of disease in the Americas, and here a good place to start would be Alfred Crosby's seminal work, *The Columbian Exchange: Biological and Cultural Consequences of 1492* (Westport, VA, 1972). Kenneth Kiple has done work on Africans and diseases. See his edited volume *The African Exchange: Toward a Biological History of Black People* (Durham, NC, 1988) and *The Caribbean Slave: A Biological History* (Cambridge, 1984). See also vol. 26 in this series by Kenneth Kiple and Stephen Beck, *Biological Consequences of the European Expansion, 1450–1800* (forthcoming 1997.) On the spread of sugar cane, see Sidney Mintz, *Sweetness and Power* (New York, 1985).

[9] Alfred W. Crosby, *Ecological Imperialism: The Biological Expansion of Europe, 900–1900* (Cambridge, 1986).

[10] For a brief analysis of some of the less obvious New World plant resources whose use has been expanded, sustained, revived or reduced since colonial times, see Joseph Ewan, 'Plant Resources in Colonial America', *Environmental Review* II, no. 1 (1977), pp. 44–56.

Much work remains to be done on the question of species introductions within Europe itself. An interesting taste of the possibilities is provided by John Sheail's, 'Rabbits and Agriculture in post-Medieval England', *Journal of Historical Geography* IV, no. 4 (1978) pp. 343–55. Much has been made of the consequences of the introduction of rabbits to regions where they encounter few predators, see for example Joseph M. Powell, *An Historical Geography of Modern Australia: The Restive Fringe* (Cambridge, 1988). Sheail reminds us that rabbits were introduced to Britain itself from the Mediterranean regions in the twelfth century, in an early effort at agricultural improvement. The selective grazing of rabbits resulted in significant – and undesirable – alteration of the plantscape, and caused tensions between warren-keepers and neighbouring farmers.

the movements of innovative technology, and the spread of elements of the unique American crop complexes provided an ideal subject of study. The diffusion of language provided one ready theoretical model for the task, and indeed language proved to be one of the means by which the diffusionists traced the movement of foods. In the 1950s French scholar Roland Portères traced the spread of maize in Africa by studying modification of terms for traditional grains such as *dura* or *masi*, while in an important article on American foods in China Ping-ti Ho studied Chinese local histories and botanical treatises to trace the first mention of peanuts, sweet potatoes and maize, cataloging the terms by which these crops were first known.[11]

The study of plant diffusion remains a rich vein of historical study. The Columbian Quincentennial, for example, prompted the publication of an edition of *Les cahiers d'Outre-mer* devoted to 'Les plantes américaines à la conquête du monde'.[12] It also served as an occasion for a major exhibition at the Smithsonian Museum of Natural History, 'Seeds of Change'.[13] One of the most common themes of such studies remains the question of routes of diffusion. In this volume, we see a recent example of that approach in Jean Andrews' 'Diffusion of Mesoamerican Food Complex to Southeastern Europe'.[14] Like the work of Portères and Ho, part of what makes Andrews' work interesting is the emphasis on diffusion by non-European agents. Andrews asserts that American foods reached Central Europe not by way of traders from the Iberian peninsula, but from the Ottomans by way of Asia (see chapter 1 below).[14]

---

[11] Roland Portères, 'L'introduction du mais en Afrique', *Journal d'agriculture tropicale et de botanique appliquée* II (Paris, 1995), pp. 221–31. Ping-ti Ho, 'The Introduction of American Food Plants into China', *American Anthropologist* LVII (1955) pp. 191–201. Few have followed the lead of Ho in Asia. Philip Huang has referred to Ho in his own sweeping studies of the economies of North China. His work on the introduction of Asian cotton provides a good model for the study of how crop diffusion can thoroughly transform a society. Yet the biological interaction of East Asia with Europe or the Atlantic world remains largely unexamined. Philip C. C. Huang, *The Peasant Economy and Social Change in North China* (Stanford, CA, 1985); and *The Peasant Family and Rural Development in the Yangtze Delta, 1350–1988* (Stanford, CA, 1990). For further information on maize in Africa, see Marvin P. Miracle, *Maize in Tropical Africa* (Madison, WI, 1966).

[12] *Les cahiers d'Outre-mer* (Revue de géographie de Bordeaux), no. 179-180 (Juillet–Decembre 1992).

[13] See Herman J. Viola and Carolyn Margolis, eds., *Seeds of Change: A Quincentennial Commemoration* (Washington, D.C., 1991).

[14] Chapter 1 in this volume, pp. 1–11. In his study of the Mediterranean, John McNeill attributes demographic increases in the mountain regions in part to the 'improved viability and attractiveness of the mountains' thanks to a three-part 'agricultural involution'. The three components to this 'involution' were irrigation, the introduction of subtropical crops which provided a new source of income, and the introduction of American crops. He attributes the first two components to Arabs; perhaps he may now credit them for the third as well. John R. McNeill, *The Mountains of the Mediterranean World: An Environmental History* (Cambridge, 1992).

The theme of diffusion is taken up by J. E. Spencer as well, in an article on 'The Rise of Maize as a Major Crop Plant in the Philippines' (see chapter 2 below). In this case, the process of adoption is explicitly linked to technology, yielding interesting results. Spencer asserts that while the Spanish introduced maize, the crop enjoyed little popularity until Chinese traders brought the knowledge and means to process the grain into a desirable form.[15] This raises important themes of agency and social incorporation, also explored by Daniel Gade in his article, 'Landscape, System and Identity in the Post-Conquest Andes'. Gade is interested in the ways that rural people of the Central Andes incorporated European technologies into their own agropastoral systems. He argues that the unique environment of the Andean highlands and the success of the indigenous system filtered out all but a few innovations, until the ecological crisis prompted by depopulation encouraged a selective Europeanization of Andean culture (see chapter 3 below).[16]

Even with intentional introductions of new plant and animal species, the effects of diffusion could ripple outward into the host environment in complex and unexpected ways. Historians of American Plains Indian culture have long noted the profound impact of the horse, introduced by the Iberians. Peoples who had yet to encounter Europeans themselves could nonetheless experience the effects of the biological exchange, as horses revolutionized the lives of nomadic societies and placed new pressures on sedentary ones. While many historians have studied the direct impact of horses on culture, the ecological effects of this new animal remain relatively unexplored. Dan Flores begins to address this question in his article, 'Bison Ecology and Bison Diplomacy: The Southern Plains from 1800 to 1850'. The horse transformed Plains ecology, argues Flores, and this too affected human societies. The connection between the rise of the horse and the decline of the buffalo is not a simple matter of enhanced hunting technology. Only an ecological analysis will unravel the complex story of species interaction. He finds that factors such as climate and the competition for grazing lands must be taken into account, as well as the enhanced hunting capacity of humans (see chapter 4 below).[17]

Elinor G. K. Melville also applies ecological analysis to her study of 'Environmental and Social Change in the Valle del Mezquital, Mexico, 1521–1600'. She charts the transformation of this Mesoamerican landscape from a 'densely populated and complex agricultural mosaic of planted and fallow fields,

[15] Chapter 2 in this volume, pp. 13–28.

[16] Chapter 3 in this volume, pp. 29–46. This too is a special issue commemorating the Columbus Quincentennial.

[17] Chapter 4 in this volume, pp. 47–67. See also Paul H. Carlson, 'Indian Agriculture, Changing Subsistence Patterns, and the Environment on the Southern Great Plains', *Agricultural History* LXVI, no. 2 (1992), pp. 52–62.

villages, quarries, lakes, dams, woods, and native grasslands' to 'the archetype of the barren, eroded regions of Mexico' (see chapter 5 below). Indigenous farmers had exploited the land for centuries, leaving little evidence of environmental degradation, but the introduction of new diseases and new animals wrought devastating change. Livestock, especially sheep, destroyed vegetative cover and encouraged erosion, while they also introduced alien plant species and selectively destroyed indigenous grasses, allowing woody shrubs and cacti to spring up where agricultural fields and grasslands had once flourished. Again, climate change may have delivered the *coup de grace*, pushing the weakened ecosystem over the edge into arid scrublands.[18]

Flores, Melville and other historians who study the ecological effects of biological exchange are careful not to separate the ecosystems from the human societies that inhabit them, although they do argue for greater attention to the environmental context of human activity. The introduction of new organisms remained connected to technology, as the diffusionists have long stressed. Yet attention to ecology serves to highlight the uncertainties of the transfers. Melville argues that the Spanish did not act in their own long-term interest when they developed the grazing economy in the Valle del Mezquital; they proved unable to 'adapt their expectations, and their institutions, to New World realities'.

In making this argument, Melville takes issue with common opinion in what may be described as a subgenre of diffusionist studies, the history of ranching in the Americas. While she argues that European settlement could result in a transformation of agricultural land to grazing land, most historians have emphasized the use of lands for ranching that are agriculturally marginal – namely the edges of forests, tropics and deserts. In the relative absence of predators and diseases, European livestock competed very successfully against indigenous grazing animals. Ranching represented a rational economic choice for Europeans facing a shortage of labour and an abundance of land that was not worth putting into agricultural production, as did the release of swine into forests that were not worth cutting down. In the face of this apparently obvious conclusion, historians took up the question of whether American ranching methods were best understood in terms of the diffusion of Old World cultural systems, or as a response to the environments and economic constraints they encountered in the New World.[19]

---

[18] Chapter 5 in this volume, pp. 69–98. A more complete account is available in her book, *A Plague of Sheep: Environmental Consequences of the Conquest of Mexico* (Cambridge, 1994).

[19] For the 'cultural diffusion' side of the argument, as well as an excellent overview of the field, see Terry G. Jordan's *North American Cattle-Ranching Frontiers: Origins, Diffusion, and Differentiation* (Albuquerque, NM, 1993). John Solomon Otto and N.E. Anderson provide the more environmentally determinist argument in their comparative study of cattle ranching in the *llanos* of Venezuela and the flatwoods of Florida, where despite differing origins of cattle complexes (Spanish and British), the similar grazing environments forced ranchers toward strikingly similar

The European standards for establishing grazing (or hunting) versus agricultural lands did not necessarily apply to indigenous production complexes, however, and what may have seemed unproductive land to Europeans may have been highly productive for indigenous users. Historians may have been too quick to yield to the claims of European settlers on this point because of their own lack of familiarity with different ecological systems or with non-European methods of subsistence. This is the point of Jacques Barreau's study of the introduction of cattle ranching to New Caledonia in the 1850s (see chapter 6 below). Colonial interlopers failed to recognize that the lush savannas that seemed to invite grazing were really the products of a sophisticated system of swidden agriculture. Cattle became '*the* means of land alienation by Europeans', resisted in a number of rebellions that were 'ferociously repressed by the French colonial authorities'. Contrary to the idea that Europeans introduced advanced technology, 'in their new environment, the white graziers had made a big leap backward to archaic animal husbandry', with effects on indigenous agriculture that echoed the experience of Mexico's Indians in the Valle del Mezquital.[20]

To summarize the findings of historians of biological exchange, one of the most important dimensions of this exchange was the diffusion of food plants and animals and the productive systems of which they were a part. In ecological terms, this should be explained partly by the removal of biota from their pest and disease environments and by their introduction to new systems where they could compete successfully with indigenous species or occupy a niche favorable to themselves or to human cultivators. The ecological dimension is important because biological systems have their own historical dynamics outside of, as well as in conjunction with human history. Biota could spread fast and far beyond the zones of their deliberate introduction, and they could transform ecological systems in unexpected ways. Both introduced biota and the technological systems that accompanied them could be perceived as an enhancement to indigenous systems of production, but they might also fail to mesh with existing systems. This could have the effect of discouraging

---

techniques. J. S. Otto and N. E. Anderson, 'Cattle Ranching in the Venezuelan Llanos and the Florida Flatwoods: A Problem in Comparative History', *Comparative Studies in Society and History,* XXVIII, no. 4 (1986), pp. 672–83.

For further detailed description of early cattle raising techniques in North America, see also John Solomon Otto's, 'Livestock-Raising in Early South Carolina, 1670–1700: Prelude to the Rice Plantation Economy', *Agricultural History* LXI, no. 4 (1987), pp. 13–24. Richard Slatta has done outstanding work on the history of ranching societies. See his *Cowboys of the Americas* (New Haven, CT, 1990) and *Gauchos and the Vanishing Frontier* (Lincoln, NE, 1983). For an exhaustive study of both introduced and indigenous pigs and their place in the American tropics, see R. A. Donkin, 'The Peccary – With Observations on the Introduction of Pigs to the New World', *Transactions of the American Philosophical Society* LXXV, pt. 5 (Philadelphia, 1985).

[20] Chapter 6 in this volume, pp. 99–111.

propagation of the introduced species, or it could force changes to existing cultural systems in order to adapt to opportunity or to environmental changes wrought by the new species or complex. Ultimately, the effects of biological diffusion can be a challenge to describe even where there is a very severe and obvious change, in part because it is difficult to reconstruct the systems that existed before the introductions occurred.

The process of biological exchange is clearly central to any analysis of the process of European expansion, and nowhere is this more true than in the history of agriculture. The concept of diffusionism provides a powerful tool for linking biology and culture, although it has yielded some ground to the growing and immensely promising field of ecological history, where the emphasis is on change in place rather than on movement of a particular organism or technology across space. Part of the importance of an ecological approach is that it provides a means for evaluating the impact of diffusion, and this in turn has highlighted the centrality of culture and power as critical elements of biological exchange. Incorporation of new biota could be the rational choice of receiving cultures, as was the case with the horse in the Americas or of various American crops in other parts of the world; or it could be accidental, as with the introduction of diseases and pests. Yet for the most part, the biological exchange that accompanied European expansion was a part of the process of incorporation of new realms into the world system, and this often linked biological exchange to exploitation.[21]

---

[21] There remain many pressing questions unanswered and relatively unexplored in regard to even the fairly simple yet momentous question of plant exchange. The question of the effects of American foods on African population, for example, has been raised in the context of the demographic debate on slavery. Here I refer to the extensive literature on West African population dynamics prompted by Philip Curtin's *The Atlantic Slave Trade: A Census* (Madison, WI, 1969). Curtin himself suggested that the introduction of manioc and maize may have balanced the scales in favour of population growth despite the losses incurred because of the slave trade, but the question has been complicated by the role of the slave trade itself, of war, and of climate. See for example James F. Searing, *West African Slavery and Atlantic Commerce: The Senegal River Valley, 1700–1860* (Cambridge, 1993). See also George E. Brooks, *Landlords and Strangers: Ecology, Society, and Trade in Western Africa, 1000–1630* (Boulder, CO, 1993) or his article 'A Provisional Historical Schema for Western Africa Based on Seven Climate Periods (*ca.* 9000 BC to the 19th Century)', *Cahiers d'études africaines* XXVI (1-2), nos. 101–102 (1986), pp. 43–62.

Brooks argues that population increases in West Africa were due to a wet period that lasted from about 1500 to 1630. An ensuing dry period created a 'push' of drought, famine and warfare that coincided with the 'pull' of the growing European demand for slaves. Brooks observes that 'there are a number of issues concerning the commercialization of agricultural and sylvan commodities during the latter part of the *ca.* 1630–1860 dry period that merit systematic study with respect to climate and ecological factors', including the effects of timber cutting at a time when forest growth was inhibited, as well as the consequences of changes in land use and the redistribution of flora and fauna, and their relationship to disease. He also notes that the relationship between climate and introduction of commercial crops remains unstudied.

Land and resource exploitation

Although it rarely stands as a line between wilderness and civilization, the expanding zone of European contact, conquest and settlement can be profitably studied as a kind of ecological frontier. Europeans sought to expand their own systems of land and resource use at the expense of indigenous systems and users. The struggle over the means and methods of exploiting a resource became especially acute when the resource was finite. Europeans often mobilized their own technology in order to assume absolute control over critical resources. Looking at South Africa, Leonard Guelke and Robert Shell find that water can provide an especially stark example of the process in their article 'Landscape of Conquest: Frontier Water Alienation and Khoikhoi Strategies of Survival, 1652–1780'. By seizing control of water, they argue, settlers insidiously but effectively

---

The theme of the issue of *Cahiers d'études africaines* containing the Brooks article is 'Milieux, histoire, historiographie' and it contains other articles of interest, although they do not directly address the question of European influence. See especially Nicoué Lodjou Goyibor, 'Écologie et histoire: les origines de la savane du Bénin'; Kenneth Swindell, 'Population and Agriculture in the Sokoto-Rima Basin of Northwest Nigeria: A Study of Political Intervention, Adaptation and Change, 1800–1980'; Palo B. Eyzaguirre, 'The Ecology of Swidden Agriculture and Agrarian History in São Tomé'; and for a subject very rarely broached, Jean-Pierre Chauveau, 'Une histoire maritime africaine est-elle possible? Historiographie et histoire de la navigation et de la pêche africaines à la côte occidentale depuis le XVe siècle'.

Timothy Weiskel suggests an intriguing avenue of ecological study of the relationship between slavery and environment in West Africa in his article 'Toward an Archeology of Colonialism', in ed., Donald Worster, *The Ends of the Earth* (Cambridge, 1989). He suggests that the slave trade itself forced an intensification of agriculture as peoples accustomed to extensive forms of production were forced to change their settlement patterns and seek refuge in fortified 'nucleated defensive settlements'. He offers no particular evidence for this speculation, however, and he suggests that 'migration, rather than intensification seems to have been the solution frequently adopted'.

This ongoing discussion of the demographic effects of the incorporation of West Africa into the Atlantic system is not merely an academic one, as it has been used as a framework for discussing current population issues in Africa. Timothy Weiskel, for example, argues that the slave trade encouraged high fertility both as a response to conflict and because of the benefits of new food crops. According to Weiskel, this set a pattern that was further encouraged by commercial agriculture which 'put a premium on families that could mobilize large numbers of dependents in order to increase their household production', and by the further catastrophes of the period of colonial conquest. Timothy C. Weiskel, 'Vicious Circles: African Demographic History as a Warning', *Harvard International Review* XVI, no 4 (1994), pp. 12–16.

It should also be noted that even the exchange from the Americas to Europe remains a field ripe for study. Although the impact of particular crops, such as the potato, is well known, a systematic and holistic analysis of the relationships among biological exchange, agricultural change, demography and industrialization has, to my knowledge, yet to be attempted. Still, the demographic effects of biological exchange have perhaps been better studied for Europe than for any other part of the world. William Langer provides a nice demographic synopsis in his 'American Foods and Europe's Population Growth, 1750-1850', *Journal of Social History* VIII, no. 1 (1975), pp. 51–60.

undermined the Khoikhoi pastoral economy in a piecemeal fashion, 'one spring or *spruijt* at a time' (see chapter 7 below).[22]

European expansion inevitably involved ecological transformation. Quite often, the purpose of expansion was simply and directly exploitative. Shawn Miller chronicles the fate of one resource in his article, 'Fuelwood in Colonial Brazil: The Economic and Social Consequences of Fuel Depletion for the Bahian Recôncavo, 1549–1820' (see chapter 8 below). Miller describes the critical linkage between forests and sugar plantations. 'With rarely a pause,' he notes, 'the fires of the sugar mills burned as many as nine months of the year, more than twenty hours of the day and six days of the week, as long as there was fuel to feed them'. Fuelwood and the labour to harvest and transport it constituted a significant cost of production, and control over supplies became a growing preoccupation of planters as local forests dwindled and transportation costs rose. The sugar industry did not bring about deforestation single-handedly; the ancillary industries of rum distillery and brick and tile production also called for firewood, as did other forms of industry from lime production to tanneries. Rural and urban residents alike required fuel wood for domestic use, and shipbuilding and lumber exports also placed pressure on the forests.[23]

David O. Percy provides further depth to the story of forest exploitation with his North American study, 'Ax or Plow? Significant Colonial Landscape Alteration Rates in the Maryland and Virginia Tidewater' (see chapter 9 below). European settlers were excited by the fertility of the soil, but they found the best land covered with hardwood forest. Settlers reasoned philosophically that good land would naturally be exploited by big plants, but this presented them with the daunting task of clearing the forest when labour was very hard to find. Percy argues that early farmers responded to this challenge by adapting indigenous swidden agriculture to their own purposes. They learned to girdle trees rather than chop them down, and they practiced a very long rotation by abandoning exhausted fields to return to forest. Yet Europeans could not adopt indigenous methods successfully because the settler population grew too rapidly, and the demands of the market economy proved too great a temptation. In the end, settlers were forced to return to European methods of husbandry if they hoped to conserve the land for farming.[24]

While Percy suggests that indigenous swidden agriculture may have provided the basis for the American settlers' tendency toward 'land mining' (exhausting the soil and then moving on), James J. Parsons suggests that the European system is inherently exploitative rather than self-sustaining. His subject is a seminal event in the history of European expansion: the Spanish occupation of the Canary Islands in the Atlantic. The Spanish presence spelled the doom of the island

[22] Chapter 7 in this volume, pp. 113–134.
[23] Chapter 8 in this volume, pp. 135–158.
[24] Chapter 9 in this volume, pp. 159–167.

ecosystems – including their indigenous inhabitants, the Guanches – as the Spanish set about putting the land to their own uses. Parsons takes a long view of how the Spanish made use of finite island resources. In 'Human Influences on the Pine and Laurel Forests of the Canary Islands', Parsons also pinpoints forests, the foundation of the island ecology, as an important key to comprehending European cultural patterns (see chapter 10 below).[25]

In the Canaries, the Spanish sought to emulate Portuguese success in Madeira by encouraging the establishment of sugar plantations on the island. This industry was appropriate to only limited portions of the islands, so settler society became polarized between those who produced commodities and those who cobbled together a subsistence culture based on European livestock and both European and American crops. Forests played an important role for all Canary Islanders: they supplied fuel, timber, pitch and forage. Fuel-hungry sugar mills put heavy pressure on the forests, as did a rising population that cleared forests for farmland. Canary Islanders recognized the crucial importance of forests. For one thing, they firmly believed that the forests were essential for capturing rainwater. They needed the forest to live, but they needed to destroy the forest to earn a living. Despite efforts to conserve the forests through regulation, Canary Islanders could not stop the exploitation.

The history of the Canaries represent something of a departure for Parsons, whose main interest has been the effect of Europeans on the tropical forest systems of the Americas. He has observed that the Caribbean and the Americas present the largest area on earth suitable to tropical rain forest. While much of this rainforest survived into the twentieth century, the Caribbean saw most of its once-extensive forest land replaced by scrub and savanna.[26]

Indigenous land use: evaluating the effects of European expansion

To understand the European impact fully, historians must find a way to measure the colonized environment against the indigenous one, and describe the process

[25] Chapter 10 in this volume, pp. 169–187. The story of the encounter between the Guanche and the Spanish is recounted in Crosby's *Ecological Imperialism*.

[26] That area includes 'almost the entire Amazon basin, the Guiana highlands and coasts, the foothills and lower slopes of the eastern Colombian and the adjacent Venezuelan Andes, and the southern half of the Lake Maracaibo depression. It includes a significant part of the lower Magdalena valley, as well as the Sinú and Atrato valleys and the entire Pacific coast of Colombia, reaching southward to Esmeraldas, Ecuador. It encompasses most of the east coast of Central America, the Yucatán Peninsula and northward along the lower flanks of the Mexican highlands to beyond Vera Cruz. Finally, it also includes several localized high rainfall pockets on the west coast of Central America and Mexico...as well as most of the islands of the Caribbean'. James J. Parsons, 'The Changing Nature of New World Tropical Forests Since European Colonization', in ed., William M. Denevan, *Hispanic Lands and Peoples: Selected Writings of James J. Parsons* (Boulder, CO, 1989).

by which Europeans transformed the landscape. For many regions, this must be predicated by some understanding of tropical ecological systems, and the differences between temperate and tropical systems. Unlike temperate ecological systems where many nutrients are stored in soil and dead matter, tropical systems store nutrients in living organisms where they are quickly and efficiently recycled. Although indigenous Americans developed the means to store nutrients in germ plasm (chiefly maize), many tropical people followed the natural pattern, relying on the living ecosystem rather than storage.

The indigenous peoples of the Caribbean Islands did not rely on maize, and they provide a good example of tropical subsistence methods prior to European arrival. Arawak subsistence was based on the *conuco* system (a *conuco* is a cultivated plot) of shifting cultivation of starch- and sugar-rich foods in kitchen and other types of gardens, supplemented by hunting, fishing, and the gathering of wild plants and animals. Arawaks, like people in other tropical parts of the world, favored root crops or those that reproduced from cuttings, rather than seeds. Starchy tubers were often grown in mounds for drainage. Land clearance was accomplished through girdling, and the felled wood was burned to provide nutrients to the crops. Arawaks planted both the tall-growing, hardy and highly productive manioc, and sweet potatoes, which provided ground cover and reduced erosion and leaching. A cleared patch might last as long as two decades, but it eventually gave way to the returning forest. As the land became less suitable to cultivation, it might be converted to a 'hunting garden' of fruit trees planted to attract game. Thus, for Arawaks as for many forest peoples, the line between cultivation and wild gathering was not a sharp divide. Arawaks harvested not only forest game (mammals were relatively limited on the islands), but like other people of the Caribbean they enjoyed a rich harvest of sea turtles as well.[27]

This means of subsistence was highly productive and supported populations which may have been comparable to those of today. Population estimates vary, but they are high. Hispaniola alone may have supported a population of anywhere from 3 million to 8 million people.[28] Parsons heeds Columbus' description of the land as 'a vast and well peopled garden "as fully cultivated as the countryside around Cordoba"'.[29]

Throughout the American tropics, whether the dominant crop was maize or manioc, some system of shifting cultivation was practiced and allowed for great population densities. In the pre-Columbian era, people favoured the lowlands of

---

[27] David Watts has published a comprehensive and exemplary study of the transformation of the Caribbean in his *The West Indies: Patterns of Development, Culture and Environmental Change Since 1492* (Cambridge, 1987).

[28] The low estimate is from Watts, the high one from Parsons.

[29] Parsons, 'The Changing Nature of New World Tropical Forests Since European Colonization' op. cit.

alluvial or coastal plains, where they undoubtedly had a significant effect on the ecology. This was especially true on the climatic fringe of the tropics, where burning could open land for cultivation over a relatively long period. Parsons estimates a population in lowland South America of at least 15 million, while Mexico and Central America added another eight to ten million. Forest colonizers in the twentieth century have uncovered evidence of surprisingly widespread cultivation in earlier eras.

Growing recognition of the carrying capacity of tropical lands, and of the sophistication of indigenous peoples in exploiting that capacity, has led necessarily to a reassessment of the extent of environmental modification by indigenous peoples prior to European colonization. The most immediate impact of European presence in the American tropics may have been the return of brushlands and ultimately of forest to the former savannas of the coastal lowlands as the indigenous populations died off, as the labour of survivors was appropriated to serve the demands of Europeans, and as free-ranging European livestock destroyed native crops. European farming methods brought a preference for the better soils of the highlands and the relatively dry Pacific coast (where soils were not immediately leached of their nutrients), so that population patterns after European colonization were significantly different from those of the pre-Columbian era.[30]

The inhospitality of the tropics to European systems of agriculture could leave tropical zones in the possession of their indigenous peoples long after other areas had been colonized. In the Yucatán Peninsula, for example, limestone bedrock, thin soil and relatively low seasonal rainfall discouraged agriculture when better opportunities beckoned elsewhere. Ranching and beekeeping were the only really viable Hispanic industries there until a rising colonial population encouraged maize and rice production.[31]

Using the land

The importation of African slaves, like the adoption of indigenous techniques, introduced some degree of expertise in exploiting what was for many Europeans a thoroughly strange environment. Judith Carney provides an analysis of the role of Africans in Carolina rice farming. In 'From Hands to Tutors: African Expertise in the South Carolina Rice Economy', she compares West African methods of

---

[30] Many Latin American historians have observed that colonists would go to great lengths to grow familiar foods, even where the terrain was not appropriate. For a good case study of European efforts to transplant their agriculture to the New World, see Robert G. Keith, *Conquest and Agrarian Change: The Emergence of the Hacienda System on the Peruvian Coast* (Cambridge, MA, 1976).

[31] Robert W. Patch, 'Agrarian Change in Eighteenth Century Yucatán', *Hispanic American Historical Review* LXV, no. 1 (1985), pp. 21–49.

cultivation to those employed in North America and finds striking similarities between them. She emphasizes the hydraulic complexity of the tidewater system developed by slaves from the Upper Guinea coast. Taking up the issue of cultural diffusion, she suggests that historical approaches to technology transfer must incorporate the dynamics of negotiation, as slaves likely bartered their knowledge in exchange for better conditions (see chapter 11 below).[32]

As Europeans learned to use the land, the resurgence of forest that marked the earliest years of European colonization of the tropics gave way to forest clearance and exploitation, especially in the areas where sugar plantations were established. As in North America, the forests of the Caribbean islands became sharply diminished. Yet in the Americas as a whole the tropical forest may still have experienced something of a renaissance in former savanna lands, at least until the late nineteenth century when the markets for both cattle and tropical fruits provided new incentives for colonization.

Europeans eventually developed some sophistication in their methods for dealing with soil loss and the leaching of nutrients as a result of the loss of natural vegetative cover in the tropical climate, but this could raise the costs of cultivation.[33] The problems and costs of intensive management, along with the

---

[32] Chapter 11 in this volume, pp. 189–218. See also Peter H. Wood, *Black Majority: Negroes in Colonial South Carolina from 1670 through the Stono Rebellion* (New York, 1974).On the theme of cross-cultural exchange of expertise, see also Gary B. Nash, *Red, White and Black: The People of Early North America*, 3rd edn (Englewood Cliffs, NJ, 1992).

[33] David Watts provides the best coverage on this subject. See also Richard K. Ormrod, 'The Evolution of Soil Management Practices in Early Jamaican Sugar Planting', *Journal of Historical Geography* V, no. 2 (1979), pp. 157–70. Ormrod, like Watts, describes a shift from a 'pioneer' pattern of land use to intensification by the early 1700s. Ormrod directly addresses the question of technological diffusion, arguing that Jamaican planters borrowed techniques directly from Britain and that they 'did not consider the two agricultural environments to be fundamentally different'.

Carville Earle suggests a slightly different approach to the question of the ecology of plantation agriculture. Studying the Chesapeake colonies, he too takes an interest in diffusion, but particularly in the relationship between the diffusion of agricultural innovations – such as the plantation system – and long waves of economic growth. In contrast to Ormrod, Earle sees innovation as a localized response to both environmental and economic conditions. Early tobacco culture went through two important periods of innovation during the 'long-wave bad times' of the 1630s and 1680s, he argues; the first decline in prices forced an increase in productivity through the practice of plant topping and to better processing. The second forced a more profound shift in cultivation practices because traditional methods of clearing new land proved too expensive while manuring hurt the quality of the tobacco. Growers were forced toward a system of rotation and diversification which superficially resembled the English system, yet it was really 'something of an ethnic amalgam … borrowing elements from Indian, Afro-American and European sources'. Eighteenth-century Europeans, he argues, showed little understanding of the Chesapeake system: 'Its unkempt fields littered with dying trees and stumps and hummocked by mini-excavation pits … its ragged old fields in various stages of succession; and its ramshackle tobacco houses in various states of decay. To be sure, the tobacco landscape after 1700 was not pretty, but it was highly

emergence of severe pest problems, were of critical importance as a causal factor in the decline of colonial tropical plantation systems. Yet they have received very little attention from historians. This is true for temperate as well as tropical environments, and there is clearly a great deal more to be done in the ecological history of commercial agriculture, and especially in the ways that the ecology of farm systems shaped land use patterns over time.

One broad foray into historical agroecology can be found in Robert MacCameron's 'Environmental Change in Colonial New Mexico'. In this case, MacCameron takes a semi-arid environment as the object of study, and he seeks to demonstrate that while commercial agriculture certainly caused a dramatic transformation of the environment, subsistence agriculture could also induce ecological change (see chapter 12 below). New Mexico remained a relatively undeveloped commercial backwater, yet livestock altered grasslands while the introduction of the metal axe enabled both the Spanish and the Pueblo to harvest timber to the point of deforestation. European landholding systems also stamped their pattern on the land, and Spanish methods of irrigation intensified soil problems that had already developed under the Pueblo. Despite these changes, the effects of agriculture remained tied to its interaction with climate cycles through much of the colonial period, until the consolidation of Spanish control at the end of the eighteenth century induced a level of immigration that tipped the balance of ecological power decisively into the hands of settlers.[34]

As Europeans exercised their power, they sought to regulate access to the resources that they exploited. This meant achieving hegemony over land that was often contested by its indigenous inhabitants, but it also meant controlling the behavior of Europeans themselves. Systems of regulation became entailed in the technology of exploitation, as Europeans developed greater ecological understanding of the resources they used. We have seen that indigenous culture could be incorporated into European systems, but Europeans were even more likely to develop regulatory methods based strictly upon their own cultural systems

functional in economy and ecology'. Carville Earle, 'The Myth of the Southern Soil Miner', in ed., Donald Worster, *The Ends of the Earth: Perspectives on Modern Environmental History* (Cambridge, 1989).

Reform is also discussed by J. H. Galloway, 'Agricultural Reform and the Enlightenment in Late Colonial Brazil', *Agricultural History* LIII, no. 4 (1979), pp. 763–79. Galloway observes that Brazilian reformers pointed to the intensification of agriculture in the West Indies as an appropriate model for Brazilian sugar growers. Galloway notes that these theorists did not understand the pressures – especially deforestation – which had pushed West Indian planters to new methods. On the other hand, he finds merit in their larger economic critique of the dangers of monoculture and their search for appropriate crops to diversify the Brazilian economy. Like Earle, Galloway suggests that the desire to innovate was a response to a declining economy in the late 1700s.

[34] Chapter 12 in this volume, pp. 219–241.

and goals. European science sometimes played a role in this process, although some historians assert that the name of Science was invoked more frequently than its empirical methods.[35] However they developed, European methods of resource regulation fell eventually under the rubric of Conservation, and historians have barely begun to examine this regulatory dimension to resource exploitation.

Conservation

Although forests and agricultural land constituted major attractions for Europeans, they also ventured forth into the world in order to exploit other kinds of resources. European mastery of the seas translated into robust industries in whaling, fisheries and sealing, all topics that await revisiting by historians with an ecological and world historical perspective. Chesley Sanger has struck out in an ecological direction with his study of the Right whale harvest in the seventeenth through nineteenth centuries, 'Saw Several Finners but No Whales'. Sanger uses the accounts of whalers themselves to address the relationship between the industry and the 'habits and character of the resource' (see chapter 13 below). Besides its high yield of oil and whalebone, the Right whale proved the most susceptible to exploitation because of its slow movement and its need to surface two to three times per hour to breathe. By contrast, the swiftness of the 'finner', or norqual, defeated the technology of the fishermen. The population of Black Right whales, found in the Bay of Biscay and off the coast of Labrador, was depleted by the end of the seventeenth century. Whalers moved on to Greenland in pursuit of whale oil and baleen. Sanger uses British logbooks to estimate the migratory routes of the Greenland Right whale, a species exploited so heavily that it became the first whale to be officially protected.[36]

The fur trade also played an important role in European expansion. Like the Right whale, the populations of heavily exploited fur bearing animals such as the beaver could be depleted as trappers raced to meet the insatiable demand of the world market. In Western Canada, for example, the intense competition between rival fur companies had ravaged the ecosystem by the end of the eighteenth century. The collapse of the population of fur bearers prompted Indians to migrate out of areas under the control of the Hudson's Bay Company. Arthur Ray, a leading historian of the North American fur trade, addresses European efforts to regulate this overexploited animal resource in his 1975 study, 'The Conservation Schemes of the Hudson's Bay Company, 1821–50'.

Ray uses the regulatory efforts of the Hudson's Bay Company as an opportunity to examine the relationship between the Company and indigenous

---

[35] Madhav Gadgil and Ramachandra Guha make the distinction between scientific claims and scientific methods in their history of British forestry in India, *This Fissured Land: An Ecological History of India* (Berkeley, CA, 1992.)

[36] Chapter 13 in this volume, pp. 243–270.

trappers. Taking up a theme later explored by other historians, most notably Calvin Martin,[37] Ray suggests that Indians as well as Métis proved to be uninterested in husbanding fur resources according to conservation schemes worked out by the company. Concepts of land tenure and resource ownership, as well as religious beliefs, clashed with European cultural approaches to resource use. Above all, a social structure of competing rival bands and families, as well as rival fur companies, made conservation of an easily appropriated resource all but impossible. The opportunity to pursue a conservation strategy became possible only with the merger of Hudson's Bay Company with one of its rivals, North West Companies. Secure in its monopoly, the Company tried to shift trappers to new areas and new prey, but the interests of the trappers did not necessarily coincide with the Company goal of achieving sustainable yields. Nor could some company traders accept the harsh measures proposed by their superiors to drive Indians out of damaged areas. In the end, the beaver was saved not by conservation, but by the shifting currents of European fashion (see chapter 14 below).[38]

Peter Boomgaard assesses the conservation tactics of another charter company in his article, 'Forest Management and Exploitation in Colonial Java, 1677–1897'. The Dutch East India Company exploited the teak forests for shipbuilding, construction, furniture-making and fuel, and its first task was to negotiate with Javanese rulers to secure adequate supplies. Boomgaard finds that Dutch efforts to promote sustained yields were minimal, as a rotational theory of clearcutting was eventually suggested but never pursued. Instead, the VOC attempted to conserve existing resources, an effort that put its agents in opposition to indigenous customs such as annual burning. Because the Javanese regarded the practice as essential to their own livelihoods in the forest, Dutch officials were forced into forbearance until they achieved enough political control to ban burning in 1857. The consolidation of Dutch control could also be found in the

---

[37] Calvin Martin, *Keepers of the Game: Indian-Animal Relationships and the Fur Trade* (Berkeley, CA, 1978). See also Shepard Krech, ed., *Indians, Animals and the Fur Trade: A Critique of Keepers of the Game* (Atlanta, GA, 1981). William Cronon also addresses indigenous cultural attitudes toward fur bearing animals in a chapter entitled 'Commodities of the Hunt' in his *Changes in the Land: Indians, Colonists, and the Ecology of New England* (1983).

[38] Chapter 14 in this volume, pp. 271–290. See also Arthur J. Ray and Donald B. Freeman, *'Give Us Good Measure': An Economic Analysis of Relations Between the Indians and the Hudson's Bay Company before 1763* (Toronto, 1978). James R. Gibson devotes a chapter to the impact of the fur trade on the indigenous peoples of the Northwest Coast and in the Pacific in his book, *Otter Skins, Boston Ships and China Goods: The Maritime Fur Trade of the Northwest Coast, 1785–1841* (Seattle, 1992). The role of fashion is explored in more depth by Lorne Hammond, 'Marketing Wildlife: The Hudson's Bay Company and the Pacific Northwest, 1821–49', *Forest and Conservation History* XXXVII, no. 1 (1993), pp. 14–25. Hammond sees the HBC squeezed between two extremely volatile realms: that of fashion, on one end, and wildlife demographics on the other.

establishment of teak plantations and bans on tree felling in forests under Dutch governance. Yet the fate of the forests became a secondary consideration as a plantation economy took hold, and Indonesian forests fell under the same pressures as those of the West Indies. At the same time, the Dutch succeeded in replacing indigenous forest labour systems with a commercial timber economy, managed by a national forest service (see chapter 15 below).[39]

While Boomgaard makes little of the political or ideological implications of Dutch forest management practices in Indonesia, these are the central questions for Richard Grove in his article, 'Conserving Eden: The (European) East India Companies and their Environmental Policies on St. Helena, Mauritius and in Western India, 1660–1854'. For Grove, deforestation is directly linked to the integration of these regions into the global market economy, and thus to the exploitative ideology of imperialism. Yet he challenges those who would leave the interaction of economy and ideology unexamined. Empire also tied colonies to the ideologies and politics of their home countries, where the effects of environmental degradation and the struggle over control of resources had created complex and innovative currents of conservationist thought. Small but influential cadres of scientists and naturalists subjected colonial practices to both social and scientific critique, apart from and even in opposition to the kind of economic logic that prompted conservationist efforts by the likes of the Hudson's Bay Company.

The European association of Nature with Paradise 'provided a philosophical and quasi-theological basis for an interventionist response to environmental destruction'. Tropical islands held a peculiar place in this discourse, as the emergence of a genre of desert island literature, as well as the obvious and immediate effects they suffered from environmental degradation, turned them into a utopian metaphor for the world. Ironically, the very fact of absolute imperial control made colonies accessible to forms of social experimentation – including the imposition of conservationist regulations by the state – that were not possible in Europe itself. State motivations for carrying out such reforms represented a complex amalgam of commercial, political and social interests that were part and parcel of the phenomenon of imperialism (see chapter 16 below).[40]

[39] Chapter 15 in this volume, pp. 291–311.

[40] Chapter 16 in this volume, pp. 313–346. See also Nancy Lee Peluso, 'The History of State Forest Management in Colonial Java', *Forest and Conservation History* XXXV, no. 2 (1991), pp. 65–75. Like Grove, Peluso emphasizes the importance of conservation ideology to imperial Dutch practices: 'By the end of the colonial period, Dutch foresters were using an international ideology of conservation and state stewardship to legitimate their control of land, tree species, and labour in Java's forests'. Control of land differentiated colonial rule from pre-colonial rulers who 'typically concentrated more on controlling populations and products … than on controlling access to the land itself'.

While most historians are drawn (quite rightly) to chronicling the encounter of cultures on the meeting ground of a particular landscape, Grove reminds us that conflict occurs not only between cultures, but within them. The commodification of land and resources served as both a reason and means for European expansion, yet the ecological effects of the market economy proved as challenging and destabilizing to Europeans themselves as to other cultures pulled into the European sphere of influence.

Like many of the articles in this volume, Grove's analysis suggests the need for a more holistic and ecological view of European expansion than either the models of technological diffusionism or world systems theory can accomodate on their own. There can be no doubt that the ecological effects of European activity were profound, but they did not necessarily go in one direction. Our awareness of the far-reaching effects of the industrial revolution tends to obscure the less linear dynamics of earlier centuries of encounter. The effects of biological exchange make the many directions of change more clear, and the changing ecologies of particular places remind us that nature's own complex causal role in history cannot be ignored, especially in a time when no civilization could even pretend to step outside the cycles of nature's domain. The meanings that people made of ecological change were no less complex; it would be easy to assume that Europeans lauded progress toward civilization and order, yet no less an exploiter and civilizer than Columbus admired the agricultural industry of the Arawaks and worried that deforestation might bring the West Indies to the sorry state Spain had made of the Canaries. Europeans were stunned by a world of abundance and diversity that they had never imagined – and that we shall never know.

Select Additional Bibiography

Cronon, William, *Changes in the Land: Indians, Colonists and the Ecology of New England* (New York, 1983).

Cuddihy, Linda W., and Stone, Charles P., *Alteration of Native Hawaiian Vegetation: Effects of Humans, Their Activities and Introductions* (Honolulu, 1990).

Crosby Alfred, and Reilly, Kevin, *Germs, Seeds and Animals: Studies in Ecological History* (Armonk, NY, 1994).

Dean, Warren, *With Broadax and Firebrand: The Destruction of the Brazilian Atlantic Forest* (Berkeley, CA 1995).

Denevan, William M., ed., *The Native Population of the Americas in 1492* (Madison, WI 1992).

Glacken, C., *Traces on the Rhodian Shore: Attitudes to Nature from Classical Times to 1800* (Berkeley, CA 1967).

Grove, Richard, *Green Imperialism* (Cambridge, 1994).
Kimber, Clarissa Therese, *Martinique Revisited: The Changing Plant Geographies of a West Indian Island* (College Station, TX, 1988).
Lines, William, *Taming the Great Southland* (Berkeley, CA 1991).
MacKenzie, John, ed., *Imperialism and the Natural World* (Manchester, 1990).
Merchant, Carolyn, *Ecological Revolutions: Nature, Gender and Science in New England* (1989).
Perlin, John, *A Forest Journey: The Role of Wood in the Development of Civilization* (Cambridge MA, 1989).
Pyne, Stephen J., *Burning Bush: A Fire History of Australia* (New York, 1991).
Pyne, Stephen J., *Fire in America: A Cultural History of Wildland and Rural Fire* (Princeton, NJ 1982).
Richards, John F., and Tucker, Richard P., *Global Deforestation and the Nineteenth Century World* (Durham, NC 1983).
Sauer, Carl O., *Agricultural Origins and Dispersals: The Domestication of Animals and Foodstuffs* (Cambridge, MA, 1952).
Solbrig Otto Thomas, and Solbrig, Dorothy J., *So Shall You Reap: Farming and Crops in Human Affairs* (Washington, DC, 1994).
Wallerstein, Immanuel, *The Modern World System: Capitalist Agriculture and the Origins of the European World Economy in the 16th Century* (New York, 1972).
White, Richard, *The Roots of Dependency: Subsistence, Environment, and Social Change among the Choctaws, Pawnees, and Navajos* (Lincoln, NB 1983).
Williams, Michael, *Americans and their Forests: A Historical Geography* (Cambridge, 1989).
Wolf, Eric, *Europe and the People Without History* (Berkeley, CA 1982).
Worster, Donald, *Nature's Economy: A History of Ecological Ideas* (Cambridge, 1977).
Viola Herman, and Margolis, Carolyn, eds., *Seeds of Change: A Quincentennial Commemoration* (Washington, DC, 1991).

1

Diffusion of MesoAmerican Food Complex to Southeastern Europe

Jean Andrews

THE Columbian quincentennial celebration has brought an intensified interest in the related diffusion of ideas, adaptive strategies, material culture, and domesticated plants of the New World. One crucial element that achieved a widened distribution as a result of the post-Columbian exchange was the traditional Mesoamerican food complex of maize, beans, squash, and peppers (*Capsicum*), to which the turkey might be added. Oddly, the Ottoman Turkish Empire, especially Anatolia, rather than Iberia became a center of diversity for squashes, pumpkins, popcorn, and possibly other American crops, which presents the puzzling Anatolian mystery (Anderson 1958).

My own work, focused on the spread of the five domesticated capsicums—*Capsicum annuum* var. *annuum* Linne, *Capsicum chinense* Jacquin, *Capsicum frutescens* Linne, *Capsicum pubescens* Ruiz & Pavon, and *Capsicum baccatum pendulum* (Willdenow) Eshbaugh—suggests that peppers diffused as part of this complex, that the spread to the Old World was far more complicated than is usually assumed, and that the circuitous routes by which the complex reached Anatolia and southeastern Europe largely bypassed the western Mediterranean. My findings also suggest, improbably, that the Portuguese and Turks were far more influential than the Spaniards in the diffusion of the Mesoamerican plant complex, even though the source lay in the Spanish colonies and the complex was discovered by Columbus on several voyages, probably including the first. I was led to these conclusions by the initially troubling fact that the prevalent pepper brought to their Atlantic islands and India by the Portuguese was the Mexican-derived C. *annuum* var. *annuum* rather than the South American-West Indian-Brazilian C. *chinense*, popularly called *ají*.

What is generally agreed is that the diffusion of *Capsicum* and the related complex occurred with great rapidity. Their spread to Africa and Asia occurred in such a short time that centuries later Europeans thought they had originated in the Orient. Nicholas J. Jacquin in 1776 named a new *Capsicum* species *chinense* because he thought it had originated in the Orient. In 1542 the first European illustrations of peppers were published in a German herbal. Such books listed plants and their medical properties and were written for the most part by medical doctors. The herbal by Leonhart Fuchs (1543) proves that peppers were known in central Europe no more than a half-century after the first Columbian voyage. All appear to be the Mesoamerican *C. annuum* var. *annuum*, which implies an even more rapid diffusion that almost certainly began before the conquest of Mexico by Cortés between 1519 and 1521. Moreover, though published in 1542, the herbal may have been written as early as 1538, which allows less than two decades for the diffusion from Mexico (Sauer 1969, 148). The herbal included illustrations and descriptions of the Mesoamerican squash, beans, and maize.

Pre-Columbian Diffusion and Columbian Discovery

This troubling narrow time frame rests on false assumptions concerning the pre-Columbian distribution of *C. annuum* var. *annuum*. I propose that Columbus encountered the plant perhaps several times. On New Year's Day 1493, near Navidad, Española, he recorded a momentous circumstance in his journal: the pepper that the natives used as spice was more abundant and valuable than either black pepper or melegueta (grains of paradise) peppers. The three peppers have no botanical relationship with one another (Fig. 1). Melegueta pepper, of the ginger family, is native to Guinea, Africa, and was known in Venice during the thirteenth century as a less expensive substitute for black pepper, which is of Indian origin. Melegueta was also known as Guinea or ginnie pepper. Soon after the Columbian discovery, several American chillies of the cayenne type became established in the Portuguese colonies of Guinea and were likewise called ginnie pepper, while in the Brazilian colonies tiny local capsicums were referred to as malaguetas by the African slaves transported from Guinea. All this made for a confusing situation.

Columbus left a recommendation that people who were to remain on Española collect as much of the local capsicum pepper as possible (Morison 1963, 142). Two weeks later, near Samana Bay at the other end of the island, Columbus continued to record the crops he was seeing, one of which was the native pepper. The account of that voyage by the royal historian Peter Martyr is of relevance. He not only wrote of Mesoamerican maize being grown in the islands in 1492 but also mentioned two types of peppers. "The *sweet pepper* [my emphasis] is called *boniato*, and the hot pepper is called *canibal* [apparently ají], meaning sharp and strong" after the characteristics of cannibals (d'Anghiera 1965, 532–533; Andrews 1984, 4). Today the *C. chinense* of the West Indies are known for their pungency, and it is improbable that the reference is to a sweet pepper of that species. It must refer instead to a Mesoamerican *C. annuum* var. *annuum*, which already had sweet varieties at that time.

On his fourth voyage, Columbus landed on the Honduran and Nicaraguan shores, where he encountered maize, roots, and "victuals like they eat

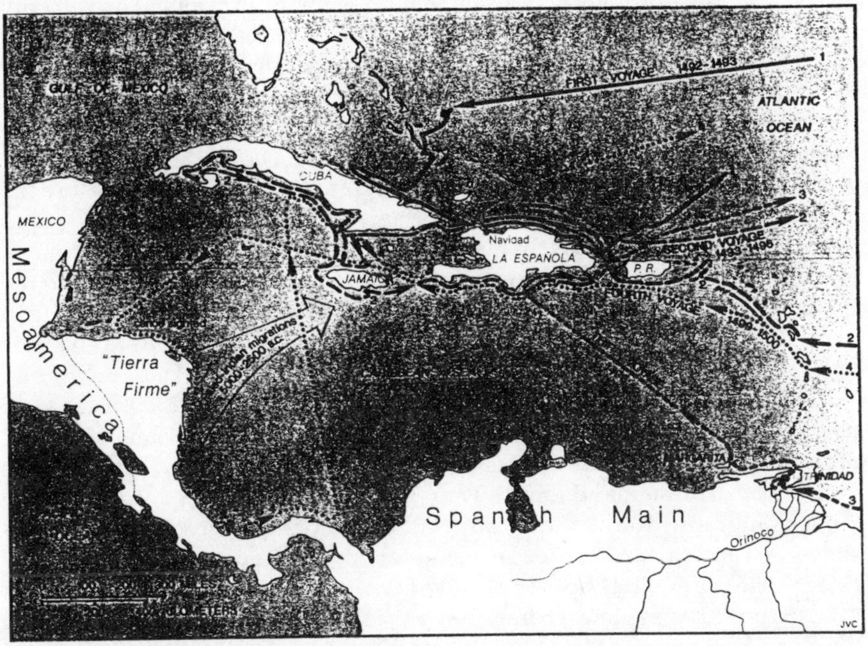

FIG. 1.—Pre-Columbian migration path from mainland to Caribbean islands and the routes of the four Columbian voyages.

in Española," along with woven cotton fabrics. In fact, there was such a wide selection of indigenous goods aboard a native trading vessel that Columbus exclaimed "Thanks to God, that he has given us a sample of all the things of that land without danger or fatigue to our people" (Columbus 1947, 274). Capsicums were probably included, as Columbus commented that all the natives knew red pepper. If so, it would almost certainly have been *C. annuum* var. *annuum*.

Later, in 1502, when the Spaniards arrived in Panama, they found maize, beans, and perhaps peppers, which had gradually diffused from Mexico through Central America (Columbus 1947, 296). From there the pre-Columbian route of diffusion had led eastward across northern South America and thence northward into the Antilles, perhaps borne in part by the Arawak and Carib Indians during their migrations (Sauer 1966, 54). Concurrently, following the same overland trail, domesticated turkeys had reached at least as far south as Costa Rica. With the immediate post-Columbian help of the Spaniards these foods quickly went from the isthmus to all of the islands.

It is my deduction that the Spaniards probably found *C. annuum* var. *annuum* in some of the islands, specifically Española, and certainly found it in Central America by 1502, and from there quickly spread it throughout the West Indies. It is highly possible that Mesoamerican peppers had already reached some parts of the West Indies at the time of the first Columbian voyage by traveling as the handmaiden to maize over pre-Columbian trade routes along the Central American corridor, across northern South America, and northward through the lesser Antilles to Española (Sauer 1966). Though highly probable, the assertion by W. C. Sturtevant (1961, 71) that *C. annuum* was being cultivated in the gardens of Taino Indians on Española before the arrival of Columbus is questionable, because the only cited source is a paper by Heiser and Smith (1953) that recognized only *C. annuum* and *C. frutescens.* Not until 1957 did botanists recognize *C. chinense,* which was brought to the West Indies during Arawak migration from South America and which had long been considered the only species growing on those islands at the time of the Columbian discovery (Heiser and Smith 1957).

Perhaps the diffusion occurred even earlier—5000 to 2500 B.C.—with paleo-Indians from Mesoamerica across the mid-Caribbean island chain; by birds, their natural dispersal agents; or through both agencies (Cruxent and Rouse 1969; Millman and Emery 1986; Watts 1987). *C. annuum* was available to those migrating hunters and gatherers, as it had been cultivated by humans since 5000 B.C. in the Tehuacán valley of Mexico and probably elsewhere (MacNeish 1967). They spread rapidly after the contact on the mainland in 1502. Supportive evidence comes not only from the chroniclers Martyr and Chanca (1870) but also from Fray Bartolomé de las Casas (1967), who recorded some botanical observations in *Apologética Historia,* which is assumed to have been drafted on Española between 1526 and 1529 (Fernandez 1971, 84). In Seville, some twenty-five years after making the first draft, Las Casas separated the two sections and enriched them with numerous interpolations (1967, xxxiv). He remarked that the peppers he saw were "like those already known in all Spain" (Las Casas 1967, 58), and it would be very significant to know when he was referring to them growing in Spain—at the time of his first sighting, at the time of his first draft, or at the time of the final version. During his second stay, 1508–1515, in the West Indies, he farmed a grant of land on the Arimao River, Española, between 1513 and 1514. I tend to think that he made his agricultural observations during that sojourn, which was well before the discovery of Mexico, an interpretation that strengthens my Mesoamerican trade thesis.

Two of the three capsicums that Las Casas describes in Española fit the illustrations in the German herbal. He had seen two kinds of peppers being cultivated in the West Indies (Las Casas 1967, 58). One ají, which is now the name for all capsicums in the Dominican Republic, was long, red, and finger shaped; the second was globular like a cherry and more pungent than the first type. A third was a wild pepper that bore very small fruits. The first of Las Casas's descriptions fits the cayenne-type *C. annuum* var. *annuum,* figured as "Langer Indianischer Pfeffer" in the German herbal, rather than any *C. chinense.*

Portuguese Trade Corridors

Why might that Mesoamerican complex have taken earlier and deeper root in Anatolia? The answer seems to lie with the Portuguese, who acquired peppers and other items from the Spanish Main despite the mutual trade exclusion that prevailed.

Few Spanish trading ships actually came to the Antilles in the crucial early years, and in the first half of the sixteenth century exchange between Seville and the New World was very limited. In contrast, trade between Portugal and the New World was much greater. Although the Treaty of Tordesillas assigned most of the New World to Spain, the elimination of Portugal from the Spanish arena remained more theoretical than real in the early sixteenth century. The paucity of Spanish shipping allowed the Portuguese surreptitiously to enter the region with their African slaves and other trade goods (Watts 1987). That illicit trade was aided by Spanish subjects in the New World, not from inherent disloyalty but from dire necessity (Means 1935, 61). Moreover, relatively cordial relations existed between Spain and Portugal in the late fifteenth and early sixteenth centuries. As a result, communication existed between Lisbon, Seville, and Barcelona (Walsh 1939) that could have contributed to the rapid diffusion of *Capsicum* and other Mesoamerican domesticates.

Another form of communication that could have transferred New World seed from Spain to Portugal soon after its introduction to Spain was the trade in small grains. At the time of the conquest and throughout the period of rapid diffusion of New World economic plants, 1492 to 1550, bread grain was deficient, and Portugal depended on a well-established trade with Spain and North Africa for its cereal supply (Pounds 1979, 63). Surely the traders would have been aware of the new intertilled symbiont—squash, beans, maize, and its attendant peppers.

The Portuguese, then, could have acquired *C. annuum* var. *annuum* and the other seeds either in Iberian ports or in the Spanish Main after initiation of the African slave trade (Watts 1987). Reaching Lisbon by one or the other of those routes, the next crucial link in the diffusion of *Capsicum*, ultimately even for Europe itself, led improbably from Portugal to the Atlantic islands, Africa, and India (Miracle 1967). The pepper and perhaps the turkey almost certainly accompanied maize, beans, and squash on that route. Evidence suggests that the Portuguese started growing maize and peppers in the Azores and Madeira as well as Guinea and Angola very early.

The herbalist John Gerard (1974, 292) described the capsicums in England and called them "ginnie pepper." He added that the peppers introduced to Spain and Italy had come from "foreign countries as Ginnie, India, and those parts into Spain and Italy." Soon the term ginnie pepper would become one of the several inappropriate names for *C. annuum* var. *annuum*. Maize reputedly arrived in Cape Verde, São Tomé, and Príncipe as early as 1502 (Jeffreys 1953, 966; 1954, 193; 1975, 35), the same year Columbus began his

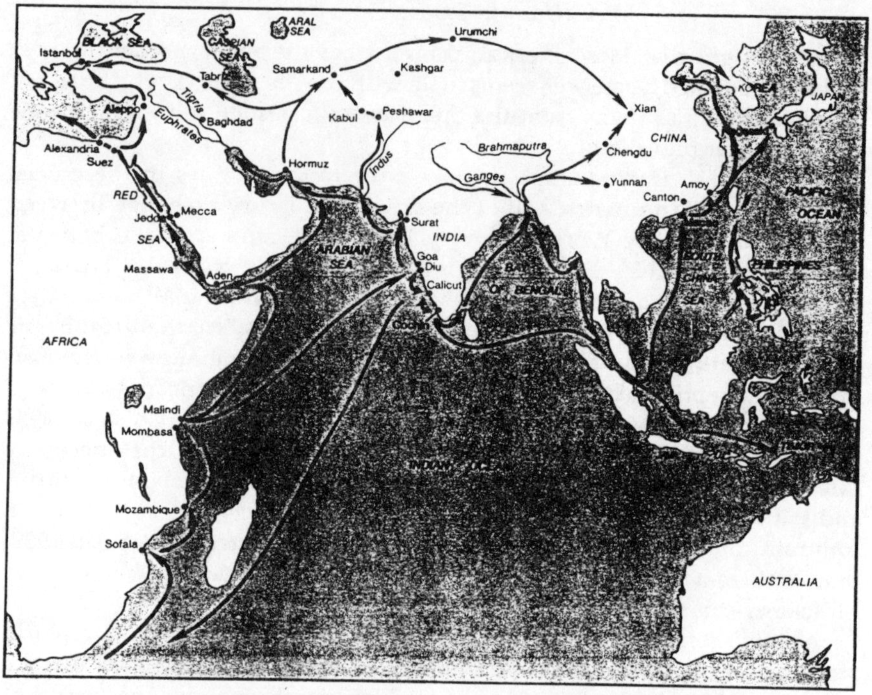

FIG. 2.—Diffusion of *Capsicum* through Asia.

fourth voyage, which took him to the mainland of Mesoamerica for the first time. If domesticated *C. annuum* var. *annuum* peppers went with maize to the Portuguese Atlantic islands at that time, the Portuguese would have had to acquire them from an earlier Spanish West Indian source, which implies that peppers were being cultivated on Española at the time of the discovery. From Portuguese Africa, capsicums soon reached the colonies in India, probably accompanied by the maize-squash-bean complex. Writing in 1576, the Flemish botanist Matthias de Lobel observed that capsicums had been brought to Goa and Calicut at a very early date. From that observation George Watt (1889) declared, "There can be no doubt that the Portuguese very possibly began exporting them in competition with black pepper (Piper nigrum)" (Fig. 2).

A Portuguese official in India from 1500 to 1516 reported that an abundance of *milho grosso* (maize) had been exported from Gujarat (Barbosa 1918). The preference for the Mesoamerican peppers caused any earlier introduction to India of other chilli-pepper species such as *C. chinense* to be replaced. The new spice was welcomed by Indian cooks who, accustomed to pungent black pepper and biting ginger, produced hot, spicy foods. The Mesoamerican pepper provided more heat with less grinding and expense. It grew readily and fruited abundantly in a sympathetic environment. The easily cultivated and naturalizing *C. annuum* var. *annuum* was a welcome addition to the native spices, whose restrictive cultural requirements and high costs put them in a luxury category. Into the curries they went.

Remarkably soon thereafter, the Mesoamerican food complex sailed the sea-lanes to Malacca and Indonesia with coasting Chinese, Gujarati, and Arabic traders. Within a brief period, considering communications and travel time in the sixteenth century, those exotic edibles were also added to the baggage carried so laboriously from the Gangetic delta at the Bay of Bengal through Burma and Chengdu in Szechuan (Ho 1955; Gode 1960, 290). Yet another possible route from the Indian Ocean began at Portuguese Diu and Surat on the Gulf of Cambay, went inland over a low divide to tributaries of the Ganges, then up the Brahmaputra River, and across the Himalayas to Szechuan. Moreover, another possible avenue to China was accessible: Portuguese controlled the mouth of the Indus River, which led to the Himalayan silk routes. These new foods melded into the gardens and cuisines of China, India, Indonesia, and other areas of the Far East, which were mainly vegetarian.

From Africa and India to Anatolia

The diffusion of Mesoamerican foodstuffs from India to the Orient is reasonably well known and noncontroversial. Less orthodox is my proposal that the eastern Mediterranean, Balkans, and even parts of central and western Europe also received the Mesoamerican complex from India and East Africa (Boxer 1969). The route from India to Europe most likely followed well-established ancient sea-lanes. Coasting from Goa, Diu, Surat, or Hormuz to the Persian Gulf or in convoys across the Indian Ocean to the Red Sea, Mesoamerican peppers, maize, beans, and cucurbits joined the oriental spices on two ancient medieval trade routes, the Aleppo and the Alexandria routes, long used by Turks, Arabs, and other Muslims for the lucrative trade from monsoon Asia to the Levant. They also could have traveled from the Indian Ocean up the Indus River to Afghan Kabul, meeting the historic course followed by Marco Polo, then westward along the toilsome way through Persia to Turkey. Most likely, the Arabs were the first middlemen in the diffusion to Europe who passed the Mesoamerican complex to the Ottoman Turks, although it should be remembered that Portuguese traders were very active at least as far as Hormuz at the entrance to the Persian Gulf and Massawa on the Red Sea, if not beyond. Hormuz was a port open to every type of immigrant, every form of commerce, and every kind of smuggling, whether by Venetians, Armenians, Turks, or Portuguese renegades who departed in astounding numbers for Turkey, where their knowledge of the East Indies was an important asset in the clandestine trade. Through Hormuz the best of India reached Venice (Braudel 1976, 564).

A Turkish document, written between 1498 and 1513, mentions a New World plant, the common bean, for the first time. By 1539 maize was already playing a pivotal role in the Turkish Empire, and squash was known (Parry and others 1976, 89; Johnson 1981, 130). True, these Mesoamerican foodstuffs, including peppers, could have reached Turkey from Spain via the Ottoman contacts with exiled Spanish Moors or with expelled Iberian Jews, who distributed them throughout North Africa all the way to Egypt. However, in view of the nature of trade and the extent of warfare in that crucial period, these are not the most likely scenarios. It is also conceivable that the trade in firearms and gun flints between Spain and the Turks accomplished incidental *Capsicum* diffusion, but I am not inclined to accept such a plot (Witthoft 1966).

Much more plausible, I propose, is that capsicums, together with turkeys, squash, maize, and beans, arrived in Turkey from Portuguese West African colonies by way of India. During that transfer, there could have been an introduction to Ethiopia as a result of the establishment of the first (1520–1526) or second (1541) Portuguese embassy to Massawa, Abyssina. Maize was probably introduced there by Turks via Harrar and by Portuguese sources via Massawa (Wright 1949, 80). European terms such as Egyptian grain, Turkish wheat, Arabian grain, and granoturko for maize; Turkish peppers for capsicums; pomo di Moro for tomatoes; and turkey for the turkey bird in diverse languages suggest the importance of the Ottomans in the European diffusion, while use of dinde for turkey, wheat of India for maize in French, and Indian peppers reveals the more remote intermediary source of the Mesoamerican complex (Stoianovich 1966). Too, the medieval spice market, still in Istanbul, has been known throughout its history as the Egyptian market, and in the old days most of the spice sold there traveled overland from the Red Sea to Aleppo and thence to Antioch. The trading vessels from the west coast of India crossed to the Red Sea where their precious cargoes were transferred to camel caravans, whose caravansaries along the route still stand as monuments to that ancient spice trade.

After the Portuguese, the Ottoman Turks were probably more responsible than any other group of people for the distribution of Mesoamerican foodstuffs (Fig. 3). In the aftermath of forays to Hormuz and beyond, the Turkish armies could easily have brought peppers with them along the medieval trade routes across Asia Minor to the Black Sea and into Hungary, which they conquered in 1526. A short time after the first Turkish siege of Vienna in that year, peppers were recorded in central Europe. The Turks, like others later, probably also recognized the value of Mesoamerican maize as livestock feed, and they had maize grown by peasants in garrison gardens in the expanding Ottoman Empire to feed the large numbers of animals required to transport the huge armies and supplies. Maize later came into use as a necessity food for troops and peasants alike. As a result it became quickly established in newly conquered Turkish territories, including Greece and elsewhere in the Balkans, perhaps as early as the 1520s. After the Turkish armies eventually departed, the peasants there continued to grow these crops

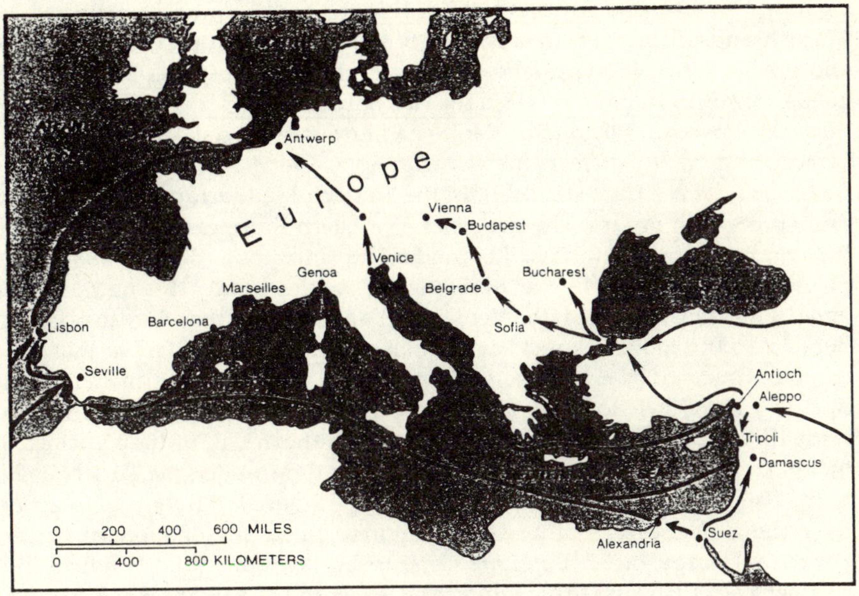

FIG. 3.—Diffusion of *Capsicum* through Europe.

in their own gardens, not only because they were more productive than native crops but also because they were still largely unknown to the noble landowners and hence untaxed. *Capsicum* peppers also traveled in the military baggage train and were grown in the same gardens as the traditional Turkish staples (Stoianovich 1966).

The diffused Mesoamerican crop-poultry complex is even now more deeply accepted in former Turkish-ruled southeastern Europe than anywhere else on the continent. Examples include the flocks of turkeys on the Peloponnesus, the use of paprika in Hungarian dishes, the South Slav attention to maize, and the role of maize in Balkan folklore since the eighteenth century (Stoianovich 1966). The Andean potato eventually prevailed in the cooler lands of northern and western Europe, but the Balkans owe far more to Mesoamerica. As Sauer (1969, 151) noted, "Cultivated seed plants originating in the New World are more significant in the eastern end of the Mediterranean and in Italy than they are in Spain, and seem to have been so as far back as there is knowledge of them."

Nor did the eastern-based diffusion halt at the borders of the Ottoman Empire. Often aided by Venetian traders, capsicums soon spread into central and western Europe. German herbalists, as previously noted, acquired peppers by about 1540, apparently through the Turks, and capsicums reached England by 1548. Most often they appeared in northern and western Europe under names such as Turkish, Calicut, or India pepper.

Italy represented a diffusionary divide in the Mediterranean area between Turkish and Spanish influences. Venice depended on the Turks for spices and wheat, while Florence and Genoa relied on the Iberians for imported goods. Communications between the two halves of the sea was slow. During the crucial period 1492 to 1542, Venice had few commercial connections with Lisbon, and trade between the eastern and western basin of the Mediterranean was at a virtual standstill. In the western Mediterranean, Genoa had trade links with Bruges. The two poles of western European commerce were Italy and the Low Countries, then under Hapsburg rule (Braudel 1982, 419).

Although the Turks were a presence in southern Italy during the 1500s and dealt extensively with the Venetians, I do not think that they introduced peppers to the Italian peninsula or acquired peppers from Italy at that time or place. Instead, sweet peppers and tomatoes came to western Italy from Spain by 1535 (Oviedo 1950). The historical record of western Mediterranean trade linked Seville and Barcelona with the southern half of Italy, including Sicily. Too, the pungent peppers of the cayenne type, which the Turks carried to the Balkans, were not the sweet-pepper type favored in Italy, and selection over the centuries has reduced the original pungency of modern Balkan paprika. A large, sweet *Capsicum* came to be favored by both Italian and Spanish cooks. The usage of peppers and spices in the Balkans is more Indian in nature, while that of Italy and Spain is characteristically Mediterranean.

Conclusion

The Mesoamerican crop complex of maize, beans, squash, and peppers was present in the West Indies when Columbus arrived, was acquired by the Portuguese about 1500, and diffused by way of the Portuguese Atlantic islands, Angola, Mozambique, India, and the Ottoman Empire to the Balkans by the 1530s. In the Balkans the complex has received its most profound European diversity and acceptance, apparently in company with the turkey. Improbably, people of the eastern, not western, Mediterranean display the greatest imprint of the Mesoamerican agricultural and culinary heritage, and the Portuguese were the most influential dispersers of that food complex. Part of the reason could be the low status of agricultural work in Spain at that time, as evidenced by the aristocratic indifference Spaniards exhibited to that activity, in contrast with Portuguese enthusiasm for highly specialized horticulture and a particular gift for agricultural botany.

References

Anderson, E. 1958. Anatolian mystery. *Landscape* (spring):14–16.

Andrews, J. 1984. Peppers: the domesticated capsicums. Austin: University of Texas Press.

d'Anghiera, P. M. 1965. Decadas del Nuevo Mundo, por Pedro Martir de Angleria, primer cronista de Indias, vol. 2. México, D.F.: Jose Porrúa e Hijos.

Barbosa. D. 1918. Book of Duarte Barbosa: an account of countries bordering on the Indian Ocean and their inhabitants [1518], vol. 1. London: Hakluyt Society.

Boxer, C. R. 1969. Portuguese seaborne empire: 1415–1825. London: Hutchinson.

Braudel, F. 1976. Mediterranean and the world in the age of Philip II. 2 vols. New York: Harper and Row.

———. 1982. Structures of everyday life. New York: Harper and Row.
Chanca, D. A. 1870. Letter to the municipal council of the city of Seville, Spain [1497]. Select letters of Christopher Columbus, ed. R. H. Major, 19-71. 2d ed. London: Hakluyt Society.
Columbus, F. 1947. Vida del almirante don Cristóbal Colón [1571]. México: Fondo de Cultura Económica.
Cruxent, J. M., and I. Rouse. 1969. Early man in the West Indies. *Scientific American* 221(5):42–69.
Fernandez, M. G. 1971. Life of Las Casas. Bartolome de las Casas in history, eds. J. Friede and B. Keen, 67–125. DeKalb, Ill.: Northern Illinois Press.
Fuchs, L. 1543. New kreuterbuch (De historia stirpium [1542]). Basel: Isingrin.
Gerard, J. 1974. Herball or general historie of plants [1597]. Norwood, N.J.: Theatrum Orbis.
Gode, P. K. 1960. History of maize (maka) in India between A.D. 1500-1900. *Studies in Indian Cultural History* 2(32):283-294.
Heiser, C. B., and P. G. Smith. 1953. Cultivated *Capsicum* peppers. *Economic Botany* 7(3):214-227.
———. 1957. New species of *Capsicum* from South America. *Brittonia* 10:194-201.
Ho, P. T. 1955. Introduction of American food plants into China. *American Anthropologist* 55:191–201.
Jeffreys, M. D. W. 1953. Pre-columbian maize to Africa. London: Macmillan Journals. 172(4386): 965-966.
———. 1954. Maize names. *Uganda Journal* 18(2):192-194.
———. 1975. Pre-Columbian maize in the Old World: an examination of Portuguese sources. Gastronomy: the anthropology of food and food habits, ed. M. L. Arnott, 23-66. The Hague: Mouton.
Johnson, M. 1981. North Balkan food, past and present. Natural and regional styles of cookery, proceedings, Oxford symposium, 122-133. London: Prospect Books.
Las Casas, B. de. 1967. Apologética historia sumaria, 1520-1561, vol. 1, ed. E. O'Gorman. México, D.F.: Universidad Nacional Autónoma de México.
MacNeish, R. S. 1967. Summary of subsistence. Prehistory of the Tehuacan valley, vol. 1: environment and subsistence, ed. D. S. Byers, 290-309. Austin: University of Texas Press.
Means, P. A. 1935. Spanish Main: focus on envy 1492-1700. New York: Charles Scribner's Sons.
Millman, J. D., and K. O. Emery. 1986. *Science* 162:1122.
Miracle, M. P. 1967. Agriculture in the Congo basin: tradition and change in African rural economics. Madison: University of Wisconsin Press.
Morison, S. E. 1963. Journals and other documents on the life and voyages of Christopher Columbus. New York: Heritage Press.
Oviedo y Valdes, G. F. de. 1950. Sumario de la historia natural de las Indias, ed. J. Miranda. México, D.F.: Fondo de Cultura Económica.
Parry, V. J., H. Inalcik, A. N. Kurat, and J. S. Bromley. 1976. History of the Ottoman Empire to 1730. Cambridge history of Islam and new Cambridge modern history, vol. 2, ed. M. A. Cook, 62-103. Cambridge, U.K.: Cambridge University Press.
Pounds, N. J. G. 1979. Historical geography of Europe 1500-1840. Cambridge, U.K.: Cambridge University Press.
Sauer, C. O. 1966. Early Spanish Main. Berkeley and Los Angeles: University of California Press.
———. 1969. Seeds, spades, hearths, and herds. Cambridge, Mass.: MIT Press.
Stoianovich, T. 1966. Le mais dans les Balkans. *Annales: Economies Societes Civilizations* 21(2):1026–1040.
Sturtevant, W. C. 1961. Taino agriculture. Evolution of horticulture systems in native South America: causes and consequences, ed. J. W. Ibert, 69-82. Caracas: Sociedad de Ciencias Naturales LaSalle.
Walsh, W. T. 1939. Isabella of Spain. London: Steel and Ward.
Watt, G. 1889. Dictionary of the economic products of India, vol. 2. Delhi: Cosmo [1972 reprint edition].
Watts, D. 1987. West Indies: patterns of development, culture and environmental change since 1492. Cambridge, U.K.: Cambridge University Press.
Witthoft, J. 1966. History of gunflints. *Pennsylvania Archaeologist* 36:1-49.
Wright, A. C. A. 1949. Maize names as indicators of economic contacts. *Uganda Journal* 13:61.

2

The Rise of Maize as a Major Crop Plant in the Philippines

J.E. Spencer

The domestication and diffusion of crop plants as a major theme in cultural geography has several sub-themes of consuming interest. Among the aspects that have attracted serious attention are the places of domestication, the varying techniques and procedures, the times and routes of diffusion, the appearance and breeding of divergent varieties, and the roles of newly accepted plants in altering cropping combinations in particular regions. There is still another sub-theme that often has been touched upon lightly but which has not received the attention requisite to a full understanding of the evolutionary development of regional agricultural economies. This is the sequence of developmental occurrences taking place in a particular culture region after a crop plant reaches that area by some pattern of diffusion. There often are problems of integration of new crops into cropping systems, of dietary or textile-use acceptance, and time intervals during which acceptances are worked out. On the contrary, certain crop plants catch on with amazing rapidity and soon become so integrated into habituated use as to cause their new hosts to react that they have always had them.

We know rather little about the processes of assessment of potential value and acceptance of new plants by the recipients thereof. What is the perception of the new plant and its product? How does acceptance of a new plant come about? Wherein does the acceptance lie, with the producer of the crop who may be able to grow it to maturity, or with the consumer who must determine how and when to use it? How long does acceptance take, and what changes in regional technology must occur to promote the final acceptance of an introduced crop plant? If we knew the full answers to some of these kinds of questions it might be that the modern processes of upgrading traditional agricultural economies could be greatly improved.

The literatures of geography, economic history, ethnology and regional economy make many passing references to the times and sources of new crop plant introductions in many parts of the world. The later regional importance of particular introduced crop plants is well treated, but there often is a time gap between the date of introduction and the date at which the introduced crop plant assumes an important role in agricultural economy. It is to the developments during this time gap that this paper is directed for one particular crop plant in one geographical region, maize in the Philippines.

Maize today holds the second ranking position, behind rice, among the food staples of the Philippines. The agricultural landscapes of the Philippines are not totally mixed ones, with both rice and maize being grown in some variable proportion by farmers everywhere, but there are regional concentrations of rice landscapes and maize landscapes. Although almost all contemporary Filipinos eat a little of both rice and maize, as food staples there is regional dominance by one or the other. There are Filipinos whose staple food is rice, and who eat maize only as a green vegetable or a snack food, and there are Filipinos whose staple food is maize in the form of "corn rice", coarsely ground maize that is cooked in the same way that rice is cooked. In the dominantly rice-growing and rice-eating regions maize is distinctly a minor crop treated as a green vegetable, and in the maize-growing and maize-eating regions rice is often scantily grown but is an interregional import eaten only at particular times by the maize-eaters or by urban classes who are in-migrants from rice-consuming regions. The increase in maize production in the twentieth century has been very marked, and relatively greater than the increase in rice production, so that maize has been plentiful and rice normally has been deficient in supply. Maize is the cheaper grain in price as a result of the long-term consumer preference for rice in many local regions in the Islands, but it is likely that many Filipinos have somewhat unwillingly increased their tolerance of maize as a dietary staple. Recent government campaigns to increase the acceptance of maize may be currently enlarging the population of maize consumers.

Stated generally, the primary "corn rice" consuming region lies in is the central Philippine Islands of Cebu, Negros, Bohol, Masbate, northern and southern Leyte, and the northern coast and upland interior sections of Mindanao, including most of the smaller islands within the central Visayan zone. As the expansion of commercial maize production has grown in the present century, the Cagayan Valley of northeastern Luzon has become an important producer of surplus maize that is shipped to the Visayan food-deficit zone. And as the settlement of Mindanao has proceeded both the Cotabato Valley of southwestern Mindanao and the Davao Gulf hinterland in southeastern Mindanao have become important commercial producers supplying the Visayan food-deficit zone. The Zamboanaga Peninsula and interior Mindanao are both producers of surplus maize that moves toward the Visayan Islands and are also regions in which many recent settlers from the Visayas belong to the "corn rice" dietary group.

The introduction of maize

The position is here taken that maize was not present in the Philippines in the pre-Columbian era, specificially meaning prior to the arrival of Ferdinand Magellan in the Philippines in 1521.[1] The actual date of the first introduction of maize in uncertain, then, as will be seen in the following paragraphs. The term maize is employed in this paper, but in English in the Philippines the term corn is almost universally used, whereas *mais* is the vernacular language term in almost all indigenous Philippine languages.

It is very doubtful that Magellan's ships initially loaded stocks of maize on setting out for the Philippines, since they were outfitted in Spain. It is possible that some

[1] The negative stand here taken does not argue that maize was not present in southern Europe, Africa or western Asia in the pre-Columbian period. This older, standard view was well presented in I. H. Burkhill, *A dictionary of the economic products of the Malay Peninsula* (Kuala Lumpur, Malaysia 1935 and 1966) 2327–34 in the repaged 1966 edition; C. O. Sauer, Maize into Europe *Acts 34th International Congress of Americanists* (Vienna 1960) 777–88, made out a good linguistic case for the pre-Columbian presence of maize in the Old World including the Philippines; M. D. W. Jeffreys, Pre-Columbian maize in Asia, pp. 376–400 of C. L. Riley, J. C. Kelley, C. W. Pennington, and R. L. Rands (Eds), *Man across the sea, problems of pre-Columbian contacts* (Austin, Texas 1971), added to the linguistic case; I, however, accept the positions stated (for the Philippines at least) by E. K. Reed, Commentary: section I, pp. 106–11 and H. G. Baker, Commentary: section III, pp. 428–44 of Riley, *et al.*, *op. cit.*, that the case for pre-Columbian maize in the Old World is not incontrovertibly proven

maize stocks were acquired on the South American coast where supplies were taken on, although maize was not so specified. Certainly the food stocks were in poor shape as the expedition arrived in the Philippines.[1] After sighting the island of Samar, the expedition touched at several islands, and made various local contacts in which there were exchanges of food and presents prior to the major landing on the east coast of Cebu Island, where there existed a periodic regional trade centre for the central Philippines; in the nineteenth century this centre has grown into the city of Cebu. After Magellan was killed in a skirmish on nearby Mactan Island, the expedition touched at a number of other islands before leaving the Philippines.[2] The members of Magellan's expedition spent little time ashore at the various landfalls, and remained at no single location long enough to plant and harvest a maize crop. There is no absolute evidence, positive or negative, regarding the release of any seed maize at any of the points of contact, but the circumstantial evidence is strongly against such an introduction of maize to the Philippines.

A second expedition from Spain in 1525 sailed under Juan Gofre de Loaisa, but the flagship of the seven-ship fleet was the only one to make a landfall in northeastern Mindanao in May 1526.[3] A local scarcity of food at the point of landfall, plus hostility displayed by the inhabitants, caused the flagship to set sail for Cebu, but contrary winds carried the Spaniards southward, via the Sarangani Islands, to the Moluccas in December 1526, where they remained until rescued. Available records yield almost no information on what they did between May and December, and no information as to whether the fleet carried any maize.

Alvardo de Saavedra Ceron commanded a three-ship expedition despatched by Cortez from Mexico in October 1527 to find de Loaisa.[4] Only the flagship of this

[1] D. D. Brand, Geographical exploration by the Spaniards, pp. 109–44 of H. R. Friis (Ed.), *The Pacific Basin, a history of its geographical exploration* (New York 1967) detailed the food stocks supposedly loaded in Spain; A. Pigafetta, First voyage around the world, vols 33 and 34, ref. to vol. 33, p. 27 of E. H. Blair and J. A. Robertson (Eds), *The Philippines, 1493–1898* (55 vols, Cleveland 1903–9) listed some of the supplies taken on in South America but did not include maize. Not all versions of Pigafetta read alike in translation and, in addition to the above version, I have used P. S. Paige (trans.), *The voyage of Magellan, the journal of Antonio Pigafetta* (Englewood Cliffs 1969); see also A. M. Molina, *The Philippines through the centuries* (Manila 1960) **1**, 32–4

[2] The names of islands touched at, and the precise details of Magellan's activities, vary slightly in different translations and different secondary accounts. E. G. Bourne, Historical introduction, vol. 1, pp. 322–28 of Blair and Robertson, *op. cit.*, cited only some names and certain details of the voyage; C. Benitez, *History of the Philippines, economic, social, political* (Boston 1926), employed modern place-names and cited particular details as interpreted from the rendition of Pigafetta in Blair and Robertson, *op. cit.*; Brand, *op. cit.*, using a number of published and manuscript sources, provided the longest list of places visited; Molina, *op. cit.*, 33–47, derived still a different version from manuscript and published sources

[3] For the several Spanish expeditions to the Philippines during the two and a half decades following Magellan's expedition I have depended on secondary sources for accounts of the various activities; Benitez, *op. cit.*, 44; E. H. Blair and J. A. Robertson, Introduction, vol. 1, pp. 30–1, Preface, vol 2, pp. 11–12 and Expedition of Garcia de Loaisa, 1525–6, vol. 2, pp. 25–35 of Blair and Robertson, *op. cit.*; Brand, *op. cit.*, 119–21; Molina, *op. cit.*, vol. 1, pp. 48–9; J. Montero y Vidal, *Historia general de Filipinas* (3 vols, Madrid 1887–95), vol. 1, pp. 22–4

[4] Benitez, *op. cit.*, 44–5; E. H. Blair and J. A. Robertson, Voyage of Alvaro de Saavedra, 1527–1528, vol 2, pp. 36–43 of Blair and Robertson, *op. cit.*; Brand, *op. cit.*, 121; Molina, *op. cit.*, vol. 1, 49; Montero y Vidal, *op. cit.*, vol. 1, 25–6

fleet survived, and it touched a landfall on the eastern coast of Mindanao. Here the ship was repaired, provisioned, some survivors of the de Loaisa expedition were picked up, and the flagship was able to track de Loaisa to the Moluccas. The records do not yield acceptable evidence concerning any distribution of maize in the Philippines, but the expedition outfitted in Mexico, and it is likely that some maize was carried as a food staple; it is just possible that barter exchange did release some maize on the Mindanao coast.

On 1st November 1542, Ruy Lopez de Villalobos left Mexico with a fleet of six ships, 360 men and supplies, under instructions to settle in, colonise, trade with and Christianise the Western Islands.[1] Touching first at some point in northern Mindanao in February 1543, where the fleet spent a month, the ships were then driven southward by storms and made a landing on one of the Sarangani Islands, just south of Mindanao, very short of food. Contacts were unfriendly and there were numerous skirmishes between the Spaniards and the local Filipinos, during which each side looted the other of sundry possessions. The Villalobos expedition abandoned Sarangani under protest from the Portuguese, spent almost two months on the southern Mindanao coast short of food and, driven by storms, also ended up in the Moluccas. The record seems to indicate that from the Filipinos the Spanish got porcelains, metal bells, musk, amber, aromatic resins and gold dust; we are not told what they lost. Outfitted in Mexico to settle permanently in the new land, the expedition carried seeds stocks of numerous crop plants, and did make plantings of maize on Sarangani.[2] It is possible that despite the food shortages, the seeds stocks remained intact and that some maize got into southern Filipino hands from bartered or looted stores, but there is no specific evidence of this fact.

Portuguese contacts in the East Indies were quite widespread by 1543, and native interregional traffic between the southern Philippines and several parts of the East Indies was rather regular during the sixteenth century. It is entirely possible that small amounts of seed maize may have reached the southern and central Philippines by the 1540s, via Portuguese sources, entirely independently of any Spanish introduction, although conventional historiography credits the Spanish with the introduction of the crop plant.[3]

[1] Benitez, *op. cit.*, 45; E. H. Blair and J. A. Robertson, Expedition of Ruy Lopez de Villalobos, 1541–46, vol. 2, pp. 45–73 of Blair and Robertson, *op. cit.*; Brand, *op. cit.*, 122–3; Montero y Vidal, *op. cit.*, vol. 1, 26–8

[2] Published records differ concerning the success of the plantings. Benitez, *op. cit.*, 45, without citing a source, stated: "Here they stayed long enough to plant maize, which yielded an abundant crop", whereas E. H. Blair and J. A. Robertson, "Expedition of Ruy Lopez" *op. cit.*, 69, quoted G. de Escalante Alvarado, *Relaciom de viaje que hizo desde Nueva Espana a las islas del poniente Ruy Lopez Villalobos* (Lisbon 1548): "Which was done twice, but it did not come up"

[3] Most of the older histories, such as Benitez, *op. cit.*, Blair and Robertson, *op. cit.*, and others make flat statements to this effect; J. L. Phelan, *The hispanization of the Philippines* (Madison 1959) accepted Spanish introduction of maize; R. R. Reed, Corn, pp. 242–67 of R. E. Huke (Ed.), *Shadows on the land, an economic geography of the Philippines* (Manila 1963), assumed that any permanent introduction of maize owed to the Legazpi expedition, still be referred to in this paper; see D. D. Brand, Geographical exploration by the Portuguese, pp. 145–50 of Friis, *op. cit.*, for a short account of Portuguese travels in and around the Moluccas; for Portuguese trade and traffic movements in the South China Sea and neighbouring waters; see J. C. van Leur, *Indonesian trade and society* (The Hague 1955), and M. A. P. Meilink Roelofsz, *Asian trade and European influences in the Indonesian Archipelago between 1500 and about 1630* (The Hague 1962)

The first successful Spanish settlement contact with the Philippines did not occur until 1565.[1] On 21st November 1564, Miguel Lopez de Legazpi sailed from Mexico with five ships, just under 400 men and settlement supplies. Food stores were taken on in the Marianas, but the expedition was rather short of food staples when it arrived off Cebu Island in February 1565. Demonstrated hostility caused Legazpi to reconnoitre other parts of the Visayan Islands region, and landings were made on several islands before acquired intelligence directed Legazpi back to Cebu in late April 1565. Since the expedition had outfitted in Mexico and was stocked for permanent settlement, it is almost certain that maize was among the food stores and the seed stocks carried aboard the fleet, and it is entirely possible that some maize may have been distributed via gift or barter at one or more of the points touched, although no specific mention of gifts of maize affords details of such exchanges.

Following a militant exhibition off Cebu, Legazpi made a landing that initiated contacts with the local population, exchanges of presents and a small flow of food supplies to the Spanish. The forceful taking of the settlement came next, followed by a drying up of the inter-island flow of goods, so that food shortages promptly developed. Cebu had long been a regional trade centre but, although the east coast of the island was well populated, the island was not a region productive of notable food surpluses, and it depended significantly on food imports. Legazpi soon temporarily moved his headquarters to southeastern Panay Island, since that sector of Panay was a producer of surplus rice and other foods. In 1568 headquarters were returned to Cebu after two more ships arrived from Mexico with orders to do so.[2]

By 1570 the Spanish had carried out many inter-island trips of exploration, and had touched at many of the Visayan Islands in the central Philippines, and had undertaken many ceremonial exchanges of presents. By late 1571 the Spanish had taken control of Manila and had made that city their chief settlement. The shores of Manila Bay comprised a zone of relatively dense population, the hinterland had a good agricultural potential, and it afforded food surpluses in variety, and here the Spanish also met Chinese traders in numbers, affording them access to the trade of the East. Persuasive efforts and militant gestures brought Spanish contact with, and native acquiescence to the Spanish presence in, most of western lowland Luzon by 1576.[3]

The available evidence therefore suggests that any significant Spanish introduction of maize probably occurred during the late 1560s and that by the mid-1570s the extensive Spanish contact with the central Philippines could either have disseminated maize rather widely or facilitated its planting in a few places only.

[1] Benitez, *op. cit.*, 45–6; M. L. de Legazpi, Relation of the voyage to the Philippines, vol. 2, pp. 196–216 of Blair and Robertson, *op. cit.*, Brand, *op. cit.*, 129–30; Molina, *op. cit.*, vol. 1, 54–60; Montero y Vidal, *op. cit.*, vol. 1, 29–43

[2] Benitez, *op. cit.*, 46–53; E. H. Blair and J. A. Robertson, Expedition of Miguel Lopez de Lagazpi, 1564–68, vol. 2, pp. 114–23 and 131–60 of Blair and Robertson, *op. cit.*; N. P. Cushner, Legazpi, 1564–1572, *Philippine Studies* **13** (1965) 163–206. A. de Mirandaola, Letter to Felipe II, 8th June 1569, vol. 3, pp. 33–43 of Blair and Robertson, *op. cit.*; Molina, *op. cit.*, vol. 1, 60–3

[3] Benitez, *op. cit.*, 50–3; Molina, *op. cit.*, 63–7; Montero y Vidal, *op. cit.*, vol. 1, 44–77; A. de Morga, *Sucesos de las islas Filipinas*, 1609 (trans.) J. S. Cummins (Cambridge, Eng. 1971) 55–77; H. Riquel, Foundation of the city of Manila, 19th June 1572, vol. 3, pp. 173–4 of Blair and Robertson, *op. cit.*; W. L. Schurz, *The Manila galleon* (New York 1939)

The evidence for Spanish introduction and dissemination of the seed stock is circumstantial rather than categorical, but there must be little doubt that it occurred. There is no categorical evidence on immediate Filipino reaction to maize, and none concerning the rates and patterns of secondary diffusion among Filipinos themselves for the sixteenth century.

Early agricultural conditions and dietary patterns in the Philippines

Once established in the Philippines, the Spanish rarely carried on agricultural production themselves but depended on Filipino crop-growers and Chinese market-gardeners for their food supplies. Where they could do so the Spanish levied food stocks from local Filipino populations. After about 1590, around Manila, in addition to such local levies, the Spanish depended significantly on Chinese fishermen, market-gardeners, and general farmers for a share of the whole range of foodstuffs needed to supply the city.[1]

Rice was the preferred crop staple of the lowland littoral dwellers, grown either as a permanent wet-field, lowland crop or as a dry-land crop under shifting cultivation. The local native rice dietary normally was augmented by fishery products from shore waters and by small amounts of dooryard garden crops of fruits and vegetables:

> Their ordinary food is cooked rice which has been pounded in wooden mortars, is known as *morisqueta* and is the ordinary bread in all the country; boiled fish is quite abundant and there is also pork, venison, and the meat of wild buffaloes, which they call carabaos. They prefer their meat and fish to be high and smelling bad. They also eat cooked camotes (which are sweet potatoes), beans, *quilites*, and other vegetables; also all kinds of guavas, pineapples, custard apples, oranges of many different varieties and other kinds of fruits and vegetables with which the country is abundantly provided.[2]

Quilites refer to the amaranths: *Morisqueta* was rice boiled dry enough that it stuck together in crusted pieces, remained edible for a period of days, and was often carried as food by travellers, so that the Spanish frequently referred to it as bread. Note that this 1609 report does not mention maize, but that it does include several other New World plants, the guavas, pineapples and custard apples, which de Morga mentions as though they were native rather than introduced. The *camotes* had, of course, preceded the Spanish, but a fairly large number of New World fruits had already spread with great rapidity throughout the Philippines

[1] A. de Morga, Report on conditions in the Philippines, 8th June 1598, vol. 10, pp. 75–102 of Blair and Robertson, *op. cit.*, 86–8; de Morga, "Sucesos" *op. cit.*, 310; Phelan, *op. cit.*, 108–16; G. de San Augustin, San Augustin's letter on the Filipinos, 1720, vol. 40, pp. 183–295 of Blair and Robertson, *op. cit.*, 292–3; Governor-General F. de Sande, Relation of the Filipinas Islands, 7th June 1576, vol. 4, pp. 21–97 of Blair and Robertson, *op. cit.*, 81: "The provinces in these islands that would be profitable to settle are those that can maintain the Spaniards and can provide them with food"; Y. de Santibanez, Letter from the archbishop of Manila to Felipe II, 24th June 1598, vol. 10, pp. 141–52 of Blair and Robertson, *op. cit.* 149–50

[2] de Morga, "Sucesos" *op. cit.*, 251

and came into common use without facing significant problems. The hunting of wild game often was seasonal and was complementary to the total dietary.

In those areas not conducive to rice production millet was grown as a substitute crop staple, but yields were relatively low and did not provide adequate subsistence, so that such areas often were rice importers. Early Spanish references to scarce food supplies in a given locality may possibly reflect disinterest in foods available but the urgent needs of the Spanish suggest, rather, low totals of permanently resident inhabitants and little or inadequate development of rice cultivation, with small amounts of millet being the only grain regularly available. References to abundant food supplies suggest local sedentary population concentrations, edaphic conditions satisfactory to wet-field rice production, and the presence of ample dooryard gardens productive of sugar-cane, coconuts, a variety of fruits and vegetables, and ample numbers of fowls, pigs and water buffaloes.

Pigafetta reported a variety of staple food crops grown in the Philippines at the time of Magellan's visit in 1521.[1] The next explicit records of food crops pertain to the early years of the Legazpi settlement period, 1573 and 1574. The first of these stated:

> Rice is the main article of food in these islands. In a few of them people gather enough of it to last them the whole year. In most of the islands, during the greater part of the year, they live on millet, *borona*, roasted bananas, certain roots resembling sweet potatoes and called *oropisa*, as well as on yams (yunames) and *camotes*, whose leaves they also eat, boiled. They eat Castilian fowls and pork. In a few of the islands inhabited by Moros, some goats are raised; but there are so few of them that wherever fifteen of twenty Spaniards arrive, no goats will be seen for the next two or three years. The cocoa-palm offers the greatest means of sustenance to the natives, for they obtain from it wine, fruit, oil, and vinegar. These people eat many kinds of herbs which grow both on the land and in the sea. Some of these herbs have been used by our people as articles of food. The scarcity of all kinds of food here is such that—with all that is brought continually from all these islands, in three frigates, one *patache*, and all the other native boats that could be obtained—each soldier or captain could only receive (as his rations) each week two *almudes* of unwinnowed rice—which, when winnowed, yielded no more than three cuartillos. This ration was accomplished by nothing else, neither meat nor fish. . . . The inhabitants of the coast are fishermen who barter their fish and buy from those living inland, who till the soil, the above-named foods. They eat all kinds of shell-fish and slimy plants that grow at the bottom of the sea.[2]

The second record, dated 1574, reads:

> At this point he enumerates the kinds of food used by the natives—"namely rice, millet, borona (a grain, also called *mijo*, resembling Indian corn),

[1] Pigafetta, *op. cit.*, vol. 33, 133 and 187, recorded rice, millet, panicum and sorghum as produced in the Visayan Islands and on Cebu, though Sauer, *op. cit.*, believed that Pigafetta himself also recorded maize in his original manuscript. In reporting food supplies provided to the Spanish after the Magellan expedition left Cebu there is no other reference to maize, but rice and millet are mentioned

[2] D. de Artieda, Relation of the western islands called Filipinas, 1573, vol. 3, pp. 190–208 of Blair and Robertson, *op. cit.*, 201–2

Castilian fowls, buffaloes, swine, and goats. They have wines of many kinds: brandy, made from palm-wine (which is obtained from the cocoa-nut palm, and from wild nipa palm); *pitarillos*, which are the wines made from rice, millet, and borona; and other wines made from sugar-cane. There are fragrant fruits,—large and small bananas, and *nancas*. . . . There is abundance of fish, and much game—deer, mountain boars, and excellent waterfowl."[1]

A report of 1582 commented that there were no deer left on the island of Cebu and that: "The island of Cebu produces a small quantity of rice, borona and millet and little or no cotton. . . . All are provided with fowls, swine, a few goats, beans, and a kind of root resembling the potatoes of Sancto Domingo, called by the natives *camotes*. After rice, fish is the main article of maintenance in this and the other islands. . . ."[2] Borona has often been equated to sorghum, but was identified as the millet *Setaria italica* Beauv by Merrill.[3] *Nancas* refer to the jackfruit.

Other reports of the late sixteenth and very early seventeenth centuries frequently mentioned rice, millet, borona and camotes, but none of the reports between Legazpi's arrival and the mid-seventeenth century make any reference to maize.[4]

The Europeans who arrived in the Philippines with and soon after Legazpi were an assorted lot of men only—Spanish, Portuguese, Greek, Italian, French and Flemming—who were eaters of wheat products and none of them probably had taken to maize as a regular food item.[5] As a substitute for wheat the Europeans sought rice, with which they were familiar, a fact made clear in all the early accounts dealing with the sixteenth and seventeenth centuries.

It could be concluded that most of the late sixteenth- and early seventeenth-century reports on the available native foodstuffs of the Philippines were written by Spaniards who, themselves, were neither interested in nor familiar with maize, but the complete lack of any comment on the presence, and food or beverage use, of the crop plant seems to throw doubt on its presence in the pre-Columbian period. At the very least it suggests that the crop was a very minor one not widely grown, and not at all significant to the basic dietary of any local group in the Philippines.

In the early Spanish occupancy of the Philippines many rules of operation were adopted from the Laws of the Indies developed for the New World.[6] Both civil administrators and Catholic friars made some efforts to encourage increased

[1] The second record is contained in a footnote written by E. H. Blair and J. A. Robertson to M. L. de Lagazpi, Relation of the Filipinas Islands and of the character and conditions of their inhabitants, 7th July 1569, vol. 3, pp. 54–61 of Blair and Robertson, *op. cit.*, 55–6, in which there is extracted one passage from a letter written in 1574 by A. de Mirandaola supplementing Legazpi's 1569 report, the extract complementing the earlier listing of food items given by Artieda, *op. cit.*

[2] M. de Loarca, Relation of the Filipinas Islands, June 1582, vol. 5, pp. 34–187 of Blair and Robertson, *op. cit.*, 43–5

[3] E. D. Merrill, *Dictionary of plant names of the Philippine Islands* (Manila 1903)

[4] P. Chirino, Relation of the Filipinas Islands and what has there been accomplished by the fathers of the society of Jesus, 1604, vols 12 and 13 of Blair and Robertson, *op. cit.*, and republished separately (Manila 1969) and de Morga, "Sucessos" *op. cit.*, each set down long lists of food plants without mentioning maize

[5] M. D. de Legazpi, Letter from Miguel Lopez de Legazpi to Felipe II, 25th July 1570, vol. 3, pp. 108–12 of Blair and Robertson, *op. cit.*, 111

[6] Benitez, *op. cit.*, 62–7 and 180–92; Phelan, *op. cit.*, chaps. IV–IX; R. R. Reed, Historic urbanism in the Philippines: a study of the impact of church and state *University of Manila Journal of East Asiatic Studies* **11** (1967) 1–222

agricultural production by Filipinos, but the early efforts seldom were very effective, and in some cases resulted in decreased production. Programmes of agricultural resettlement were steadily carried on, gathering thinly scattered and seasonally resident populations into permanent *pueblo* and *cabecera* sites wherein civil and religious administrative machinery, respectively, could both develop patterns of civilian rule and carry on the work of Christianising the native population.[1] This involved transforming shifting cultivation into sedentary crop gardening in areas surrounding the new settlements, but the early Spanish made few direct contributions to agricultural productivity beyond the introduction of a group of American crop plants and Old World farm animals. Repeated urging of the native population to enlarge agricultural production sought both food supplies for the Spanish and year-round stocks of food staples for the new permanent settlements.[2]

At the time of the Spanish entry into the Philippines there were many island areas having but light population totals, and a large share of the Filipinos were littoral dwellers who maintained considerable mobility. Many small islands and sections of large islands had long dry seasons and such small areas of moist soils as to be generally unsuitable for rice production. Numerous other small and large islands carried coralline soil covers too porous and dry for the growing of good rice crops by any technique. Several of the islands and island fringes in the central, Visayan, sector had both long dry seasons and coralline soils so that there millet, rather than rice, was the chief crop staple grown by Filipinos. The island of Cebu is such an island, and it apparently was a food deficit region when the Spanish came, although its location facilitated interregional trade and the physical situation of its chief east-coast settlements afforded good anchorage conditions for seasonal traders.[3]

Filipino reactions to maize

Assuming that the Spanish introduced maize into the Philippines in the late sixteenth century and may have disseminated it fairly widely by the end of the century, it appears that they did little more than mildly encourage its growth by Filipinos as part of the general urging toward increased agricultural production. The Spanish apparently did not introduce Mexican technology for rendering whole grain, ripe maize into an edible staple. There were very few European women in the Philippines during the early Spanish period and, apparently, extremely few Mexican women, women who could have introduced the appropriate techniques applicable to processing maize. The grinding of maize into flour, the making of

[1] Phelan, *op. cit.*, 4 and 46–52; Reed, "Historic" *op. cit.*, 33–71

[2] Scattered notations may be found recorded in the documents published by Blair and Robertson, *op. cit.*, on seventeenth-century Spanish encouragement of agriculture, but these chiefly are inferential rather than explicit, and most of them involve complaints about the native disinclination to produce commodities for Spanish consumption

[3] There are many scattered references in documents reproduced by Blair and Robertson, *op. cit.*, pointing to both periodic and regional food shortages. On soils and regional cultivation problems see A. Barrera, Soils and natural vegetation, pp. 53–71 of Huke, *op. cit.*, and F. L. Wernstedt and J. E. Spencer, *The Philippine island world, a physical, cultural, and regional geography* (Berkeley 1967) 69–74, 468–70, and 472–3; for a modern report on the soils of Cebu Island see A. Barrera, *Soil survey of Cebu Province*, soil report no. 17 (Manila 1954); for a discussion of dry seasons see A. Manalo, The distribution of rainfall in the Philippines, *Philippine Geographical Journal* **4** (1956) 104–67

tortillas and other edible products, and the preparation of hominy in Mexico were essentially women's work, and these skills and the tools and utensils needed did not get imported into the Philippines.[1]

Maize is a very different commodity thanrice, and it requires its own processing technology. To remove the husks from rice the Filipinos pounded the grain-in-husk in hollow-log mortars with wooden-post pestles. After the chaff was winnowed the grain was ready for the cooking. All native food commodities used in any volume by Filipinos at the Spanish entry could be cooked by boiling or steaming, and no important food commodity required grinding prior to cooking. Millet, after pounding, could be cooked into a gruel. At the Spanish entry the Filipinos possessed no grinding tools suitable to process such a bulky commodity as maize. Ripe maize cannot be made satisfactorily edible by simple boiling, in competition with rice, millet, yams, taro or green vegetables. Filipinos apparently tried pounding maize in their rice mortars but this did not prove very satisfactory.

The Spanish, not greatly interested in maize, left the acceptance of the crop plant entirely to the cultural dynamics of the Filipinos. Having no practical technology for handling ripe maize as a grain, the plant had no utility as a field crop, so that the first acceptance was only as a green vegetable, the immature ears being plucked for boiling or roasting. As a complementary green vegetable crop, maize had many competitors and could assume no more than a minor role in the dooryard garden and in the dietary system, since Filipinos did not at first like it very much.[2] Had the whole of the Philippines possessed both humid climatic conditions and soils amenable to productive rice cultivation it is doubtful that maize would ever have risen to the rank of a major crop plant, given the milling problem, since maize could not have competed with both established cultivation and dietary systems.

In China, on the other hand, there was relatively prompt initial acceptance of maize in at least Yunnan and on the southeast coast (assuming both sea and land post-Columbian introductions) both as a green vegetable and as an additional flour-producing ripe grain, since the Chinese possessed milling technologies adequate to the handling of ripe-grain maize. It is true that the rise of maize to the status of a major crop plant in China also came only slowly, since it had to complete with rice in the humid lowland, many parts of which were already landscaped for wet-field cultivation, and with several dry-field grains on lands not used for rice; in the humid landscaped lowlands maize still does not compete with rice even today. It was only when the Chinese began cultivating the hitherto unfarmed rough upper hill lands that maize demonstrated its value, over a century later.[3] However, the two cases are not identical since the Chinese had long grown grains needing milling and possessed all the necessary milling technologies.

During the seventeenth century the maize plant apparently spread fairly widely throughout the Philippines as a complementary dooryard garden food plant, and for small plantings in the *caingins* of the shifting cultivators, from both locations

[1] J. J. Delgado, *Historia sacro-profana, politica y natural de las islas del poniente lammadas filipinas* (completed in 1754 but published in Manila only in 1892) 707, stated that the Visayans (of the central Philippines) and the Tagalogs (of central, southern Luzon) made only a very poor gruel out of maize and did not make tortillas of it as was done in Mexico

[2] Phelan, *op. cit.*, 111 and note 13, p. 193; C. VanderMeer, Population patterns on the island of Cebu, the Philippines, 1500 to 1900 *Annals of the Association of American Geographers* **57** (1967)

[3] P. T. Ho, *Studies on the population of China, 1368–1953* (Cambridge, Mass. 1959) 187–9

being picked as green corn for boiling or roasting on the cob. The stalks and leaves of the plant in due time became incorporated into several nonculinary uses in addition to serving as animal fodder. Maize grew well in the Philippines, so that its agricultural production presented no agronomic problems that the Filipino farmer could not readily solve by minimal amounts of experimentation. In time Spanish encouragement was given to the growing of maize, along with programmes encouraging other patterns of production, and some new varieties were introduced from the New World. Although there is no effective evidence of the width of distribution of the crop plant it is likely that by 1700 the crop had become known and casually accepted throughout much of the Philippines as a complementary green vegetable. For most of the eighteenth century maize presumably continued to be grown in both dooryard gardens and in shifting cultivation *caingins* as a minor complementary crop in the mixed gardens of the Filipinos.[1]

Changing technology and the rise of maize

How, then, could there begin the rise of maize to its modern importance, and what changes took place to introduce "corn rice" as an acceptable substitute for true rice and millet in those areas that did not produce abundant rice crops?[2] There has been no explanation for this shift and I have so far found no reliable written record for these events satisfactory to the normal criteria employed in the annals of historical geography in documenting the introduction of technological change.[3] The explanation offered here is built largely on intangible and circumstantial evidence, but I believe it basically sound and generically correct in nature.[4]

The important role of the Chinese in Philippine economy is a fairly well-worked theme, although its depths have not been exhausted or fully documented.[5] The Spanish created, and suffered, periodic shortages and supply problems in consumer goods that resulted in difficulties for local populations, be they Spanish, Filipino or Chinese. Spanish exactions of food levies and purchases of food stocks that sometimes went unpaid for long intervals discouraged native production. Spanish policy alternately encouraged the Chinese as producers, handicraftsmen and traders

[1] In 1753, Delgado, *op. cit.*, 60, wrote that on Cebu the economy still revolved around the coconut palm and borona, with some cultivation of cotton, tobacco, cacao, sugarcane and rice; Phelan, *op. cit.*, 111, stated that during the early Spanish period: "All the available evidence suggests that maize production was not very large . . ."

[2] VanderMeer, *op. cit.*, 319, remarked: "The farmers on Cebu have long been forced to raise a dry crop as their staple food. Two hundred years ago, millet constituted the main crop. Today corn rules that agricultural landscape and provides the staple food for eighty-five per cent of the total population"

[3] Limited search in the National Library and the National Archives in Manila, both of which suffered damage in World War II, failed to uncover sources of value about the early use of maize

[4] Considerable time was spent in Cebu questioning older Filipinos and Chinese about the evolution of the "corn rice" complex, and my reconstruction of the development is a tentative piecing together of intangible threads of information as an oral history dependent on the memories of old stories

[5] A. Felix, Jr., *The Chinese in the Philippines, 1570–1898*, (2 vols Manila 1966–9); S. S. C. Liao (Ed.), *Chinese participation in Philippine culture and economy* (Manila 1964); Phelan, *op. cit.*; E. Wickberg, *The Chinese in Philippine life 1850–1898* (New Haven 1965)

and forcibly restricted their activities.[1] The periodic raids by the Islamic inhabitants of southern Mindanao and Sulu (Moros) on the Christian Filipinos often disrupted local economies, particularly in the central and southern Philippines.[2] After about 1760 Chinese were again restricted as to area of residence and in total numbers, although by this period there were beginning to be sufficiently numerous Philippinised Chinese families and Chinese Mestizo families that Spanish policies of restriction no longer could eliminate all Chinese from provincial hinterlands.[3]

There is necessary, here, a short digression dealing with wheat, wheat flour and grain milling, since milling is an issue critically related to the rise of maize in my interpretation of developments. The earliest Spanish in the Philippines depended on bread products made from wheat flour imported from Mexico. Imports of flour from Mexico continued for at least a decade after Legazpi's arrival, but Chinese and Japanese soon began bringing in both wheat and wheat flour, and the imports from Mexico ceased, but the Spanish saw no need to develop grain milling facilities in a domestic economy heavily committed to unmilled rice and serviced by Chinese.[4] During the late sixteenth century both the Chinese and the Japanese brought wheat and flour into the Philippines, but in the early seventeenth century, as Japan closed herself off from the outside world, the Japanese traders stopped coming and the Chinese took over the supply of wheat and flour, and became the chief bakers in Manila.[5]

Contrary to many historical assertions, wheat as a crop plant was introduced into the Philippines and became fairly productive, but there are the same questions around its introduction that parallel those of maize, discussed earlier. In the early seventeenth century some trials were made, and in time Spanish encouragement succeeded in getting Filipino farmers to produce wheat in particularly favourable local environments. The earliest record for actual production that I have found is

[1] Felix, *op. cit.*; Wickberg, *op. cit.*

[2] VanderMeer, *op. cit.*, commented on the Moro raids on Cebu; almost all histories of the Philippines contain reference to the effect of the Islamic raids on the Christian Filipinos

[3] J. A. Larkin, *The Pampangans, colonial society in a Philippine province* (Berkeley 1972); Wickberg, *op. cit.*; W. Draper, Plan for an expedition for the conquest of the southern Philippines, *c.* 1759, vol. 49, pp. 27–43 of Blair and Robertson, *op. cit.*, 37

[4] The earliest report I have found on imports of Chinese wheat or flour is G. de Lavezaris, Letter to Felipe II, 17th July 1574, vol. 3, pp. 272–81 of Blair and Robertson, *op. cit.*, 276, referring to wheat and barley flour; Chinese flour was reported low in quality by A. G. Garrillo in Annual income of the royal exchequer, 1584, vol. 6, pp. 47–53 of Blair and Robertson, *op. cit.*, 51, but Bishop D. de Salazar, The Chinese and the Parian at Manila, 24th June 1590, vol. 7, pp. 212–38 of Blair and Robertson, *op. cit.*, 221, reported Chinese flour then of high quality; de Morga, "Report" in 1598, *op. cit.*, 84, reported both wheat and wheat flour as brought in by Chinese, and that good quality flour and biscuits came in from Japan; by 1609, de Morga, "Sucesos" *op. cit.*, 306 and 308, reported that good wheat and flour came from China and that large amounts of good wheat flour came from Japan

[5] I have found no record of wheat milling in the Philippines, but by inference the Chinese, as the chief bakers, did their own milling, as indicated by Salazar, *op. cit.*, 227: an unsigned report of 1663, Events in Manila, 1662–3, vol. 36, pp. 218–60 of Blair and Robertson, *op. cit.*, 259, commented: "For not only does everything necessary for life come to us from China—as wheat, cloth, and earthenware—but it is the Sangleys who carry on all the crafts, and who with their traffic maintain the fortunes of the citizens . . .". Archbishop Hernando, Report of the archbishop on the bakery of Manila, 3rd August 1634, vol. 24, pp. 295–6 of Blair and Robertson, *op. cit.*, described a government controlled bakery but made no comment on its source of flour.

for 1649.[1] As late as 1860, at least, some wheat continued to be grown in a few upland areas.[2] Although some wheat was grown in the Philippines for two hundred years, the production was never adequate to the consumer demand, and there were imports of both wheat and flour until its import was forbidden in the early nineteenth century in a last effort to make the islands self-sufficient.

The wheat story differs from the maize story, however, in that rural Filipinos in the Spanish period did not grow wheat for their own home consumption, but produced it for Spanish consumption and for use in religious rituals of the Catholic Church. It must be noted, of course, that at an early date the few urban upper-class Filipinos began to eat some wheat products, that bread was sold on the streets by Chinese peddlers, and that the dietary acceptance of breadstuffs slowly increased among the urban populations, since these groups gradually adopted both Spanish breakfast habits and increasingly ate Chinese dishes in the restaurants of Manila and other growing cities.

Any wheat that was imported as grain during the early Spanish period came into the port of Manila, and indications are that all wheat grown in the islands during the era of production was shipped to Manila. Since the Filipinos neither milled grains nor ate milled food commodities in their own native systems, it must be presumed that the Chinese bakers converted the wheat into flour and initiated the grain-milling industry. Even the southern Chinese, who composed most of the Chinese population resident in the Philippines, were fully familiar with the milling of flours from such commodities as wheat, rice, beans, sorghum and other food crops. Certainly during the early Spanish period the milling must have been done on Chinese stone mills imported to Manila.

The oral record in Cebu suggests that at some point in the second half of the eighteenth century Chinese brought stone mills to Cebu from Manila, during a local food shortage on the island of Cebu, for the milling of various local products to be used by them as substitutes for the rice normally supplied by other islands.

[1] J. N. de Tavora, Letter to Felipe IV from Governor Juan Nino de Tavora, Cavite, 1st August 1629, pp. 29–42 of Blair and Robertson, *op. cit.*, 37–8, gave indication that wheat had been tried and that the governor intended to make an extensive planting, but no following report was made on the harvest; an unsigned report, Early Franciscan missions, 1649, vol. 35, pp. 278–322 of Blair and Robertson, *op. cit.*, 302, noted that wheat was being grown but that the seed-stock needed periodic replenishment; B. de Letona, Description of Filipinas islands, 1662, vol. 36, pp. 189–217 of Blair and Robertson, *op. cit.*, 201, stated: "... in some places wheat is sown and harvested"; Governor D. de Salcedo, in a letter to the crown on 16th July 1664, in *Coleccion Pastells de Madrid*, 15, 250–250v, extracted in H. de la Costa, *Readings in Philippine History* (Manila 1965) 67–8, reported having investigated the prospects and having arranged for the growing of wheat in upland Laguna and Batangas provinces, southwest of Manila; J. M. de Puga, The order of St. John of God, 1742, vol. 47, pp. 161–229 of Blair and Robertson, *op. cit.*, 185, referred both to the import of and the domestic production of wheat; Draper, *op. cit.*, 38, noted very fine wheat produced in Panay Island in the 1750s

[2] R. MacMicking, *Recollection of Manila and the Philippines during 1848, 1849 & 1850* (Calcutta 1851, republ. Manila 1967) 189, reported wheat being grown in the Ilocos provinces of northwest Luzon, in Tayabas province in eastern Luzon (now Quezon province), and in Laguna province south of Manila, and stated that the import of wheat was forbidden; the 1967 republication of MacMicking included an appendix of four letters by the British vice-consul in Iloilo, Panay, in the Visayan zone, in which the first letter, Loney to Farren, 12th April 1857, PRO F. O. 72/927, 226, commented: "wheat, which grows freely in the more elevated districts of the island, and of which 1,125 bags were sent from Iloilo and Antique during 1856 ..." and another letter, Loney to Farren, 10th July 1861, PRO F. O. 72/1017, 256, listed only a small amount of wheat shipped in 1860

The oral record also suggests that the mill-stones were adjusted to variable degrees of coarseness, for milling different commodities, and that some corn was milled into several different kinds of "meal". One result of this experimental milling was the production of a very coarsely ground maize having particles about the size of rice grains. When this coarse meal was cooked by steaming methods common to Chinese culinary practice the maize particles did not stick together as a mush as when maize is finely ground, but resembled properly cooked rice in texture. Chinese began substituting this coarsely ground maize for rice in their own dietary. A new food commodity was thus created, one that today is known as "corn rice" in English reference. Coarsely ground maize of uniform quality was offered for sale in Chinese shops and, since it was more satisfactory than maize pounded in wooden mortars, Filipinos began buying it in times of local food shortages. Filipinos were then largely subsistence crop growers not yet able or habituated to buy all their food supplies, but they possessed no facilities for converting whole-grain maize into a satisfactory substitute.

At this point, probably the late eighteenth century, the Chinese began two significant practices. Initially Cebu Chinese began to mill maize into "corn rice" for Filipino producers of maize, a practice that began the process of turning the Chinese into the leading commercial grain millers of the Philippines in the contemporary period. The second practice was the importing of small mill-stones that could be grooved coarsely to break whole-grain maize into the new commodity "corn rice". Such small mills were sold cheaply, were light enough to be transported to any rural locality, were easily operated by hand, and were gradually acquired by rural Filipino families who could then mill at home those amounts of "corn rice" periodically needed.[1] And with a cheap home-milling facility the maize plant took on a whole new value as a crop plant in those localities having dry soils, such as Cebu, that traditionally had been unable to grow good crops of rice.[2]

The rise of maize as a major crop plant

In consequence of the availability of a cheap home-milling facility there began the increased planting of maize on Cebu Island as a subsistence food crop that could be processed directly into a food staple as needed. I believe it was the availability of the milling device that stimulated the planting of maize, since the yield of maize was greater than that of millet on the dry-soil areas that could not

[1] In a letter "Father Jose Maria Clotet to the Reverend Father Rector of Ateneo Municipal", 11th May 1889, written at Talisayan, Bukidnon province, Mindanao, vol. 43, pp. 288–309 of Blair and Robertson, *op. cit.*, 302–3, is the comment: "Scarcely will one find a house of the Buquidnon where there are not one or at times more small mills for grinding maize. These are made of two very hard stone cylinders. The inner is fixed on a wooden upright, while the upper is movable, and has an orifice in its center through which the maize is poured. The circular movement by which the grain is crushed is produced by a handle securely fastened to one side of the movable cylinder." This statement could probably have been made concerning almost any rural household in the central Visayan islands or northern Mindanao at the time

[2] Draper, *op. cit.*, 34, for the late 1750s for Cebu noted: "The productions of this Island are Borona, a small Grain like Millet which is the chief food of the Common People as rice is scarce, Tobacco, Abacca & Cotton of which two they make Cloth", suggesting that at that date maize had not yet begun its real expansion as a staple food; see also the comment in the early 1750s in footnote 1, p. 11 by Delgado, *op. cit.*

grow rice, and I believe it was this increased food supply that supported the relatively rapid growth of the population of Cebu and other areas in the Visayan region after 1800. The liberalised policies of government under Governor Basco y Vargas in the late eighteenth century must have aided patterns of development, but such policies in themselves were not responsible for the technological development that started Visayans on the road to being producers of maize and consumers of "corn rice".

Maize began to replace millet and borona as field crops and food staples in the rural Cebuano dietary system. I believe this shift began to occur somewhat prior to 1800, and that it occurred progressively in the first two decades after 1800, although dating from the oral record must remain unprecise. I have not been able to plot the specific chronology of the geographical spread of corn rice among the separate islands of the Visayan region, but I believe it a domestic occurrence that largely escaped the notice of observers, writers and commentators since it was a rural development affecting home subsistence patterns. Foreign visitors were not numerous, as hinterland travellers, prior to the middle of the nineteenth century, but there is no comment about millet and borona as crop plants in the latter half of the century.[1] Maize as an increasingly important domestic crop plant was not of interest to the growing body of international traders that visited the Philippines, since their attentions were focused on sugar, tobacco and abaca.[2]

The spread beyond Cebu of maize production as a crop staple, and the acceptance of "corn rice" as a daily food, took place rapidly during the early nineteenth century, both by direct diffusion by emigrants from Cebu settling over Visayan Islands and the northern coastal fringes of Mindanao, and by stimulus diffusion. The spread of maize as a crop staple continued in the settling of northern Mindanao during the late nineteenth century, and it has been notable in southern Mindanao since World War II. Once the innovation became widely accepted, "corn rice" turned into a regional dietary staple throughout the Visayan Islands and their fringes, and the growing of maize as a primary crop became an ecologically sound basis for the expansion of agricultural settlement in a region in which rice locally was often not a successful crop.

The final stages of the development of maize as an important crop relate to the steady expansion in the acreage of maize production in both the southern sectors of the Philippines and in other local regions within the islands. Beginning in the nineteenth century, the growth of the Visayan Islands population created a food-deficit zone, bringing about increasing interregional trade in maize, and the expansion of commercial maize agriculture in any part of the Philippines in which

[1] J. M. de Zuniga, *An historical view of the Philippine Islands* (Manila 1803), in the John Maver translation (republished Manila 1966) p. 6, still mentioned millet as a complementary crop but, p. 8, did not list maize in a list of crops introduced by the Spanish; Zuniga, however, was writing administrative history and his descriptive and economic notations were not critical to his account

[2] For example, J. Bowring, *A visit to the Philippine islands* (London 1859) mentioned maize as introduced from the New World but never mentioned it again, since his chief concerns were with the export of sugar and tobacco; MacMicking, *op. cit.*, never mentioned maize at all in his extensive comments on products potentially valuable in foreign trade; F. Jagor, *Travels in the Philippines* (London 1875) gave considerable space to potential foreign trade crops but commented on maize only as a dooryard garden crop in several parts of the Philippines, and he noted that maize was then being grown on Cebu but that Cebu was food deficient; J. Montero y Vidal, *El archipelago Filipina y las islas Marianas, Carolinas y Palaos* (Madrid 1886) in a thirteen-page chapter on agriculture gave five lines to maize, to say only that it was grown in some provinces in which it was a substitute for rice

maize could compete with other crops.[1] Maize remains a domestic crop, since it has never significantly entered foreign trade as an export. Maize began to outrank the sweet potatoes-yams by about 1880, and to assume a strong position as the second ranking crop by about 1920. No other American crop plant has rivalled maize; manioc developed local regional importance during the nineteenth century in a few areas such as the Sulu Archipelago but in this century manioc has most commonly been consumed by people only during periods of shortage, and when other foods are plentiful or can be afforded manioc is consumed by pigs. About three-quarters of all Philippine farms grow some rice, and the acreage runs over 8,000,000 annually, whereas almost half of all farms grow maize, with a total acreage of about 5,000,000 per year. Whereas some rice is grown almost everywhere in the Philippines, grain maize is regionally concentrated in the dry-soil sectors of the islands.

Ecologically, maize is well adapted to many parts of the Philippines and perhaps by the present has spread almost to its limits. Only the uppermost highlands and the very wet lowlands lie outside its range as a significant crop. Maize does not compete seriously with lowland wet-field rice in those areas in which rice does well, but it has competed very successfully with "upland rice". Maize yields are not high in the Philippines, and there are severe problems with plant pests, but three crops may be grown per year. In those rural areas in which maize cannot compete with rice as a primary crop it now remains what it was in the earlier period, a complementary dooryard garden crop eaten as green corn on the cob. Around the fringes of the growing urban settlements this demand for green corn has expanded its production as part of the market-garden activity in the twentieth century.

In the years since World War II the Filipino dietary system has been undergoing further expansionary change, particularly among urban populations. Wheat products, today entirely imported, and mid-latitude green vegetables and fruits are expanding their patterns of consumption. The agricultural economy of the Philippines, however, continues to be one involving two primary food crop staples, rice and maize, grown in ecologically different environments and the dietary regime continues to be a dual system in which local historical and cultural preference stipulates the primary staple in the daily diet, as between rice and maize.

[1] Montero y Vidal, "El archipelago", *op. cit.*, gave statistics in which, in 1880, rice still exceeded maize by far, but those statistics also suggest that northern Luzon, the Visayan provinces, and the Mindanao provinces already displayed regionally concentrated maize production patterns relatively similar to those of the contemporary period although the 1880 acreages and production totals were much smaller than those today

3

Landscape, System, and Identity in the Post-Conquest Andes

Daniel W. Gade

Abstract. This article synthesizes the broad impact of Spanish introductions in the New World for the Central Andes (Peru, Ecuador, and Bolivia). Beginning in 1531, the Spaniards brought, from Iberia and Middle America, material elements of their culture which in time were acquired by native people through both imposition and free choice. Plants, animals, and tools were selectively integrated into native agropastoral systems and architectural elements into settlement patterns. Of the screens that filtered the array of Old World rural traits, keeping some out, permitting others to successfully pass and be adopted, the most significant were conditions that the highland environment imposed and competition from existing elements of the already well-developed Andean agricultural complex. Depopulation disrupted the native agroecosystem, and in the restructuring that followed, European goods and practices were adopted along with the indigenous. About a dozen crop introductions became important among peasants out of a total list three times that long, but European domesticated animals contributed most saliently to peasant livelihoods. These Old World biotic contributions juxtaposed with the native elements into a complex that crystallized between 1550–1650. With house types, building materials, and settlement pattern, the two traditions melded. Much of the Central Andes since then has changed relatively little.

Key Words: cultural ecology, cultural landscape, agropastoralism, crops, animals, tools, house types, settlement, Andes, Peru, Bolivia, Ecuador.

FROM the time Spaniards unleashed their brand of European civilization on the lands across the sea, the native peoples of the Western Hemisphere had to deal with a jarring array of new realities. Only those pieces of Spanish cultural baggage that were adaptable to New World conditions survived their passage through the screen of time, distance, and competition. To varying degrees, the rural landscapes of the Americas are juxtapositions of selected elements of the Old World tradition and those indigenous to the hemisphere. In some places the introduced has largely displaced the native American components. In still other regions, the material bases of everyday life manifest a felicitous convergence in which the two are merged as one.

These folkways are the stuff of a particularistic geography in touch with visible landscapes as complex compositions of multiple influences. They evoke reflections on both the origin and spread of contrasting forms, as well as on the integration of elements into a cultural-ecological system that attests to the adaptive choices of a group. Form and function of these elements are intertwined concerns of a geographical approach to the outcome of the cultural encounter that started with Columbus five hundred years ago. The availability of Spanish material culture in the New World depended on either its physical transfer or replication in place (Foster 1960; Arguedas 1968; Alcina Franch 1986). Successful diffusion required both adaptability to new environments and human receptivity (Denevan 1983).

The manifold geographical diversity of Hispanic America almost requires that a discussion of impacts on landscape and system be confined to one specific setting. The Central Andean Highlands are chosen to examine the process and outcome of Spanish introductions to rural life in the New World. In this region, pre-Columbian land use had evolved into a sustainable sedentary existence that was no less stable than the farming-herding complex across the Atlantic. Unlike most of the lands of tropical latitudes, European elements flourished in this mountain redoubt that bore certain broad resemblances to the homeland. Spaniards were, in all cases, the initial agents of diffusion, largely to meet their own needs and desires, but the focus here is on the transfer of these introduced elements of material culture to the native people of the Andes.

An Andean focus also places contemporary agropastoralism and settlement there as a double heritage. Peasant life and livelihood in the Andes have often been viewed as a straight and pure line from the Incas. Exaltation of the indigenous tended to place Western influence as a negative force or, at best, an epiphenomenon. The notion of a continuity with the indigenous past holds a compelling fascination for many members of postindustrial societies who are, moreover, able to marshall broad sympathy outside the realm for a folk often viewed as oppressed. Dispossession and death were grim chapters in post-Conquest Andean history, and European subjugation brought with it the ruin of some of the intricate adjustments that people had made with their montane environment. But the inhabitants of upland South America also profited from this impingement, however rude it was, to sustain themselves by learning new techniques of increased efficiency; accepting new crops that added new textures, flavors, and nutrition to their diets; and adopting domesticated animals that, in addition to their uses, greatly increased the energy flow to humans from non cultivated lands. Indians were never as culturally hermetic as some of the ethnographic and travel literature purport them to have been.[1]

The Central Andean Stage

Highland Ecuador, Peru, and Bolivia form a supranational unit whose coherence is more than the physiographic outcome of the Cenozoic collision of tectonic plates. Cultural groups in that part of the tropical Andes reached a higher level of sophistication than in other mountain regions of the world before the modern era. More than five thousand years of sedentary living gave rise to a series of regional polities of whom the Inca are best known. An elaborate state ideology, official religion, policy of cultural transplantation, and a road system with Cuzco at its center consolidated an exceptionally fractured territory. Regional continuity persisted after the Spanish Conquest for almost two hundred years, when the Viceroyalty of Peru with its capital in Lima placed the Central Andes under one administrative unit. Since then, the politically regionalized highlands have continued to share a cultural configuration of locally organized societies of conservative peasants long noted for their frugality, poverty, and often passive resistance to the modernizing assumptions emanating from their capital cities.

Andean land use has been marked by the extravagant verticality which telescopes climates and compresses ecological belts like few other places on earth (Troll 1931). Such diversity in short distances has encouraged a single group to manage several zones. Tuber cultivation and camelid herding above 3700 m contrasted with the temperate valleys and basins where the staple crop, maize, was grown on irrigated and regularly fertilized land. Warm depressions etched below ca. 2500 m produced tender crops such as capsicum pepper, cherimoya and lucma; below ca. 1800 m, coca grew. Periodic movement of a single community among different production zones promoted economic self-sufficiency, though trade between different groups also occurred.

Andean people were proficient in manipulating their mountain habitat for sustainable food production. Earthworks, whose construction was plausibly prompted by population increase that started long before the Inca, had water management as their raison d'être (Donkin 1979; Guillet 1987). More than a million hectares in the Central Andes were fashioned into irrigated agricultural terraces, most of them in Peru (Masson 1984). Stone-faced bench terraces found in valleys at elevations suitable for maize cultivation were the most elaborate of these cascading structures (Denevan 1987). Aside from terraces, hillsides also underwent ridging and furrowing in places whose mark on the land lingers even after they have been long since abandoned (Schjellerup 1985). Raised fields, identified by scientists only in the 1960s, were built on flat land in various places from Northern Highland Ecuador and Central Peru to the Lake Titicaca Basin and beyond (Smith, et al. 1968; Knapp and Denevan 1985; Knapp 1988). Recent experiments to reconstruct their ancient use highlight their considerable food productivity. The ridges improved drainage, and the standing water between them may have fixed nutrients and moderated temperatures for the crops grown there; organic matter that accumulated in the ditches was periodically spread over the mounds as fertilizer.

Pathways of Hispanic Diffusion

Francisco Pizarro and his men arrived on the coast of Peru in 1531 and conquered the Incas within a year. More than any other part of South America, the Central Andes with their mineral wealth, valleys of dense populations, and high productive capacity appealed to sixteenth-century Spaniards. Four decades intervened between Columbus's landfall and the first European incursions into this zone, and during that time the Spaniards had become self-confident in their ability to dominate whomever and whatever they encountered in the New World. This aplomb of the hidalgo mentality, combined with firearms and horses, resulted in one of the most sudden and traumatic conquests of any high civilization recorded in world history. The acculturation to European ways, however, lagged far behind the political hegemony imposed on the land and native people.

Most of those Spaniards who came directly to the broadly defined land called Peru in the first years after the Conquest were from Andalusia, Castile, and Extremadura (Pérez de Barradas 1941). Emigrants from eastern and northern Spain were far fewer, and fewest of all were Catalans. In 1535, of the 56 people who embarked for Peru from Seville, 65 percent of them were plateau dwellers from Old and New Castile and Extremadura (Bermúdez Plata 1942). The eightfold increase in movement to Peru a quarter of a century later did not notably change the provenance. Of the 444 people listed as going to Peru in 1560, most came originally from the Meseta (Romera Iruela and Galbis Diez 1980). Spaniards from the plateau may have had a special affinity for the Andes as a colonial destination, especially if they had knowledge of the contrast between the cool, malaria-free highlands as compared to the general unhealthiness of the lowlands. Those colonists already present in the New World when Pizarro conquered Peru had sounder information on which to make a decision to go to the Andes. The Spanish settlements on the Pacific coast from Mexico to Panama contributed many migrants in the first decade after the Conquest.

Reaching the Andes from Spain necessitated much more time and effort than going to the Antilles. The journey from Seville to Callao commonly took five months or more and required crossing the Isthmus of Panama (Fig. 1). The trip across this strip of land from the Caribbean port at Nombre de Dios (and after 1597 at Portobelo) to the Pacific side required from four to fourteen days. Once in Panama City, delays often occurred, for the minimum sixty-day trip to Callao could only be undertaken during the six months of the year when winds favored the southward sail along the west coast of South America. Little of the precious space on board the smallish ships was available to bring the paraphernalia of rural life from Spain directly to Peru. Bulky tools, millstones, plant cuttings, domesticated animals, and the necessary food and water for their transfer were low-value cargo. As much as the segmented sea voyages, the Panama Isthmus hindered the direct flow of European artifacts to western South America. The high cost of moving anything across those ten leagues of forest and swamp could only be justified by its high monetary or exceptional sentimental value. The difficult traverse also took its pathogenic toll on people, beasts, and plants.

Not surprisingly, most material introductions of Spanish culture to the Andes did not come directly from Spain. Western Nicaragua and Mexico were the major early sources of cows, pigs, goats, horses, donkeys, sheep, and chickens, and of most plants that had earlier been taken from Hispaniola and Jamaica (Borah 1954, 84, 86). In spite of long distances, however, from whatever source, by 1540 the bulk of Spanish material culture had landed in Western South America, and by 1555 this dominion was reproducing most of its own plants, animals, and farming implements. That relative rapidity suggests how a limited number of introductions, even just one, was all that was required to launch a diffusionary spiral with historic con-

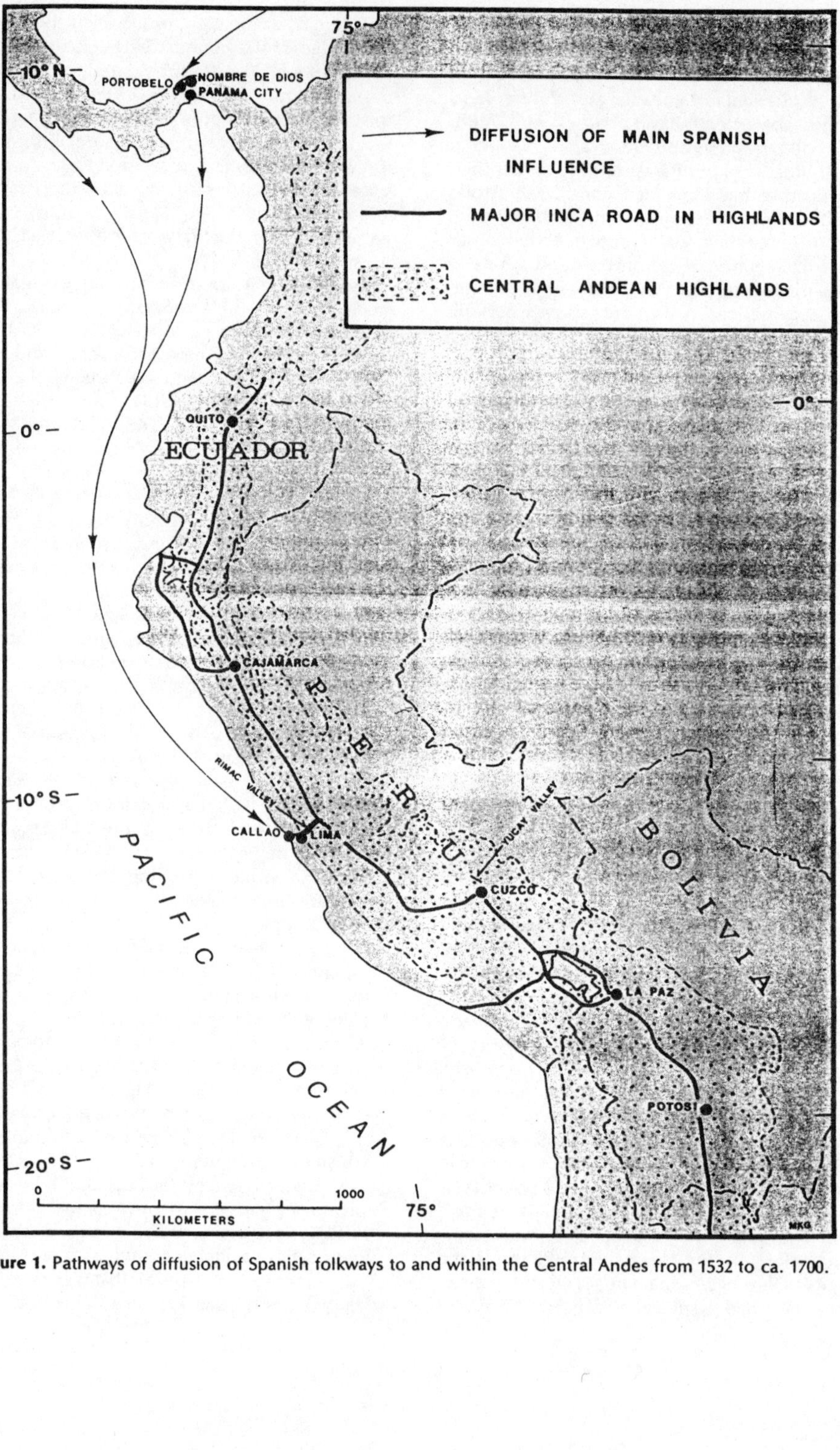

Figure 1. Pathways of diffusion of Spanish folkways to and within the Central Andes from 1532 to ca. 1700.

sequences. Callao was the main port of entry for introductions that eventually diffused to the entire Central Andean region, both north and south over the Inca road system.

Agricultural innovations first reached the irrigated farms established by Spaniards in the Rimac Valley around Lima, and from there spread to nearby valleys. Beginning in 1535, this one small coastal oasis became the seat of Spanish administration in western South America and a major focus of Spanish colonization. Though the coastal region was rainless and lay only 10° from the equator, its anomalous coolness (19°C yearly average) and water for irrigation permitted the successful cultivation of many temperate crops and most tropical plants as well. In that setting, seeds and cuttings multiplied, as did offspring of the denizens of the Iberian barnyard; they did not farther north in the hot, wet Guayas Basin. Transportation access also explained why the coastal valleys around Lima served as the critical staging area for subsequent diffusion of Spanish material culture to the highlands that loomed only a few kilometers to the east over an existing Inca road.

Spaniards moving to the Sierra were much more likely to have picked up seeds, cuttings, and animals in Lima than to have brought them from Spain as part of their personal effects. Once in the highlands, many Spaniards could fashion from local materials tools and building styles they had known in Spain. Over the next several decades, these introductions spread throughout a large part of the highlands. They comprised a combination of particularly useful elements that tended to be accepted together. By the 1590s, this bundle of Spanish rural traits had evolved into an established complex which remained in place, due to custom, isolation, and inertia, for the next four centuries.

Contact, Acceptance, and Refusal

Reconstruction of the elusive process of entrance, rejection or acceptance, and spread of European traits into Andean life cannot be sketched in detail.[2] Even an intensive historical analysis of one limited highland zone near Lima does not pinpoint these changes (Spalding 1984). Spaniards originally brought their material culture for themselves and their compatriots, who in 1570 numbered between 10,000–15,000 in the highlands outside the mining areas (López de Velasco 1971). Subsequent deployment of Old World elements to Indians, who, at that time, numbered about 1,000,000 in the region, depended on three sets of agents: clerics, headmen, and colonists. Virtually all missionaries were Spaniards who belonged to religious orders with designated territories. In the Indian communities they served, their uncontested moral authority carried over to the mundane content of rural life. Priestly directives, blandishments, and threats bent the normally submissive native population to their will. Indians of high rank called *curacas* served as intermediaries for the Spanish authorities to collect tribute and enforce compliance. In addition to his civil authority, the *curaca* was an innovatory exemplar, for he was the native person schooled in Spanish ways and most likely to first receive European introductions.

Spanish colonists, disdaining manual labor, depended on a docile supply of native workers for agricultural tasks. In the *encomienda* system, the Crown obligated the Indians to work for the conquerors; later, as land was taken over as private property, peons farmed and herded for the estate owners as well as for themselves in a system of servitude that continued in force past the mid-twentieth century.

In those nonvolitional settings, Indians rapidly learned to use the tools, grow the crops, eat the foods, and raise the animals brought by their European overlords. Embracing these strange things was another matter: like many peasants everywhere, native Andean people initially were distrustful of innovations. Gradually they experimented with the available European elements while maintaining their own. In the 1570s–80s, Indians had accepted without coercion a series of Old World items in locales closest to Spanish towns. Only later did native communities in isolated districts integrate pieces of the new order into their agroecosystem and living patterns. By the early seventeenth century, enough had been accepted to modify, though not transform, the cultural landscape of the Andes outside the grassy heights called the *puna* and its more humid facies known as the *páramo*.

This particular schedule might not have been achieved as it was if Indian populations had not dropped so precipitiously after the Conquest. From the estimated 12–14,000,000 people in the coast and highlands of Peru, Bolivia, and Ecuador in 1520, disease caused the population to decline by 90 percent by 1620 (Cook 1981). This, combined with dislocations due to forced mine work, led to less intensive land use. The fact that indigenous people of the Andes suffered egregiously at the hands of the Europeans has become a dominant theme of the Old World-New World encounter. Scholarly discussion of Spanish impact has often focused on its disadvantages, disharmony, or disorganization (Dollfus 1982; Usselmann 1987). Their status as putative victims has overshadowed the ingenious ability of Andean peasants to manipulate elements of Spanish culture and incorporate them into their livelihoods.

Post-Conquest Crop Accretions

Broad thermal similarities made it environmentally feasible to transfer much of the content of Iberian agriculture, which followed a Mediterranean rhythm of winter cropping of annuals, to an Andean setting most suitable for high-sun cultivation. Altitudes mitigated the heat of a tropical latitude to the extent that seasonal frost occurrence above 2500 m truncated the growing season as it did in most of Spain at sea level. Furthermore, a sharply defined rainfall regime, which set a cyclical rhythm for plowing, planting, and harvest, also occurred in the Central Andes. Although the warmest parts of the year in the two areas were calendrically reversed, the rainy months and the vegetative cycle of the staple annual crops occurred in the same months in both Iberia and the Andean Highlands. Harvest festivals to honor St. Isidore, patron saint of farmers in Spain and the Andes, took place in the month of May in both of these disparate regions far removed from each other.

The Old World crops that passed into the agroecosystems of the native peasantry in the early colonial period met the tests of usefulness, environmental fit, and niche competition. Within a century after the Conquest, Andean peasants had effectively integrated a dozen plants and peripherally accepted a dozen others, together comprising less than half of the total number of plants brought by the Spaniards (Gade 1975). Spanish colonists had a list of accepted crops about twice that of the native peasantry's introductions. Some plants—saffron, endive, artichoke, and hazelnut among others—grown in Spain may not have reached the Andes at all (Herrera 1970; Oliveros de Castro 1968).

Among the grain crops, only wheat and barley became widely accepted in the Andes. From the time of their arrival, Spaniards were determined to have wheat. By the 1540s, they were growing it in the highlands, and in the 1550s, it was abundant enough for the price to fall to that of other staples. To Spaniards, wheat bread was essential to a civilized existence. Its cultural importance as much as its nutritional value caused the Iberians to include wheat among the tribute items required from Indians (Cook 1975). Perhaps because of this imposition, native people in some locales were growing it as one of their own food staples by the late sixteenth century ("Reparto de tierras en 1595" 1957). Indians parched and triturated wheat kernels to make a gruel as they did with their native grains, but almost everywhere they learned to prefer bread when they could get it. Wheat straw and stubble were fed to the growing number of Indian-owned livestock (Fig. 2). In terms of land use, hillsides of closely sown wheat were less vulnerable to soil erosion than were those planted with row crops. Normally planted and harvested at least a month after maize, wheat cultivation did not directly conflict with other labor commitments in the

Figure 2. Wheat harvest using a sickle, with animals grazing on the stubble, near Sangarará, Peru (3600 m).

agricultural schedule.

Indians adopted barley for its grain, straw, and stubble. Cut green, it was sold to Spaniards as high-quality fodder for horses and mules. Barley's main advantage over wheat was its ability to grow in drier zones, somewhat salty soils, and lower temperatures. As with wheat, the introduction and acceptance of barley by Indian communities expanded nonirrigated agriculture. Both crops fit well into sectorial fallowing on communal lands, a pre-Conquest practice that survived through the Colonial period. After a variable period (four, seven, or ten years) of fallow, the cropping cycle of a sector started with the potato. Either during the second, third, or fourth year of the cycle, that same plot was used for wheat and/or barley (Orlove and Godoy 1986). Neither oats, rye, millet, or rice emerged as even minor crops among Spanish colonists or Indian peasants in the highlands. In Iberia oats and rye were typically planted where wheat would not yield well, whether in cool rainy zones or on poor or exhausted soils. The Andes had quinoa and its even hardier relative known as cañihua for the marginally productive lands.

Mediterranean perennials adapted to summer drought did not bear much, if any, fruit in the Andes with its high-sun rainfall. Almond, carob, pistachio, jujube, or the olive did not yield satisfactorily enough under Andean conditions to become accepted tree crops. If the olive had successfully fructified, its adoption still would have been unlikely: Indians had no agricultural tradition of oil crops or culinary tradition that included frying. The grapevine was in a somewhat different category among Mediterranean crops. Spaniards in the Andes grew it with mixed success; among Indians, grapes were rarely grown and, in any case, not for making wine, though they were fond of its Dionysian effects. To Spaniards wine enhanced sociability; to Indians it offered inebriation until supplanted in the nineteenth century by brandy and then rum. Though chicha making from maize persisted without interruption, a craving for stronger alcoholic beverages emboldened native people to enter the money economy controlled by Spaniards and mestizos. A technological barrier also explained an early Spanish monopoly on growing sugarcane. Although sugarcane was taken to the warm valleys of the Sierra from the Coast by the 1540s, only Europeans had the technology needed to process the juice into white sugar and distilled spirits. Indians later grew sugarcane for its sweet stalk and juice to make a fermented beverage and hardened molasses. Artisanal wooden presses moved by oxen or by hand were replicated from a Spanish design to extract the juice, but the sugarcane product they valued most was beyond their means to fabricate.

Several Andean domesticated plants declined in the face of competition from Old World introductions; for example, wheat expansion brought the contraction of quinoa, the latter a crop that required multiple washings to remove a bitter principle. Still quinoa and cañihua continued to dominate at high elevations where European small grains failed to mature their seeds. Broad bean (*haba*), in Spain a crop of the cool Mediterranean winter, adapted well to the zone between 3300–3800 m, much the same niche as that of tarwi, a native lupine. Though high yielding and disease resistant, the advantages of this native domesticate did not compensate for the multiple washings needed to remove the toxic substances in its seeds.

Turnips, leeks, parsnips, salsify, and beets did not catch on among highland Indians. Yet several native root crops nevertheless declined: oca, ullucu, añu (mashua), and maca eventually lost agronomic space to another native, the potato, especially at elevations below 3500 m. Greater food choice and security as a result of Old World introductions marginalized the small tubers, but the main beneficiary was the more efficient potato that was also regarded by Andean people as better tasting.

Vegetables, condiments, and fruits of Old World origin added to the food possibilities of the post-Conquest Andes. Native potherbs were quinoa and amaranth leaves, and those of certain field weeds collectively called *yuyo;* their use has persisted. Of the many European greens, only cabbage gained marginal Indian approval, whereas lettuce, eggplant, celery, cardoon, and orach scarcely were known or appreciated.[3] Onions, garlic, coriander, basil, and other herbs were embraced by native people, but did not replace capsicum pepper, which remains the most appreciated condiment to this day. Anise, a Mediterranean herb with a strong flavoring in its seed-like fruits, has been cultivated sparingly but continuously since its sixteenth-century introduction.

Almost a dozen sweet fruits of the warm valleys were domesticated in or near the Andes, but none of them could compare with the food value of the banana or the refreshing acidity of the orange. In the temperate valleys, trees of apple, medlar, quince, pear, plum, and peach that the Spaniards brought gained some currency among Indians, who were enjoined to plant them where the local climate permitted (Saravia Viejo 1989, II:258). But these rosaceous fruit trees produced rather poorly in their transplanted Andean setting, compared to the capuli cherry, which the Spaniards brought to the Andes from Middle America in the sixteenth century.

Assimilation of European Livestock

The sharp population decline in the Central Andes in the first century after the Conquest may have hastened native adoption of European livestock. In the production of food calories, animal husbandry normally requires less human labor than does crop growing. It was more feasible to pasture animals on the weeds of abandoned terraced or ridged fields than to cultivate and repair them (Caillavet 1989, 122–24). Small children enlisted to watch grazing animals freed elders for other tasks.

The Spanish animal inventory offered Andean folk several advantages and had a greater overall impact than did the corresponding plant list. The range of useful beasts increased subsistence security by providing a form of living capital that could be kept in reserve until butchered, traded or sold, at which time quadrupeds could move under their own power to local markets. Livestock also served as a food reservoir in case drought, flood, or frost destroyed standing crops. Andean acceptance of Old World herbivores was encouraged by the institution of common lands that continued after the Conquest (Godoy 1991). Spaniards were accustomed to this shared resource surrounding most villages in Iberia (Vassberg 1984); they viewed collective land use in the New World Indies as appropriate for herds of privately-owned animals. Unlike its fine-tuned seasonality in late Medieval Spain, transhumance never broadly emerged in the Andes, although localized movements of animals are not unknown (Stewart et al. 1976). Livestock of Eurasian origin also provided the main organic fertilizers after the Conquest when the regular transport of guano from the coastal islands to highland maize plots stopped with the decline of llama trains. In Spain at the time of the Conquest, fields were typically manured by livestock grazing on the stubble, which became one of the ways in which livestock raising was part of a larger agricultural system (Butzer 1988).

European beasts of burden were generally superior to the llama. Though cheaper to maintain, capable of surviving on dry native grasses and brackish water, and more sure-footed in rugged terrain than a mule, the llama carried only one fourth the weight, had less stamina, and was harder to drive. Mules became important in the Andes because they had no real competition from oxcarts. The rugged terrain, absence of suitable bridges, and the strong tendency for native people to porter their belongings did not favor cart transport of the kind used in medieval Castile (Ringrose 1970). In the early Colonial period, most mule breeding and selling was in the hands of Spaniards and mestizos, although Indians were engaged as muledrivers. For most of the sixteenth century, Indians were forbidden to ride horses, as riding was a prerogative of Spaniards, though they did use them as pack animals. Donkeys better fit the needs of a hardscrabble peasantry. Hardy and economical to keep, they carried harvested crops, firewood, and fertilizer over irregular terrain. By the end of the Colonial period, equines had largely supplanted llamas as pack animals in Ecuador, northern Peru, and southernmost Bolivia, as well as elsewhere at elevations below 3000 m.

The sheep was the single European introduction most valuable to native people. First kept on Spanish estates, by the 1590s they formed large flocks (Egaña 1966). At least as early as 1560, Indians were keeping sheep; by the seventeenth century, they had well integrated sheep into the native economy and land use. Softer yet greasier than llama fiber, sheep's wool was woven for the warm homespun clothing needed in the frigid Andean nights. Greasiness is an advantage for water-resistant garments. Ovines yielded mutton, tallow, dung, and in a few places, milk. Sheep, with no set lambing period in these tropical latitudes, had lower mortality and higher fertility than did the alpaca. Below 3500 m, sheep displaced the camelids, but they also prevailed above that elevation and elsewhere if Spanish influence was strong enough (Fig. 3). For example, on the *puna* of Tarma in the eighteenth century, camelids were no longer mentioned as livestock on either haciendas or in native communities (Arellano Hoffmann 1988).

Cattle proved to be less suitable for Indian needs than were sheep, especially at high elevations and on steep terrain. Satisfactory fodder during the dry season was often in short supply. A cow or bull represented a major investment whose loss through accident, disease, or theft created economic hardship. Milk, one of the major reasons for keeping cows elsewhere in the world, was not originally part of the Andean diet, though some Indians did learn to make cheese. When butchered, cattle provided meat, and cow dung had use as a fuel and fertilizer. Where peasants adopted the plow, oxen were the only draft animals, for unlike in Spain, cows, horses, mules, or donkeys were not used for this purpose in the Andes.

Innate resourcefulness of the goat and pig stood these creatures in good stead. The goat's Andean niche was in warm dry valleys where a market developed for wineskins (*odres*) used to carry liquids, especially alcohol. In these places, the goat's browsing habits made creative use of land that was not feasible to irrigate. Introduced with Pizarro in 1531, the pig became a component of Andean rural life for the twin rewards of meat and lard (Gade 1986). The key to its ready acceptance was the foraging ability of the long-legged, narrow-snouted Iberian breed. Lacking the acorns that they fed on in Spain, the Iberian pig in the Andes became a village forager of garbage, gleanings, and—in the absence of a latrine tradition—excrement.

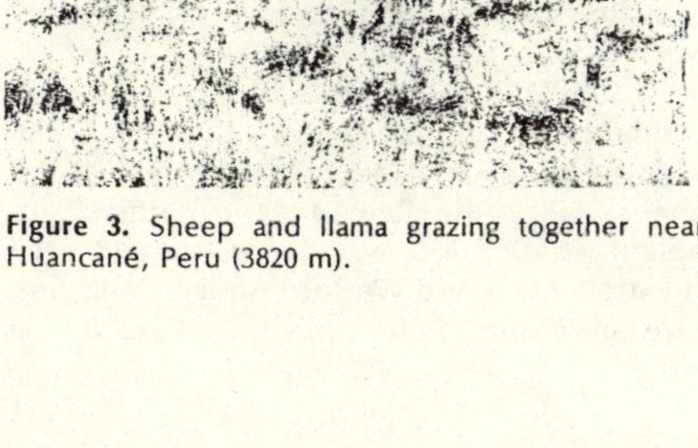

Figure 3. Sheep and llama grazing together near Huancané, Peru (3820 m).

Live chickens and eggs were sixteenth-century tribute items, so that their initial adoption may not have involved free choice. Ability to forage weed seeds and dooryard insects was critical in their acceptance into the household economy. But live hens and eggs, when sold, provided market income for peasant women. The meat, characteristically tough and therefore stewed, was eaten on special occasions, and the fat had medicinal uses. The chicken partially displaced the Muscovy duck, a native Andean domesticate that is a less efficient egg or meat producer. Other fowl introductions were generally unsuccessful. Old World species of ducks and geese, and the turkey brought to the Central Andes from Mesoamerica after the Conquest, seldom became part of the barnyard menagerie. Iberians also brought domestic pigeons to the Andes to be used for food and fertilizer, but only in a few places of Spanish influence did peasants take to raising them.

The Old World rabbit and the New World guinea pig occupied similar niches: small but prolific, both herbivores yield valuable protein. Rabbits (*conejos de Castilla*) were the more efficient meat producers, but Indians stuck to their guinea pigs (*conejillos de Indias*), which had the ability to thrive with little care in the dark recesses of the dwelling (Gade 1967). Andean affinity for their squeaking rodents went deeply into Andean tradition as a featured food at events marking one's life cycle and in folk curing rites.

Old World livestock contributed heavily to native livelihood, but they also had their negative effects. The livestock trampled and obliterated many abandoned ridged fields and weakened the facing of agricultural terraces. Their sharp hooves and heavy weight damaged the traditional bridges woven of plant fibers (Gade 1972). Roving flocks pulverized the soil surface

Figure 4. Work teams near Yauyos, Peru in 1942, plowing bench terraces for maize cultivation with the *chaquitaclla* as women sow on prepared land. (A. Guillén, photographer. Abraham Guillén collection, Latin American Library, Tulane University. Reproduced with permission.)

on slopes, provoking gully and sheet erosion during the torrential downpours of the rainy season. Need for additional pasture accelerated removal of some of the few woodlands still remaining after the Conquest. Livestock intrusion into fields of standing crops was a frequent cause of social discontent and violence within and between communities.

Rural Technology

Indian agriculturalists accepted the European plow without giving up their native spade-like equivalent, the *chaquitaclla,* a curved piece of wood with a bronze blade or a fire-hardened wooden point, and a footrest to provide leverage (Gade and Rios 1972) (Fig. 4). Beginning in the latter half of the sixteenth century, the oxen-pulled plow gave an alternative to the *chaquitaclla* on valley floors, first where Spaniards took over the land and later in indigenous communities. Native acceptance was arguably accelerated by Viceroy Toledo's edict of 1575 mandating a plow and oxen for each Indian agglomeration. Acceptance of the ard brought with it a radically new idea in Andean agriculture: the collective cultivation of certain crops as a field unit rather than individual attention imparted to each plant by the work of a mattock.

Of the four basic kinds of plows used in sixteenth-century Spain, only the Mediterranean scratch plow (*arado dental*) spread to the Andes. Since the Roman period, this lightweight wooden tool was the instrument used in Extremadura and Western Andalusia to prepare the ground before planting (Caro Baroja

Figure 5. A scratch plow constructed of eucalyptus wood but with no metal plowshare, near Sucre, Bolivia (2900 m).

1949). The ard's New World success can be attributed to the simplicity of its design and perhaps the small amount of iron needed in the plowshare; even that could be dispensed with in light soils (Fig. 5). No document shows that a prototype of the scratch plow was unloaded in Peru from either Seville, Central America, or the Antilles. Shipping a bulky tool such as a plow halfway around the world was hardly necessary, since many Spanish peasants who disembarked in South America could have constructed them from memory using the wood of native trees.

The *chaquitaclla* was relegated to slopelands too steep for oxen to maneuver. It also remained useful for working heavy soils, boulder-strewn flatlands, and narrow agricultural terraces and for specific tasks such as drilling seed holes, harvesting tubers, preparing mud for making adobes, removing rocks, digging ditches, and making furrows. Likewise the mattock, a short-handled and sharp-angled tool used for weeding maize and harvesting tubers, was not abandoned when the long-handled hoe was introduced. Instead, the mattock was fitted with a metal piece the width of a shovel and called a *lampa*. But the sickle, closely associated with wheat and barley harvesting, had no native counterpart. Ripe chenopod plants are still often pulled up by their roots. The scythe, a later improvement on the sickle, was not used in the Andes.

Small seeds in the pre-Conquest Andes were threshed with sticks or poles. The flail, one of several threshing devices in Spain, was an improvement over a simple pole, but rarely seen in the colonial Andes. What did catch on was having animals, preferably equines but never llamas, trample wheat and barley on a circular floor (Fig. 6). In Spain the hooves of livestock were largely superseded by the tribulum or *trillo* (Sanz Moreno 1985). This flint or iron-studded sled pulled by a draft animal removed grain kernels from their glumes and cut the straw into small pieces for use as fodder. It was a relatively efficient device, but the absence of flint and scarcity of iron in the early decades may explain why the Spaniards did not build and use it in the Andes.

Bread-making came to the Andes in the sixteenth century. Spanish artisans brought the concept of the arch and built water-driven grist mills of that design. Sixteenth-century Spain had both the horizontal waterwheel (*molino de rodezno*) and the more efficient vertical waterwheel (*aceña*), which dominated milling in or near cities. The more primitive technology continued to operate on streams close to small settlements in rural Iberia (Escalera Reyes and Villegas Santaella 1983). Only the horizontal wheel, whose gearless simplicity required minimal skill to operate or repair, was successfully implanted in the Andes (Gade 1971). Although used and operated by Indians, the mills long remained in the hands of Spaniards or mestizos. The bread oven also was of Spanish origin;

Figure 6. Donkeys threshing barley near Layo, Peru (3900 m).

Indians had baked in earthen pits. As with mills, oven design scarcely changed over the next four centuries (Urribari de González 1967).

Hispanic Influences on Settlement

The Folk Dwelling

The possibilities for melding rather than simply juxtaposing traditions were greater with settlement features than with agropastoral elements. The basic rural Andean dwelling of rectangular form and gabled roof goes back to the Inca period. Other forms coexisted with it both before and after the Conquest, as different ethnic groups maintained their own pre-Inca construction styles that included round form, more than one room and/or story, and flat roofs. Europeans, like the Incas before them, favored rectangularity, although circular construction was not unknown in Spain. Viceroy Toledo's decree of 1570 mandated a squarish dwelling form in the Indian villages. Straight walls best accommodated the two pieces of furniture that Spaniards regarded as necessary to a civilized existence: tables to eat from and beds to sleep on (Vargas Ugarte 1951, I:373–74). In contrast to the importance the Spaniards placed on those domestic accoutrements—perhaps a cultural reaction against Moorish custom, which had neither—Andean Indians did not use furniture and even today often see no need to partition their one-story dwellings in which cooking, eating, sleeping, and crop storage occur within a single room. In Southern Spain, most rural houses were one story (Giese 1951, 588), and although two-story dwellings were known to both Incas and Spaniards, those constructed in the Colonial period reflect Spanish design.

Adobe, sod, fieldstone, and cut stone were Andean building materials before the arrival of the Spaniards. The exquisitely fashioned ashlars associated with Inca civilization were not used for vernacular buildings. Fieldstone with mud mortar continues to serve for wall construction up to the present, but in the Colonial period adobe became the most widespread building material. In both pre- and post-Conquest times, adobes were made by puddling soil of at least 15 percent clay content with wild grass as the binding agent and drying in the sun (Sutter Esquenet 1985). A Spanish introduction, the wooden mold to shape the mud, substantially improved the uniformity of the block over the irregular Inca adobes (Fig. 7). Spaniards from some parts of Spain were experienced in adobe construction, which the Moors may have taught them; the word is derived from the Arabic, *attub,* meaning brick.

Figure 7. Adobe dwellings with thatch roofs at Capacmarca, Peru (3400 m).

Another earthen wall construction of independent invention was the tapial method, which saves time and labor over adobes. Used in pre-Conquest Peru (Cobo 1956, 241), this type of wall was also known in Moorish Spain. The Spanish method of constructing the wall by using a box-like wooden frame in which to pack the earth diffused in the Andes at the expense of the indigenous method in which earth was molded around straight bamboo-like canes (Muelle 1978). Whether tapia or adobe, the Spaniards brought and sometimes imposed on the indigenous population whitewashing for protection and appearance. Fired bricks also came from the Iberian Peninsula, where they date from the Roman period, but were not much used in the rural Andes. Before the nineteenth-century arrival of eucalyptus, fuelwood in the Andes was much scarcer than it is now (Johannessen and Hastorf 1990).

The highland cottage has had a pitched roof whose steepness varied regionally. In rural zones, the indigenous method of lashing roofing timbers long prevailed over the use of jointed beams, which was a Spanish introduction. Roofing materials also showed cultural differences. Thatch of locally available grasses was the only real roofing material before the Spaniards came. Baked tiles made of clay, used first by Spaniards on their own buildings, gradually spread to the countryside in the late Colonial period. Grass thatch, though free for the taking, requires some skill to install properly, normally lasts less than ten years, and shelters vermin. The conquerors tried to impose the concept of a proper lockable door on native dwellings for security and privacy (Jiménez de la Espada 1965, I:97). At least the wooden door set in a wooden frame was more successfully implanted than were windows. Inca dwellings sometimes had a small gable opening to let in light, whereas Spanish design favored windows below the roof line. The interior wall niche, another design detail of the Inca vernacular house, does not appear in native dwellings of the Colonial period or later.

Nucleation and Dispersion

At the time of the Spanish Conquest and for four decades thereafter, most Andean people lived near their fields. In 1570 the Spanish viceroy, Francisco de Toledo, ordered native people to group in compact agricultural villages (*reducciones*) to facilitate catechization, tribute collection, and work assignments (*mita*). Reordering the Andean settlement pattern rested on a fundamental cultural assumption of the Iberian conquerors that agglomeration was another of the elements essential for civilized living. In New Castile, Extremadura, and Andalusia of the fifteenth and sixteenth centuries, peasants lived in nucleated settlements. Towns, not the countrysides, controlled and directed agriculture, which placed the rural focus in Spain on *urbanitas* not *rusticitas* (Blok and Driessen 1984). The daily interactions in this arrangement fostered an urban way of life, not by functional diversity, which was nonexistent in these villages, but by providing the opportunities to acquire the civilized virtues that have been the mark of Mediterranean existence for more than two millenia.

The Spanish attempt to implant the values of *polis* and *civitas* in the isolated valleys and plateaus of the High Andes evoked the agrotowns of the Iberian Peninsula. Nucleation was perceived as a fundamental prerequisite for learning European values of cleanliness and order, as well as the more subtle human qualities of finesse, vivacity, and the art of pleasing. A sixteenth-century Spanish jurist in Peru, Juan de Matienzo (1967, 48), expressed that attitude when he wrote of native people that "they cannot be Christianized nor even men without being together in towns . . . ." The gridiron layout of these tight clusters of farmers around an open square paralleled the accepted design for Spanish colonists. The inspiration for right-angled settlements is uncertain, for most Spanish towns or cities had no such layout. A notable exception and possibly the model for Spanish America was Santafé, near Granada. Founded by Queen Isabella in 1491 with a gridiron regularity of streets, Santafé was the headquarters for the siege of the last redoubt of Moorish control in Europe.

In the Andes, the Mediterranean concept of nucleation failed to achieve its goals, for the agro-village did not retain the indigenous farming population (Gade and Escobar 1982). A farmer could commute not much more than four kilometers between his/her village and fields. Even before the end of the sixteenth century, Indians departed *reducción* villages and reestablished themselves near farm plots. They also dispersed to remote, inhospitable *puna* zones to escape their oppressors or to avoid social conflict with other families (Favre 1975; Houdart 1980). Nucleation made working the land more difficult and exacerbated tensions of living near strangers that contradicted their kin-based concepts of proper social organization.

A 1604 decree from Madrid commanding Indians to live in the *reducción* was rescinded only fourteen years later when the Spaniards realized the agglomerated ideal had disintegrated and could not be reconstituted (*Recopilación de leyes* 1841). Diffuse settlements of pre-Conquest pattern from isolated or semi-dispersed dwellings to loose hamlets reemerged, but without legal titles to the lands they occupied. To legitimize the claims of these reconstituted communities, Hispanic, not Inca, concepts were used: a chapel was constructed, a patron saint selected, a fiesta sponsored and, at a later time, a school was built. While the rural population of the Central Andes has partly reverted to the pre-Conquest dispersed pattern, most sixteenth-century *reducciones* survive, not as *pueblos de índios,* but as mestizo villages (Gade and Escobar 1982).

Spanish Impact in Space and Time

European ideas and materials were combined in uneven spatial patterns with those already in the Andes. In the more accessible temperate valleys and basins between 2500–3500 m elevation, the two cultures had converged by the late eighteenth century. Attracted by the abundant native population, alluvial soils, and amenable climate, Spaniards established their haciendas on the valley floors. Many elements of Old World material culture were preadapted to this elevational range much better than above or below it. To the agropastoral mélange was added the miscegenation (*mestizaje*) that gradually reformulated the ethnic composition of most temperate valleys. A notable example from Peru is the Yucay Valley (that portion of the Urubamba or Vilcanota depression between 2800–2900 m above sea level), which Spaniards repeatedly compared to the Tagus Valley around Aranjuez (Gade 1968). The benign climate, absence of malaria, and dense Indian population prompted Spaniards to appropriate it for themselves, which gradually shifted the racial and ethnic character of the rural peasantry from Indian to mestizo. Between 2500–1000 m, the European influence was even more prevalent due to the sparse indigenous populations at these generally insalubrious elevations. Spaniards, lured to the warm valleys by the possibilities of commercial sugar and coca production, used seasonal migrants and black slaves who themselves had little stake in the land (Gade 1973). Indigenous land use and settlement patterns persisted most tenaciously at elevations above 3600 m, where European crops were undependable in the freeze-prone climate of these altitudes. Though sheep were present, llamas and alpacas, because of their superior physiological adaptation to the altitude and natural forage, continued from central Peru to southern Bolivia to have an importance they had lost in the valleys below. Spaniards shunned these cold, hypoxic zones of *puna* and its wetter variant, the *páramo,* leaving them as refuges for Indians retreating from Spanish domination. Rudimentary dwellings dispersed in the pre-Conquest pattern reflected the pastoral emphasis in these lofty *punas* and *páramos*. Later the Spaniards and mestizos coveted the land for grazing, but even then not for residence.

The successful introductions of many Spanish rural elements to the Central Andes occurred rather early in the Colonial period. Some of these adoptions swept through the highlands not simply because they were available, but because they filled a need at a critical time. Catastrophic depopulation swept Andean communities after the Conquest and forced a shift from intensive cultivation to extensive agropastoralism. From the array of goods and ideas that the Spaniards brought, the native people selected those that enhanced family security in the face of the breakdown of the Inca system of food redistribution and the demographic decimation. Those Old World elements that were perceived to meet the exigencies of the period gained an entrée into the cultural-ecological system. The livestock additions fit into the gaps left in the changed agricultural situation that resulted after a community's labor force had been greatly diminished. They also enabled the peasantry to better survive the uncertainties of crop production due to drought, hail, and unseasonal freezes that periodically scourged the highlands. New crops helped to reduce the risk of food failure, but the two pre-Conquest staples, maize and potatoes, have remained the outstanding sources of food and nutrition in the region up to the present. Both provide high caloric content, adapt to a range of environmental conditions, and carry emotional, even religious, associations.

The Andean peasantry acquired a simplified version of the Spanish inventory of rural material life. Many elements were eliminated. Moreover, the Old World introductions to the Central Andes did not overwhelm the native elements as they did in the Argentine Pampa, Middle Chile, or Northern Mexico. The Spanish inventory did not displace the well-developed agricultural achievements in crop raising, pastoralism, technology, and settlement left by the Inca. Instead the two traditions, each having been altered—the Spanish, simplified, the Inca, disrupted—merged into one hybrid cultural complex. This expanded agropastoral complex, though it did not lead to well-being

Figure 8. Andean rural life juxtaposes different cultural forms and functions that themselves do not necessarily belong to different time periods. Pack llamas transport potatoes to a village of twentieth-century origin [Hector Tejada, Department of Cuzco, Peru] having buildings with colonial-style doors and balconies that, at the same time, incorporate the much more modern notion of the sheet-metal roof.

or escape from poverty, has lasted 400 years, evidence of its environmental adaptation and integration into Andean life. Andean peasants consider their Old World plants and animals to be autochthonous ("*criollo*"), so well are these heirloom varieties and antique breeds integrated into their system. Field preparation, rotation schemes, intertillage, mixed grazing, vernacular architecture, and settlement patterns continue to show features of both traditions (Fig. 8). European elements have also interlaced with the Andean in clothing, diet, folk medicine, religion, and language. These elements, introduced by 1550 and widely accepted by 1650, tended to crystallize into an established pattern that did not allow much subsequent variation. This concept of cultural crystallization that froze in time the wherewithal of peasant existence explains in part the archaic cast of contemporary Andean livelihood (Foster 1960, 227–34).

Conclusion

The Andean cultural landscape is above all a syncretism. In spite of their hermetic reputation, Andean people have always been pervious to outside influences when nonnative elements demonstrated superiority or when they filled an apparent vacuum if the risk of adoption was not too high. European introductions created new possibilities for enhancing rural livelihood. Those elements of Western material culture preadapted to Andean conditions that were of value to Andean people were incorporated into their folkways in the first half of the Colonial period without much later variation.

In this century and especially since the 1940s, the Andean peasant view of the world has been much broadened beyond the local community. Pendular migration to adjacent lowlands, military service, contact with development agencies, easier communications that include more vehicle roads as well as radio and telegraph, and growing politicization at several scales, have accelerated acceptance of parts of the national cultures emanating from the large cities.

Given the more than four centuries of Hispanic domination, it is quite remarkable that this irreparable cultural amalgam still characterizes the Central Andean realm. Only some of the diffused elements from Spain were successfully implanted into peasant usage; others lingered with a peripheral status, and a few never gained assent at all. But those European crops, animals, tools, and dwellings that did gain acceptance enriched Andean livelihoods by stretching the range of possibilities to better spread the risks imposed by nature and by offering nutritional combinations unknown before. The Central Andean cultural landscape of the late twentieth century, which owes much to both traditions, is still a time warp. Poverty and isolation have played a role in greatly slowing the pace of modernization into the Andes. But technological décalage should not obscure the fundamental accommodations that gradually emerged from the Encounter of these two earthly segments. Moreover, it should not deceive one into thinking that people themselves are artifacts of some earlier era. The Columbian Quincentennial holds a special charge to geographers to explore without deterministic preconceptions the adaptational dimension of intercontinental transfers around the world in the context of time, space, and real places.

Notes

1. Particularly influential in presenting the distorted view of contemporary Andean culture as indige-

nous was the study by Mishkin (1946), whose generalizations about Andean subsistence and social organization were largely based on ethnographic descriptions gathered in 1937–38 in the isolated high-altitude village of Kauri in the Department of Cuzco. Another author, Castro Pozo (1946), wrote a highly glossed essay that misrepresented the social structure of the rural Andean community as a kind of Inca socialism. Both the above works, published in the same volume, have influenced the formulation of research projects over the past four decades. The notion of continuity from the pre-Inca and Inca past into the present has been a compelling interest in Andean studies (e.g. Isbell 1977; Rasnake 1988). That romanticization of the native extends to a recent volume on the agricultural botany of lesser Andean crops, some of whose virtues are overrated (National Research Council 1989). "Andeanism" refers to the false notion, held by an array of scholarly, artistic, and popular observers, of a highland peasantry "outside the flow of modern history" (Starn 1991, 64).

2. Major lacunae in the documentary record make reconstruction of the rural folkways and land use in historic times, whether of Spain or the Andes, an approximation. Several published sources form the bulk of what we know. For the Iberian Peninsula, the work of Gabriel Alonso de Herrera (1970) describes the state of Castilian farming in the late fifteenth and early sixteenth centuries. For the Central Andes of the early Colonial period, the observations of chroniclers José de Acosta (1962), Pedro Cieza de Leon (1986), Bernabé Cobo (1956), Felipe Guaman Poma de Ayala (1980), and Antonio Vázquez de Espinosa (1948) provide accessible and generally reliable data. El Inca Garcilaso de la Vega (1960) has much to say on this theme, but his facts are not always accurate and must be used with reserve. The *Relaciones geográficas*, a collection of essentially local reports in response to a Crown request, is a rich source on late sixteenth-century rural life in the Andes (Jiménez de la Espada 1965).
3. Among the fifteenth-century crops in Seville were several that may have been originally brought to the Andes as cultivated plants, but which later spread mainly as weeds: purslane (*Portulaca oleracea*), sow thistle (*Sonchus* spp.), rocket (*Eruca vesicaria*), radish (*Raphanus sativus*), and turnip (*Brassica campestris*) (Aviñon 1885).

References

Acosta, Joseph de. 1962 [1590]. *Historia natural y moral de las Indias*. Mexico City: Fondo de Cultura económica.

Alcina Franch, J. 1986. *La cultura de Castilla y León en América: La cultura material*. In *Etnología y folklore en Castilla y León*, ed. L. Díaz Viana, pp. 357–69. Salamanca: Junta de Castilla y León.

Aviñon, J. 1885 [1418–19]. *Sevillana medicina*. Sevilla: Sociedad de Bibliófilos andaluces.

Arellano Hoffmann, C. 1988. Anotaciones del clima, ganado y tenencia de pastos en la puna de Tarma, siglo XVIII. In *Llamichos y paqocheros: pastores de llamas y alpacas*, ed. J. A. Flores Ochoa, pp. 77–84. Cuzco: CONCYTEC-CEAC-Editorial UNSAAC.

Arguedas, J. M. 1968. *Las comunidades de España y del Perú*. Lima: Universidad Nacional Mayor de San Marcos.

Bermúdez Plata, C. 1942. *Catálogo de pasajeros de Indias*. Sevilla: Consejo Superior de Investigaciones científicas.

Blok, A., and Driessen, H. 1984. Mediterranean agro-towns as a form of cultural dominance with special reference to Sicily and Andalusia. *Ethnologia Europaea* 14:111–24.

Borah, W. 1954. *Early colonial trade and navigation between Mexico and Peru*. Ibero-Americana 38. Berkeley: University of California Press.

Butzer, K. 1988. Cattle and sheep from old to New Spain: Historical antecedents. *Annals of the Association of American Geographers* 78:29–56.

Caillavet, C. 1989. Las técnicas agrarias autóctonas y la remodelación colonial del paisaje en los Andes septentrionales (siglo XVI). In *Ciencia, vida y espacio en Iberoamérica*, ed. J. L. Peset, pp. 109–26. Madrid: Consejo Superior de Investigaciones científicas.

Caro Baroja, J. 1949. Los arados españoles (sus tipos y reparticiones). *Revista de Dialectología y Tradiciones populares* 5:3–96.

Castro Pozo, H. 1946. Social and economico-political evolution of the communities of central Peru. In *Handbook of South American Indians*, vol. 2, ed. J. Steward, pp. 483–500. Bureau of American Ethnology Bulletin 143. Washington: Smithsonian Institution.

Cieza de León, P. 1986. *Crónica del Perú*. 3 vols. Lima: Fondo Editorial P. Universidad Católica del Perú.

Cobo, B. 1956 [1653]. *Obras de Bernabé Cobo*. Madrid: Biblioteca de Autores Españoles.

Cook, N., ed. 1975. *Tasa de la visita general de Francisco de Toledo*. Lima: Universidad Mayor de San Marcos.

———. 1981. *Demographic collapse: Indian Peru, 1520–1620*. New York: Cambridge University Press.

Denevan, W. 1983. Adaptation, variation and cultural geography. *The Professional Geographer* 35:399–407.

———. 1987. Terrace abandonment in the Colca Valley, Peru. In *Pre-Hispanic agricultural fields in the Andean region* (part 1), ed. W. Denevan, K. Mathewson, and G. Knapp, pp. 1–43. International Series 359 (i). Oxford: British Archaeological Reports.

Dollfus, O. 1982. Development of land-use patterns in the Central Andes. *Mountain Research and Development* 2:39–48.

Donkin, R. 1979. *Agricultural terracing in the ab-*

original New World. Tucson: University of Arizona Press.

Egaña, A., ed. 1966. *Monumenta peruviana (1586–1591)*, vol. 4. Rome: Apud "Monumenta Historica Soc. Iesu."

Escalera Reyes, J., and Villegas Santaella, A. 1983. *Molinos y panaderías tradicionales*. Madrid: Editora Nacional.

Favre, H. 1975. Le peuplement et la colonisation agricole de la steppe dans le Pérou central. *Annales de Géographie* 84:415–40.

Foster, G. 1960. *Culture and conquest: America's Spanish heritage*. Chicago: Quadrangle Books.

Gade, D. 1967. The guinea pig in Andean folk culture. *Geographical Review* 57:213–24.

———. 1968. Aranjuez of the New World. *Américas* 20:12–19.

———. 1971. Grist milling with the horizontal waterwheel in the Central Andes. *Technology and Culture* 12:43–51.

———. 1972. Bridge types in the Central Andes. *Annals of the Association of American Geographers* 62:94–109.

———. 1973. Environment and disease in the land use and settlement of Apurimac Department, Peru. *Geoforum* 16:37–45.

———. 1975. *Plants, man and the land in the Vilcanota Valley of Peru*. Biogeographica 6. The Hague: W. Junk B. V. Publishers.

———. 1986. The Iberian pig in the Central Andes. *Journal of Cultural Geography* 7:35–50.

——— **and Escobar, M.** 1982. Village settlement and the colonial legacy in southern Peru. *Geographical Review* 72:430–49.

——— **and Rios, R.** 1972. Chaquitaclla: The native footplough and its persistence in central Andean agriculture. *Tools and Tillage* 2:3–15.

Garcilaso de la Vega, I. 1960. *Comentarios reales de los Incas*. Cuzco: Ediciones de la Universidad Nacional del Cuzco.

Giese, W. 1951. Los tipos de casa de la Península Ibérica. *Revista de Dialectología y Tradiciones populares* 7:563–601.

Godoy, R. 1991. The evolution of common-field agriculture in the Andes: A hypothesis. *Comparative Studies in Society and History* 33:395–414.

Guaman Poma de Ayala, F. 1980. *El primer nueva corónica y buen gobierno*, ed. J. Murra and R. Adorno. 3 vols. Mexico City: Siglo Veintiuno.

Guillet, D. 1987. Terracing and irrigation in the Peruvian highlands. *Current Anthropology* 25:409–30.

Herrera, G. de Alonso. 1970 [1512]. *Obra de agricultura*. Madrid: Biblioteca de Autores Españoles.

Houdart, M. 1980. Un exemple de scissiparité de village dans les Andes: Le cas de Pilchaca. *Bulletin de l'Institut Français d'Etudes andines* 9:35–58.

Isbell, B. 1977. *To defend ourselves: Ecology and ritual in an Andean village*. Austin: University of Texas Press.

Jimenez de la Espada, M., ed. 1965 [1580]. *Relaciones geográficas de Indias*, 3 vols. Madrid: Biblioteca de Autores Españoles

Johannessen, S., and Hastorf, C. 1990. A history of fuel management (A.D. 500 to the present) in the Mantaro Valley, Peru. *Journal of Ethnobiology* 10:61–90.

Knapp, G. 1988. *Ecología cultural prehispánica del Ecuador*. Quito: Ediciones del Banco Central del Ecuador.

——— **and Denevan, W.** 1985. The use of wetlands in the prehistoric economy of the northern Ecuadorian Highlands. In *Prehistoric intensive agriculture in the tropics*, ed. I. Farrington, pp. 185–207. International Series 232. Oxford: British Archaeological Reports.

López de Velasco, J. 1971 [1574]. *Geografía y descripción universal de las Indias*. Madrid: Biblioteca de Autores Españoles.

Masson, L. 1984. *Las terrazas agrícolas: Una tecnología olvidada*. Lima: Banco Continental.

Matienzo, J. 1967 [1567]. *Gobierno del Perú*, ed. G. Villena Lohmann. Paris-Lima: Institut français d'Études andines.

Mishkin, B. 1946. The contemporary Quechua. In *Handbook of South American Indians*, vol. 2, ed. J. Steward, pp. 411–70. Bureau of American Ethnology Bulletin 143. Washington: Smithsonian Institution.

Muelle, J. 1978. Tecnología del barro en el Perú precolombino. In *Tecnología andina*, ed. R. Ravines, pp. 573–79. Lima: Instituto de Estudios Peruanos.

National Research Council. 1989. *Lost crops of the Incas*. Washington: National Academy Press.

Oliveros de Castro, M., and Jordana de Pozas, J. 1968. *La agricultura en tiempo de los reyes católicos*. Madrid: Instituto Nacional de Investigaciones Agronómicas.

Orlove, B., and Godoy, R. 1986. Sectorial fallowing systems in the Central Andes. *Journal of Ethnobiology* 6:169–204.

Pérez de Barradas, C. 1941. Las regiones españolas y la población de América (1509–1534). *Revista de Indias* 6:81–120.

Rasnake, R. 1988. *Domination and cultural resistance: Authority and power among an Andean people*. Durham, NC: Duke University Press.

Recopilación de leyes de los reinos de las Indias (Carlos II), 5th ed. 1841. Madrid: Boix.

Reparto de tierras en 1595. 1957. *Revista del Archivo Histórico* (Cuzco) 8:428–32.

Ringrose, D. 1970. Carting in the Hispanic world: An example of divergent development. *Hispanic American Historical Review* 50:30–51.

Romera Iruela, L., and Galbis Diez, M. 1980.

Catálogo de pasajeros a Indias, vol. 4 (1560–1566). Sevilla: Ministerio de Cultura.

Sanz Moreno, M. 1985. Una artesanía del pasado: Los trillos de Cantalejo (Segovia). *Estudios Geográficos* 46:496–502.

Sarabia Viejo, J. 1989. *Francisco de Toledo: Disposiciones gubernativas para el Virreinato del Perú 1575–1580.* 2 vols. Sevilla: Escuela de Estudios Hispano-Americanos.

Schjellerup, I. 1985. Observations on ridged fields and terracing systems in the northern highlands of Peru. *Tools and Tillage* 15:100–21.

Smith, C.; Denevan, W.; and Hamilton, P. 1968. Ancient ridged fields in the region of Lake Titicaca. *Geographical Journal* 134:353–67.

Spalding, K. 1984. *Haurochirí: An Andean society under Inca and Spanish Rule.* Stanford, CA: Stanford University Press.

Starn, O. 1991. Missing the revolution: Anthropologists and the war in Peru. *Cultural Anthropology* 6:63–91.

Stewart, N.; Belote, J.; and Belote, L. 1976. Transhumance in the Central Andes. *Annals of the Association of American Geographers* 66:377–97.

Sutter Esquenet, P. 1985. Arquitectura andina tradicional y sus problemas. *Cultura* (Quito) 7:145–214.

Troll, C. 1931. Die geographische Grundlagen der andinen Kulturen und des Inca Reiches. *Ibero-Amerikanisches Archiv* 5:258–94.

Uribarri de González, V. 1967. El pan en Ayacucho (Perú). *Revista de Dialectología y Tradiciones populares* 23:347–66.

Usselmann, P. 1987. Un acercamiento a las modificaciones del medio físico latinoamericano durante la colonización. *Bulletin de l'Institut français d'Études andines* 16:127–35.

Vargas Ugarte, R., ed. 1951. *Concilios limenses* (1551–1772). Lima: privately printed.

Vassberg, D. 1984. *Land and society in Golden Age Castile.* New York: Cambridge University Press.

Vázquez de Espinosa, A. 1948 [1613]. *Compendio y descripción de las Indias.* Smithsonian Miscellaneous Collections, vol. 109. Washington: Smithsonian Institution.

4

Bison Ecology and Bison Diplomacy: The Southern Plains from 1800–1850

Dan Flores

In bright spring light on the Great Plains of two centuries ago, governor Juan Bautista de Anza failed in the last of the three crucial tasks that his superiors had set him as part of their effort to reform New Mexico's Comanche policy. Over half a decade, Anza had followed one success with another. He had brilliantly defeated the formidable Comanche ***nomnekaht*** (war leader) Cuerno Verde in 1779, and as a consequence in 1786, he had personally fashioned the long-sought peace between New Mexico and the swelling Comanche population of the Southern Plains. His third task was to persuade the Comanches to settle in permanent villages and to farm.[1]

But the New Mexico governor found the third undertaking impossible. Observers of Plains Indian life for 250 years and committed to encouraging agriculture over hunting, the Spaniards were certain that the culture of the horse Indians was ephemeral, that the bison on which they depended were an exhaustible resource. Thus Anza pleaded with the tribes to give up the chase. The Comanches thought him unconvincing. Recently liberated by horse culture and by the teeming wildlife of the High Plains, their bands found the Arkansas River pueblo the governor built for them unendurable. They returned to the hunt with the evident expectation that their life as buffalo hunters was an endless cycle. And yet Anza proved to be a prophet. Within little more than half a century, the Comanches and other tribes of the Southern Plains were routinely suffering from starvation and complaining of shortages of bison. What had happened?[2]

Environmental historians and ethnohistorians whose interests have been environmental topics have in the two past decades been responsible for many of our most valuable recent insights into the history of native Americans since their contact with

[1] See Jacobo Loyola y Ugarte to Juan Bautista de Anza, Oct. 5, 1786, roll 10, series II, Spanish Archives of New Mexico, microfilm (New Mexico State Archives, Santa Fe).

[2] *Ibid.* One line reads: "use all your sagacity and efficiency, making evident to the [Comanche] Captains . . . that the animals they hunt with such effort at sustenance are not at base inexhaustible." See also Alfred B. Thomas, ed., *Forgotten Frontiers: A Study of the Spanish Indian Policy of Don Juan Bautista de Anza, Governor of New Mexico, 1777–1787* (Norman, 1932), 69–72, 82.

Euro-Americans.[3] Thus far, however, modern scholarship has not reevaluated the most visible historic interaction, the set piece if you will, of native American environmental history.[4] On the Great Plains of the American West during the two centuries from 1680 to 1880, almost three-dozen native American groups adopted horse-propelled, bison-hunting cultures that defined "Indianness" for white Americans and most of the world. It is the end of this process that has most captured the popular imagination: the military campaigns against and the brutal incarceration of the horse Indians, accompanied by the astonishingly rapid elimination of bison, and of an old ecology that dated back ten thousand years, at the hands of commercial hide hunters. That dramatic end, which occurred in less than fifteen years following the end of the Civil War, has by now entered American mythology. Yet our focus on the finale has obscured an examination of earlier phases that might shed new light on the historical and environmental interaction of the horse Indians and bison herds on the Plains.

In the nineteenth-century history of the Central and Southern Plains, there have long been perplexing questions that environmental history seems well suited to answer. Why were the Comanches able to replace the Apaches on the bison-rich Southern Plains? Why did the Kiowas, Cheyennes, and Arapahoes gradually shift southward into the Southern Plains between 1800 and 1825? And why, after fighting each other for two decades, did these Southern Plains peoples effect a rapprochement and alliance in the 1840s? What factors brought on such an escalation of Indian raids into Mexico and Texas in the late 1840s that the subject assumed critical importance in the Treaty of Guadalupe-Hidalgo? If the bison herds were so vast in the years before the commercial hide hunters, why were there so many reports of starving Indians on the Plains by 1850? And finally, given our standard estimates of bison numbers, why is it that the hide hunters are credited with bringing to market only some 10 million hides, including no more than 3.5 million from the Southern Plains, in the 1870s?

Apposite to all of these questions is a central issue: How successful were the horse Indians in creating a dynamic ecological equilibrium between themselves and the vast bison herds that grazed the Plains? That is, had they developed sustainable hunting practices that would maintain the herds and so permit future generations

[3] See Alfred W. Crosby, Jr., *The Columbian Exchange: Biological Consequences of 1492* (Westport, 1972); Alfred W. Crosby, Jr., *Ecological Imperialism: The Biological Expansion of Europe, 900–1900* (New York, 1986); Henry Dobyns, *Native American Historical Demography* (Bloomington, 1976); Henry Dobyns, *Their Number Become Thinned: Native American Population Dynamics in Eastern North America* (Knoxville, 1983); Calvin Martin, *Keepers of the Game: Indian-Animal Relations and the Fur Trade* (Berkeley, 1979); Richard White, *Roots of Dependency* (Lincoln, 1983); William Cronon, *Changes in the Land: Indians, Colonists, and the Ecology of New England* (New York, 1983); Paul Martin and Henry Wright, Jr., eds., *Pleistocene Extinctions: The Search for a Cause* (New Haven, 1967); and Paul Martin and Richard Klein, eds., *Quarternary Extinctions* (Tucson, 1985).

[4] See Richard White, "American Indians and the Environment," *Environmental Review*, 9 (Summer 1985), 101–3; Richard White, "Native Americans and the Environment," in *Scholars and the Indian Experience: Critical Reviews of Recent Writings in the Social Sciences*, ed. W. R. Swagerty (Bloomington, 1984), 179–204; Christopher T. Vecsey and Robert W. Venables, eds., *American Indian Environments: Ecological Issues in Native American History* (Syracuse, 1980); Richard White and William Cronon, "Ecological Change and Indian-White Relations," in *Handbook of North American Indians*, ed. William C. Sturtevant (20 vols., Washington, 1978–1989), IV, 417–29; and Donald J. Hughes, *American Indian Ecology* (El Paso, 1983).

of hunters to follow the same way of life? This is not to pose the "anachronistic question" (the term is Richard White's) of whether Indians were ecologists.[5] But how a society or a group of peoples with a shared culture makes adjustments to live within the carrying capacity of its habitat is not only a valid historical question, it may be one of the most salient questions to ask about any culture. Historians of the Plains have differed about the long-term ecological sustainability of the Indians' use of bison, particularly after the Euro-American fur trade reached the West and the tribes began hunting bison under the influence of the market economy. The standard work, Frank Roe's *The North American Buffalo*, has generally carried the debate with the argument that there is "not a shred of evidence" to indicate that the horse Indians were out of balance with the bison herds.[6] Using the new insights and methods of environmental history, it now appears possible systematically to analyze and revise our understanding of nineteenth-century history on the Great Plains. Such an approach promises to resolve some of the major questions. It can advance our understanding of when bison declined in numbers and of the intertwining roles that Indian policies—migrations, diplomacy, trade, and use of natural resources—and the growing pressures of external stimuli played in that decline. The answers are complex and offer a revision of both Plains history and western Indian ecological history.

Working our way through to them requires some digression into the large historical forces that shaped the Southern Plains over the last hundred centuries. The perspective of the *longue durée* is essential to environmental history. What transpired on the Great Plains from 1800 to 1850 is not comprehensible without taking into account the effect of the Pleistocene extinctions of ten thousand years ago, or the cycle of droughts that determined the carrying capacity for animals on the grasslands. Shallower in time than these forces but just as important to the problem are factors that stemmed from the arrival of Europeans in the New World. Trade was an ancient part of the cultural landscape of America, but the Europeans altered the patterns, the goods, and the intensity of trade. And the introduction of horses and horse culture accomplished a technological revolution for the Great Plains. The horse was the chief catalyst of an ongoing remaking of the tribal map of western America, as native American groups moved onto the Plains and incessantly shifted their ranges and alliances in response to a world where accelerating change seemed almost the only constant.

At the beginning of the nineteenth century, the dominant groups on the Southern Plains were the two major divisions of the Comanches: the Texas

[5] See White, "American Indians and the Environment," 101; and White and Cronon, "Ecological Change and Indian-White Relations."

[6] Several earlier scholars have addressed this question. For an early argument that the horse Indians overhunted bison, see William T. Hornaday, "The Extermination of the American Bison, with a Sketch of its Discovery and Life History," *Smithsonian Report,* 1887, pp. 480–90, 506. For statements of this position that offer nothing beyond anecdotal evidence, see James Malin, *History and Ecology,* ed. Robert Swierenga (Lincoln, 1984), 9, 31–54; George Hyde, *Spotted Tail's Folk* (Norman, 1961), 24; and Preston Holder, *The Hoe and the Horse on the Plains* (Lincoln, 1970), 111, 118. For the contrary view, see Frank Roe, *The North American Buffalo: A Critical Study of the Species in its Wild State* (Toronto, 1951), 500–505, 655–70; and Frank Roe, *The Indian and the Horse* (Norman, 1955), 190–91. The breadth and authority of Roe's books have given him priority in the field.

Comanches, primarily Kotsotekas, and the great New Mexico division, spread across the country from the Llano Estacado Escarpment west to the foothills of the Sangre de Cristo Mountains, and composed of Yamparika and Jupe bands that only recently had replaced the Apaches on the High Plains. The Comanches' drive to the south from their original homelands in what is now southwestern Wyoming and northwestern Colorado was a part of the original tribal adjustments to the coming of horse technology to the Great Plains. There is reason to believe that the Eastern Shoshones, from whom the Comanches were derived before achieving a different identity on the Southern Plains, were one of the first intermountain tribes of historic times to push onto the Plains. Perhaps as early as 1500 the proto-Comanches were hunting bison and using dog power to haul their mountain-adapted four-pole tipis east of the Laramie Mountains. Evidently they moved in response to a wetter time on the Central Plains and the larger bison concentrations there.[7]

These early Shoshonean hunters may not have spent more than three or four generations among the thronging Plains bison herds, for by the late seventeenth century they had been pushed back into the mountains and the sagebrush deserts by tribes newly armed with European guns moving westward from the region around the Great Lakes. If so, they were among a complex of tribes southwest of the lakes that over the next two centuries would be displaced by a massive Siouan drive to the west, an imperial expansion for domination of the prize buffalo range of the Northern Plains, and a wedge that sent ripples of tribal displacement across the Plains.[8]

Among the historic tribes, the people who became Comanches thus may have shared with the Apaches and, if linguistic arguments are correct, probably with the Kiowas the longest familiarity with a bison-hunting life-style. Pressed back toward the mountains as Shoshones, they thus turned in a different direction and emerged from the passes through the Front Range as the same people but bearing a new name given them by the Utes: Komantcia. They still lacked guns but now began their intimate association with the one animal, aside from the bison, inextricably linked with Plains life. The Comanches began acquiring horses from the Utes within a decade or so after the Pueblo Revolt of 1680 sent horses and horse culture diffusing in all directions from New Mexico. Thus were born the "hyper-Indians," as William Brandon has called the Plains people.[9]

[7] Since it utilizes long-ignored Spanish documents, on the Comanche's migration I follow Thomas Kavanagh, "Political Power and Political Organization: Comanche Politics, 1786–1875" (Ph.D. diss., University of New Mexico, 1986), rather than Ernest Wallace and E. Adamson Hoebel, *The Comanches: Lords of the South Plains* (Norman, 1951). Dimitri Boris Shimkin, *Wind River Shoshone Ethnography* (Berkeley, 1947); James A. Goss, "Basin-Plateau Shoshonean Ecological Model," in *Great Basin Cultural Ecology: A Symposium*, ed. Don D. Fowler (Reno, 1972), 123–27.

[8] Richard White, "The Winning of the West: The Expansion of the Western Sioux in the Eighteenth and Nineteenth Centuries," *Journal of American History*, 65 (Sept. 1978), 319–43.

[9] Kiowa origin myths set on the Northern Plains are at variance with the linguistic evidence, which ties them to the Tanoan speakers of the Rio Grande pueblos. Scholars are coming to believe that there is a connection between the mysterious Jumano peoples of the seventeenth- and eighteenth-century New Mexico documents and the later Kiowas. See Nancy Hickerson, "The Jumano and Trade in the Arid Southwest, 1580–1700," 1989 (in Dan Flores's possession). I am indebted to Ms. Hickerson for allowing me to examine her work. William Brandon, *Indians* (Boston, 1987), 340.

The Comanches became, along with the Sioux, the most populous and widespread of all the peoples who now began to ride onto the vast sweep of grassland to participate in the hunter's life. They began to take possession of the Southern Plains by the early 1700s. By 1800 they were in full control of all the country east of the Southern Rocky Mountains and south of the Arkansas River clear to the Texas Hill Country. Their new culture, long regarded as an ethnographic anomaly on the Plains because of its western and archaic origins, may not be unique, as older scholars had supposed it to be—at least if we believe the new Comanche revisionists. Irrespective of their degree of tribal unity, however, when they began to move onto the Southern Plains with their new horse herds, their culture was adapting in interesting ways to the wealth of resources now available to them.[10]

To the Comanches, the Southern Plains must have seemed an earthly paradise. The Pleistocene extinctions ten thousand years earlier had left dozens of grazing niches vacant on the American Great Plains. A dwarf species of bison with a higher reproductive capability than any of its ancestors evolved to flood most of those vacant niches with an enormous biomass of one grazer. In an ecological sense, bison were a weed species that had proliferated as a result of a major disturbance.[11] That disturbance still reverberated, making it easy for Spanish horses, for example, to reoccupy their old niche and rapidly spread across the Plains. Those reverberations made the horse Indians thrive on an environmental situation that has had few parallels in world history.

The dimensions of the wild bison resource on the Southern Plains, and the Great Plains in general, have been much overstated in popular literature. For one thing, pollen analysis and archaeological data indicate that for the Southern Plains there were intervals, some spanning centuries, others decades, when bison must have been almost absent. Two major times of absence occurred between 5000 and 2500 B.C. and between A.D. 500 and 1300. The archaeological levels that lack bison bones correspond to pollen data indicating droughts. The severe southwestern drought that ended early in the fourteenth century was replaced by a five hundred–year cycle of wetter and cooler conditions, and a return of bison in large numbers to the Southern Plains from their drought refugia to the east and west. This long-term pattern in the archaeological record seems to have prevailed on a smaller scale within historic times. During the nineteenth century, for example, droughts of more than five years' duration struck the Great Plains four times at roughly twenty-year in-

[10] Ernest Wallace, "The Habitat and Range of the Kiowa, Comanche and Apache Indians Before 1867," prepared for the United States Department of Justice for Case No. 257 before the Indian Claims Commission, 1959 (Southwest Collection, Texas Tech University, Lubbock). See Thomas Kavanagh, "The Comanche: Paradigmatic Anomaly or Ethnographic Fiction," *Haliksa'i*, 4 (1985), 109–28; Melburn D. Thurman, "A New Interpretation of Comanche Social Organization," *Current Anthropology*, 23 (Oct. 1982), 578–79; Daniel J. Gelo, "On a New Interpretation of Comanche Social Organization," *ibid.*, 28 (Aug.–Oct. 1987), 551–52; Melburn D. Thurman, "Reply," *ibid.*, 552–55. For the earlier position that the Comanches are atypical on the Plains, see Symmes Oliver, *Ecology and Cultural Continuity as Contributing Factors in the Social Organization of the Plains Indians* (Berkeley, 1972), 69–80.

[11] Jerry McDonald, *North American Bison: Their Classification and Evolution* (Berkeley, 1981), 250–63.

tervals, in a long-term dendrochronological pattern that seems to show a drying cycle (shorter drought-free intervals) beginning in the 1850s.[12]

More important, our popular perception of bison numbers—based on the estimates of awed nineteenth-century observers—probably sets them too high. There very likely were never 100 million or even 60 million bison on the Plains during the present climate regime because the carrying capacity of the grasslands was not so high. The best technique for determining bison carrying capacity on the Southern Plains is to extrapolate from United States census data for livestock, and the best census for the extrapolation is that of 1910, after the beef industry crashes of the 1880s had reduced animal numbers, but before the breakup of ranches and the Enlarged Homestead Act of 1909 resulted in considerable sections of the Southern Plains being broken out by farmers. Additionally, dendrochronological data seem to show that at the turn of the century rainfall on the Southern Plains was at median, between-droughts levels, rendering the census of 1910 particularly suitable as a base line for carrying capacity and animal populations.[13]

The 1910 agricultural census indicates that in the 201 counties on the Southern Plains (which covered 240,000 square miles), the nineteenth-century carrying capacity during periods of median rainfall was about 7,000,000 cattle-equivalent grazers—specifically for 1910, about 5,150,000 cattle and 1,890,000 horses and mules.[14] The bison population was almost certainly larger, since migratory grazing patterns and coevolution with the native grasses made bison as a wild species about 18 percent more efficient on the Great Plains than domestic cattle. And varying climate conditions during the nineteenth century, as I will demonstrate, noticeably affected grassland carrying capacity. The ecological reality was a dynamic cycle in which carrying capacity could swing considerably from decade to decade.[15] But if the Great Plains bovine carrying capacity of 1910 expresses a median reality, then

[12] Tom Dillehay, "Late Quarternary Bison Population Changes on the Southern Plains," *Plains Anthropologist*, 19 (Aug. 1974), 180–96; Darrell Creel et al., "A Faunal Record from West-Central Texas and Its Bearing on Late Holocene Bison Population Changes in the Southern Plains," *ibid.*, 35 (April 1990), 55–69. For Great Plains dendrochronology, see Harry Weakly, "A Tree-Ring Record of Precipitation in Western Nebraska," *Journal of Forestry*, 41 (Nov. 1943), 816–19; and Edmund Schulman, *Dendroclimatic Data from Arid America* (Tucson, 1956), 86–88. For use of meteorological data to argue that climate variability was exponentially greater on the Southern Plains than farther north, see Douglas Bamforth, *Ecology and Human Organization on the Great Plains* (New York, 1988), 74.

[13] Weakly, "Tree-Ring Record of Precipitation in Western Nebraska," 819.

[14] U.S. Department of Commerce, Bureau of the Census, *Thirteenth Census of the United States, Taken in the Year 1910*, vols. VI and VII: *Agriculture, 1909 and 1910* (Washington, 1913). My method has been to compile 1910 cattle, horse, and mule figures for the then-existing Plains counties of Texas (119), western Oklahoma (45), New Mexico (10), counties below the Arkansas River in Colorado (8), and counties in southwestern Kansas (19). The carrying capacity for a biome such as the Great Plains ought to be measured by the use of the county figures. The principal problem with this technique in the past has been overgeneralization of stock numbers through reliance on state totals. It was first used by Ernest Thompson Seton, *Life Histories of Northern Animals* (2 vols., New York, 1909), I, 259–63; and more recently by Bill Brown, "Comancheria Demography, 1805–1830," *Panhandle-Plains Historical Review*, 59 (1986), 8–12. Range management commonly assigns cows a grazing quotient of 1.0, bulls a quotient of 1.30, and horses and mules 1.25.

[15] Joseph Chapman and George Feldhamer, eds., *Wild Mammals of North America: Biology, Management, and Economics* (Baltimore, 1982), 978, 986, 1001–2. Modern bison ranchers claim that bison achieve greater land-use efficiency and larger herd size on native grass compared to cattle. The editors of the above work call for more research into this question. See also Charles Rehr, "Buffalo Population and Other Deterministic Factors in a Model of Adaptive Process on the Shortgrass Plains," *Plains Anthropologist*, 23 (Nov. 1978), 25–27.

during prehorse times the Southern Plains might have supported an average of about 8.2 million bison, the entire Great Plains perhaps 28–30 million.[16]

Although 8 million bison on the Southern Plains may not be so many as historians used to believe, to the Comanches the herds probably seemed limitless. Bison availability through horse culture caused a specialization that resulted in the loss of two-thirds of the Comanches' former plant lore and in a consequent loss of status for their women, an intriguing development that seems to have occurred to some extent among all the tribes that moved onto the Plains during the horse period.[17] As full-time bison hunters the Comanches appear to have abandoned all the old Shoshonean mechanisms, such as infanticide and polyandry, that had kept their population in line with available resources. These were replaced with such cultural mechanisms as widespread adoption of captured children and polygyny, adaptations to the Plains that were designed to keep Comanche numbers high and growing.[18] That these changes seem to have been conscious and deliberate argues, perhaps, both Comanche environmental insight and some centralized leadership and planning.

Comanche success at seizing the Southern Plains from the native groups that had held it for several hundred years was likewise the result of a conscious choice: their decision to shape their lives around bison and horses. Unlike the Comanches, many of the Apache bands had heeded the Spaniards' advice and had begun to build streamside gardening villages that became deathtraps once the Comanches located them. The Apaches' vulnerability, then, ironically stemmed from their willingness to diversify their economy. Given the overwhelming dominance of grasslands as opposed to cultivable river lands on the Plains, the specialized horse and bison culture of the Comanches exploited a greater volume of the thermodynamic energy streaming from sunlight into plants than the economies of any of their competitors—until they encountered Cheyennes and Arapahoes with a similar culture.[19]

[16] For the use of a different formula (mean potential bovine carrying capacity per acre) to arrive at similar totals see Tom McHugh, *The Time of the Buffalo* (New York, 1972), 16–17. For the reasonably convincing argument that because of climate variability and less nutritious grasses, population densities of Great Plains bison were lowest on the Southern Plains, see Bamforth, *Ecology and Human Organization on the Great Plains*, 74, 78.

[17] The figure for loss of plant lore is based on a comparison of remembered ethnobotanies for the Shoshones (172 species) and the Comanches (67 species). See Brian Spykerman, "Shoshoni Conceptualizations of Plant Relationships" (M.S. thesis, Utah State University, 1977); and Gustav Carlson and Volney Jones, "Some Notes on the Uses of Plants by the Comanche Indians," *Papers of the Michigan Academy of Science, Arts, and Letters*, 25 (1939), 517–42. On women's loss of status, see Holder, *Hoe and the Horse on the Plains*.

[18] Wallace and Hoebel, *Comanches*, 142. On the loss of Shoshone birth control mechanisms among Comanche women, see Abram Kardiner, "Analysis of Comanche Culture," in *The Psychological Frontiers of Society*, ed. Abram Kardiner et al. (New York, 1945). For a report of 500 adopted captives in a decade, see Jean Louis Berlandier, *The Indians of Texas in 1830*, ed. John C. Ewers (Washington, 1969), 119. Estimates on the Euro-American constituency of nineteenth-century Comanche bands as approaching 75% are probably too high, but 30% may not be. On Comanche adoption and captives trade, see Carl C. Rister, *Border Captives: The Traffic in Prisoners by Southern Plains Indians, 1835–1875* (Norman, 1940); and Russell Magnaghi, "The Indian Slave Trader: The Comanche, a Case Study" (Ph.D. diss., St. Louis University, 1970). Indian Agent Thomas Fitzpatrick was adamant that Comanche raids were for children, "to keep up the numbers of the tribe." See Kardiner, "Analysis of Comanche Culture," 89. On hunter-gatherer carrying capacity, see Marvin Harris and Eric Ross, *Death, Sex, and Fertility: Population Regulation in Preindustrial and Developing Societies* (New York, 1987), 23–26; and Ezra Zubrow, *Prehistoric Carrying Capacity: A Model* (Menlo Park, 1975).

[19] David Kaplan, "The Law of Cultural Dominance," in *Evolution and Culture*, ed. Marshall Sahlins and Elman

The horse-mounted Plains Indians, in other words, made very efficient use of the available energy on the Great Plains, something they seem instinctively to have recognized and exulted in. From the frequency with which the Comanches applied some version of the name "wolf" to their leaders, I suspect that they may have recognized their role as human predators and their ecological kinship with the wolf packs that like them lived off the bison herds.[20]

The Comanches were not the only people on the Southern Plains during the horse period. The New Mexicans, both Pueblo and Hispanic, continued to hunt on the wide-open Llanos, as did the prairie Caddoans, although the numbers of the latter were dwindling rapidly by 1825. The New Mexican peoples and the Caddoans of the middle Red and Brazos rivers played major trade roles for hunters on the Southern Plains, and the Comanches in particular. Although the Comanches engaged in the archetypal Plains exchange of bison products for horticultural produce and European trade goods and traded horses and mules with Anglo-American traders from Missouri, Arkansas, and Louisiana, they were not a high-volume trading people until relatively late in their history. Early experiences with American traders and disease led them to distrust trade with Euro-Americans, and only once or twice did they allow short-lived posts to be established in their country. Instead, peace with the prairie Caddoans by the 1730s and with New Mexico in 1786 sent Comanche trade both east and west, but often through Indian middlemen.[21]

In the classic, paradigmatic period between 1800 and 1850, the most interesting Southern Plains development was the cultural interaction between the Comanches and surrounding Plains Indians to the north. The Kiowas were the one of those groups most closely identified with the Comanches.

The Kiowas are and have long been an enigma. Scholars are interested in their origins because Kiowa oral tradition is at odds with the scientific evidence. The Kiowas believe that they started their journey to Rainy Mountain on the Oklahoma Plains from the north. And indeed, in the eighteenth century we find them on the Northern Plains, near the Black Hills, as one of the groups being displaced southwestward by the Siouan drive toward the buffalo range. Linguistically, however, the Kiowas are southern Indians. Their language belongs to the Tanoan group of Pueblo languages in New Mexico, and some scholars believe that the Kiowas of later history

Service (Ann Arbor, 1960), 75–82. On Apache vulnerability, see George E. Hyde, *Indians of the High Plains: From the Prehistoric Period to the Coming of Europeans* (Norman, 1975), 65, 70, 91. Other explanations include the Spanish refusal to trade guns to the Apaches and Comanche superiority at horse care. For less monocausal interpretations, see Frederick W. Rathjen, *The Texas Panhandle Frontier* (Austin, 1973), 47–48.

[20] For references to Comanche names containing the word *wolf* (rendered by Euro-Americans as *isa*, *ysa*, *esa*, or sometimes with an *sh* second syllable), see George Catlin, *Letters and Notes on the Manners, Customs, and Conditions of the North American Indians* (2 vols., New York, 1973), II, 67–69; Noel Loomis and Abraham Nasatir, eds., *Pedro Vial and the Roads to Santa Fe* (Norman, 1967), 488n22a; and Thomas, *Forgotten Frontiers*, 325–27.

[21] On Comanche trade with Anglo-Americans, see Dan Flores, ed., *Journal of an Indian Trader: Anthony Glass and the Texas Trading Frontier, 1790–1810* (College Station, 1985), esp. 3–33; and Thomas James, *Three Years among the Indians and Mexicans*, ed. Walter Douglas (St. Louis, 1916), 191–235. For the argument that the Comanches were among the earliest Plains traders, and that Comanche leadership evolved in a trade/market situation, see Kavanagh, "Political Power and Political Organization." See Charles Kenner, *A History of New Mexican–Plains Indian Relations* (Norman, 1969); and William Swagerty, "Indian Trade in the Trans-Mississippi West to 1870," in *Handbook of North American Indians*, ed. Sturtevant, IV, 351–74.

Kiowa Buffalo Hunter, by unknown Indian artist after 1875, on reservation ledger paper. By the nineteenth century, many Plains Indians had come to define themselves as bison predators, a dependency that was beginning to fail them by 1850. *Courtesy Morning Star Gallery, Santa Fe, New Mexico.*

are the same people as the Plains Jumanos of early New Mexico history, whose rancherias were associated during the 1600s and early 1700s with the headwaters of the Colorado and Concho rivers of Texas. How the Kiowas got so far north is not certainly known, but in historical times they were consummate traders, especially of horses, and since the Black Hills region was a major trade citadel they may have begun to frequent the region as traders and teachers of horse lore.[22]

Displaced by the wars for the buffalo ranges in the north, the Kiowas began to drift southward again—or perhaps, since the supply of horses was in the Southwest, simply began to stay longer on the Southern Plains. Between 1790 and 1806, they developed a rapprochement with the Comanches. Thereafter they were so closely associated with the northern Comanches that they were regarded by some as merely a Comanche band, although in many cultural details the two groups were dissimilar. Spanish and American traders and explorers of the 1820s found them camped along the two forks of the Canadian River and on the various headwater streams of the Red River.[23]

[22] See Hickerson, "Jumano and Trade in the Arid Southwest." On Kiowa origins, see also Maurice Boyd, *Kiowa Voices: Ceremonial Dance, Ritual, and Songs* (2 vols., Fort Worth, 1981); and John Wunder, *The Kiowas* (New York, 1989).

[23] Elizabeth A. H. John, "An Earlier Chapter of Kiowa History," *New Mexico Historical Review*, 60 (no. 4, 1985),

Picketing Horses, painting by Alfred Jacob Miller. Here two Euro-American traders stake horses they have chosen from the Indians' herd behind them. By reducing the carrying capacity for bison on the Great Plains and stimulating Indian/Euro-American trade, horses were a critical element in nineteenth-century Plains ecology. *Courtesy Walters Art Gallery, Baltimore.*

The other groups that increasingly began to interact with the Comanches during the 1820s and thereafter had also originated on the Northern Plains. These were the Arapahoes and the Cheyennes, who by 1825 were beginning to establish themselves on the Colorado buffalo plains from the North Platte River all the way down to the Arkansas River.

The Algonkian-speaking Arapahoes and Cheyennes had once been farmers living in earth lodges on the upper Mississippi. By the early 1700s both groups were in present North Dakota, occupying villages along the Red and Sheyenne rivers, where they first began to acquire horses, possibly from the Kiowas. Fur wars instigated by the Europeans drove them farther southwest and more and more into a Plains, bison-hunting culture, one that the women of these farming tribes probably resisted as long as possible.[24] But by the second decade of the nineteenth century the Teton Sioux wedge had made nomads and hunters of the Arapahoes and Cheyennes.

379–97. On traders' contacts with the Kiowas, see, particularly, Maxine Benson, ed., *From Pittsburgh to the Rocky Mountains: Major Stephen Long's Expedition, 1819–1820* (Golden, 1988), 327–36.

[24] Holder, *Hoe and the Horse on the Plains.* See also Bea Medicine and Pat Albers, *The Hidden Half: Studies of Plains Indian Women* (Lanham, 1983); and Katherine Weist, "Plains Indian Women: An Assessment," in *Anthropology on the Great Plains,* ed. Raymond Wood and Margot Liberty (Lincoln, 1980), 255–71.

Their search for prime buffalo grounds and for ever-larger horse herds, critical since both tribes had emerged as middlemen traders between the villagers of the Missouri and the horse reservoir to the south, first led the Cheyennes and Arapahoes west of the Black Hills, into Crow lands, and then increasingly southward along the mountain front. By 1815 the Arapahoes were becoming fixed in the minds of American traders as their own analogue on the Southern Plains; the famous trading expedition of August Pierre Chouteau and Jules De Mun that decade was designed to exploit the horse and robe trade of the Arapahoes on the Arkansas. By the 1820s, when Stephen Long's expedition and the trading party including Jacob Fowler penetrated the Southern Plains, the Arapahoes and Cheyennes were camping with the Kiowas and Comanches on the Arkansas. The Hairy Rope band of the Cheyennes, renowned for their ability to catch wild horses, was then known to be mustanging along the Cimarron River.[25]

Three factors seem to have drawn the Arapahoes and Cheyennes so far south. Unquestionably, one factor was the vast horse herds of the Comanches and Kiowas, an unending supply of horses for the trade, which by 1825 the Colorado tribes were seizing in daring raids. Another was the milder winters south of the Arkansas, which made horse pastoralism much easier. The third factor was the abnormally bountiful game of the early nineteenth-century Southern Plains, evidently the direct result of an extraordinary series of years between 1815 and 1846 when, with the exception of a minor drought in the late 1820s, rainfall south of the Arkansas was considerably above average. So lucrative was the hunting and raiding that in 1833 Charles Bent located the first of his adobe trading posts along the Arkansas, expressly to control the winter robe and summer horse trade of the Arapahoes and Cheyennes. Bent's marketing contacts were in St. Louis. Horses that Bent's traders drove to St. Louis commonly started as stock in the New Mexican Spanish settlements (and sometimes those were California horses stolen by Indians who traded them to the New Mexicans) that were stolen by the Comanches, then stolen again by Cheyenne raiders, and finally traded at Bent's or Ceran St. Vrain's posts, whence they were driven to Westport, Missouri, and sold to outfit American emigrants going to the West Coast! Unless you saw it from the wrong end, as the New Mexicans (or the horses) seem to have, it was both a profitable and a culturally stimulating economy.[26]

Thus, around 1825, the Comanches and Kiowas found themselves at war with Cheyennes, Arapahoes, and other tribes on the north. Meanwhile, the Colorado tribes opened another front in a naked effort to seize the rich buffalo range of the upper Kansas and Republican rivers from the Pawnees. These wars produced an interesting type of ecological development that appeared repeatedly across most of

[25] Joseph Jablow, *The Cheyenne in Plains Indian Trade Relations, 1795–1840* (Seattle, 1950); Donald Berthrong, *The Southern Cheyennes* (Norman, 1963), 4–21; Loomis and Nasatir, eds., *Pedro Vial and the Roads to Santa Fe,* 256–58.

[26] Alan Osburn, "Ecological Aspects of Equestrian Adaptations in Aboriginal North America," *American Anthropologist,* 85 (Sept. 1983), 563–91; Berthrong, *Southern Cheyennes,* 25–26; Jablow, *Cheyenne in Plains Indian Trade Relations,* 67; David Lavender, *Bent's Fort* (Lincoln, 1954), 141–54; George Phillips, *Chiefs and Challengers: Indian Resistance and Cooperation in Southern California* (Berkeley, 1975), 42–43; Eleanor Lawrence, "The Old Spanish Trail from Santa Fe to California" (M.A. thesis, University of California, Berkeley, 1930).

the continent. At the boundaries where warring tribes met, they left buffer zones occupied by neither side and only lightly hunted. One such buffer zone on the Southern Plains was along the region's northern perimeter, between the Arkansas and North Canadian rivers. Another was in present-day western Kansas, between the Pawnees and the main range of the Colorado tribes, and a third seems to have stretched from the forks of the Platte to the mountains. The buffer zones were important because game within them was left relatively undisturbed; they allowed the buildup of herds that might later be exploited when tribal boundaries or agreements changed.[27]

The appearance of American traders such as Bent and St. Vrain marked the Southern Plains tribes' growing immersion in a market economy increasingly tied to worldwide trade networks dominated by Euro-Americans. Like all humans, Indians had always altered their environments. But as most modern historians of Plains Indians and the western fur trade have realized, during the nineteenth century not only had the western tribes become technologically capable of pressuring their resources, but year by year they were becoming less "ecosystem people," dependent on the products of their local regions for subsistence, and increasingly tied to biospheric trade networks. Despite some speculation that the Plains tribes were experiencing ecological problems, previous scholars have not ascertained what role market hunting played in this dilemma, what combination of other factors was involved, or what the tribes attempted to do about it.[28]

The crux of the problem in studying Southern Plains Indian ecology and bison is to determine whether the Plains tribes had established a society in ecological equilibrium, one whose population did not exceed the carrying capacity of its habitat and so maintained a healthy, functioning ecology that could be sustained over the long term.[29] Answering that question involves an effort to come to grips with the factors affecting bison populations, the factors affecting Indian populations, and the cultural aspects of Plains Indians' utilization of bison. Each of the three aspects of the question presents puzzles difficult to resolve.

In modern, protected herds on the Plains, bison are a prolific species whose numbers increase by an average of 18 percent a year, assuming a normal sex ratio (51 males to 49 females) with breeding cows amounting to 35 percent of the total.[30] In other words, if the Southern Plains supported 8.2 million bison in years of median rainfall, the herds would have produced about 1.4 million calves a year. To

[27] On the intertribal buffer zones, see Berthrong, *Southern Cheyennes*, 76, 93. On their function in preserving wildlife in other ecosystems, see Harold Hickerson, "The Virginia Deer and Intertribal Buffer Zones in the Upper Mississippi Valley," *Man, Culture, and Animals*, ed. Anthony Leeds and Andrew Vayda (Washington, 1965), 43–66.

[28] Raymond Dasmann, "Future Primitive," *CoEvolution Quarterly*, 11 (1976), 26–31; David Wishart, *The Fur Trade of the American West, 1807–1840* (Lincoln, 1979); Arthur Ray, *The Fur Trade and the Indian* (Toronto, 1974); White, *Roots of Dependency*, 147–211.

[29] Zubrow, *Prehistoric Carrying Capacity*, 8–9.

[30] Chapman and Feldhamer, eds., *Wild Mammals of North America*, 980–83; Arthur Halloran, "Bison (Bovidae) Productivity on the Wichita Mountains Wildlife Refuge, Oklahoma," *Southwestern Naturalist*, 13 (May 1968), 23–26; Alisa Shull and Alan Tipton, "Effective Population Size of Bison on the Wichita Mountains Wildlife Refuge," *Conservation Biology*, 1 (May 1987), 35–41.

White Wolves Attacking Buffalo, watercolor by George Catlin, 1832. By preying on injured and older bison, and especially bison calves, wolves played a key role in holding bison population in equilibrium with the grasslands.
Courtesy Smithsonian Institution, Anthropological Archives.

maintain an ecological equilibrium with the grasses, the Plains bison's natural mortality rate also had to approach 18 percent.

Today the several protected bison herds in the western United States have a natural mortality rate, without predation, ranging between 3 and 9 percent. The Wichita Mountains herd, the only large herd left on the Southern Plains, falls midway with a 6 percent mortality rate. Despite a search for it, no inherent naturally regulating mechanism has yet been found in bison populations; thus active culling programs are needed at all the Plains bison refuges. The starvation-induced population crashes that affect ungulates such as deer were seemingly mitigated on the wild, unfenced Plains by the bison's tendency—barring any major impediments—to shift their range great distances to better pasture.[31]

[31] Data on the modern bison herds on the Great Plains are from the refuge managers and superintendents of the Wichita Mountains Wildlife Refuge, Theodore Roosevelt National Memorial Park, and Wind Cave National Park. The National Bison Refuge in Montana did not respond to my inquiries. Robert Karges to Dan Flores, March 18, 1988 (in Flores's possession); Robert Powell to Flores, Feb. 10, 1988, *ibid.*; Ernest Ortega to Flores, Feb. 11,

Determining precisely how the remaining annual mortality in the wild herds was affected is not easy, because the wolf/bison relationship on the Plains has never been studied. Judging from dozens of historical documents attesting to wolf predation of bison calves, including accounts by the Indians, wolves apparently played a critical role in Plains bison population dynamics, and not just as culling agents of diseased and old animals.[32]

Human hunters were the other source of mortality. For nine thousand years native Americans had hunted bison without exterminating them, perhaps building into their gene pool an adjustment to human predation (dwarfed size, earlier sexual maturity, and shorter gestation times, all serving to keep populations up). But there is archaeological evidence that beginning about A.D. 1450, with the advent of "mutualistic" trade between Puebloan communities recently forced by drought to relocate on the Rio Grande and a new wave of Plains hunters (probably the Athapaskan-speaking Apacheans), human pressures on the southern bison herd accelerated, evidently dramatically if the archaeological record in New Mexico is an accurate indication. That pressure would have been a function of both the size of the Indian population and the use of bison in Indian cultures. Because Plains Indians traded bison-derived goods for the produce of the horticultural villages fringing the Plains, bison would be affected by changes in human population peripheral to the Great Plains as well as on them.[33]

One attempt to estimate maximum human population size on the Southern Plains, that of Jerold Levy, fixed the upper limit at about 10,500 people. Levy argued that water would have been a more critical resource than bison in fixing a limit for Indian populations. Levy's population figures are demonstrably too low, and he lacked familiarity with the aquifer-derived drought-resistant sources of water on the Southern Plains. But his argument that water was the more critical limiting resource introduces an important element into the Plains equation.[34]

1988, *ibid.* See Graeme Caughley, "Eruption of Ungulate Populations, with Emphasis on Himalayan Thar in New Zealand," *Ecology,* 51 (Winter 1970), 53–72. This study has been widely cited in wildlife ecology as evidence that starvation rather than predation is often the key to regulating natural population eruptions. The only documentary evidence I have seen for starvation of bison on the Southern Plains is Charles Goodnight's account of seeing "millions" of starved bison along a front 25 by 100 miles between the Concho and Brazos rivers in 1867, after bison migration patterns had been disrupted by settlements. J. Evetts Haley, *Charles Goodnight, Cowman and Plainsman* (Norman, 1949), 161.

[32] The 19th-century documentary evidence assigns wolves roles as scavengers of bison killed by other agents; as cullers of weak, sick, and old animals; and as predators of bison calves. The last, I believe, best expresses the regulatory effect of wolves on Plains bison population dynamics. See Gary E. Moulton, ed., *The Journals of the Lewis and Clark Expedition* (5 vols., Lincoln, 1987), IV, 62–63; Donald Jackson and Mary Lee Spence, eds., *The Expeditions of John Charles Fremont* (5 vols., Urbana, 1970–1984), I, 190–91; Henry Boller, *Among the Indians: Four Years on the Upper Missouri, 1858–1862,* ed. Milo Milton Quaife (Lincoln, 1972), 270–71; Maria Audubon and Elliott Coues, eds., *Audubon and His Journals* (2 vols., New York, 1986), I, 49; and W. Eugene Hollon, ed., *William Bollaert's Texas* (Norman, 1989), 255. For other descriptions, see Stanley P. Young and Edward A. Goldman, *The Wolves of North America* (2 vols., New York, 1944), I, 50, 218, 224–31.

[33] Katherine Spielman, "Late Prehistoric Exchange between the Southwest and Southern Plains," *Plains Anthropologist,* 28 (Nov. 1983), 257–79. For the argument that essential plant resources from this trade ended a nutrition "bottleneck" and therefore allowed the buildup of much larger human populations on the Southern Plains, see Bamforth, *Ecology and Human Organization on the Great Plains,* 8.

[34] Jerold Levy, "Ecology of the South Plains," in *Symposium: Patterns of Land Use and Other Papers,* ed. Viola Garfield (Seattle, 1961), 18–25.

The cultural utilization of bison by horse Indians has been studied by Bill Brown. Adapting a sophisticated formula worked out first for caribou hunters in the Yukon, Brown has estimated Indian subsistence (caloric requirements plus the number of robes and hides required for domestic use) at about 47 animals per lodge per year. At an average of 8 people per lodge, that works out to almost 6 bison per person over a year's time. Brown's article is not only highly useful in getting us closer to a historic Plains equation than ever before; it is also borne out by at least one historic account. In 1821 the trader Jacob Fowler camped for several weeks with 700 lodges of Southern Plains tribes on the Arkansas River. Fowler was no ecologist; in fact, he could hardly spell. But he was a careful observer, and he wrote that the big camp was using up 100 bison a day. In other words, 700 lodges were using bison at a rate of about 52 per lodge per year, or 6.5 animals per person. These are important figures. Not only do they give us some idea of the mortality percentage that can be assigned to human hunters; by extension they help us fix a quadruped predation percentage as well.[35]

Estimates of the number of Indians on the Southern Plains during historic times are not difficult to find, but they tend to vary widely, and for good reason, as will be seen when we look closely at the historical events of the first half of the nineteenth century. Although observers' population estimates for the Comanches go as high as 30,000, six of the seven population figures for the Comanches estimated between 1786 and 1854 fall into a narrow range between 19,200 and 21,600.[36] Taken together, the number of Kiowas, Cheyennes, Arapahoes, Plains Apaches, Kiowa-Apaches, and Wichitas probably did not exceed 12,000 during that same period. Contemporaries estimated the combined number of Cheyennes and Arapahoes, for example, as 4,400 in 1838, 5,000 in 1843, and 5,200 in 1846.[37] If the historic Southern Plains hunting population reached 30,000, then human hunters would have accounted for only 195,000 bison per year if we use the estimate of 6.5 animals per person.

But another factor must have played a significant role. While quadruped predators concentrated on calves and injured or feeble animals, human hunters had different criteria. Historical documents attest to the horse Indians' preference for and success in killing two- to five-year-old bison cows, which were preferred for their meat and for their thinner, more easily processed hides and the luxurious robes

[35] Brown, "Comancheria Demography," 10–11; H. Paul Thompson, "A Technique Using Anthropological and Biological Data," *Current Anthropology*, 7 (Oct. 1966), 417–24; Jacob Fowler, *The Journal of Jacob Fowler: Narrating an Adventure from Arkansas through the Indian Territory, Oklahoma, Kansas, Colorado, and New Mexico, to the Sources of the Rio Grande del Norte*, ed. Elliott Coues (New York, 1898), 61, 63.

[36] Wallace and Hoebel, *Comanches*, 31–32. The anthropological literature tends to set Comanche population much more conservatively, often at no more than 7,000. See, for example, Bamforth, *Ecology and Human Organization on the Great Plains*, 104–14. Such low figures ignore eyewitness accounts of localized Comanche aggregations of several thousand. I have a historian's bias in favor of documentary evidence for estimating human populations; Plains observers computed village sizes relatively easily by counting the number of tents.

[37] Berthrong, *Southern Cheyennes*, 78, 92, 107. The Kiowas and Kiowa-Apaches seem to have averaged about 2,500 to 3,000 from 1825 to 1850, and the Prairie Caddoans perhaps 2,000, shrinking to 1,000 by midcentury. *Report, Commissioner of Indian Affairs, 1842*, cited in Josiah Gregg, *Commerce of the Prairies*, ed. Max Moorhead (Norman, 1964), 431–32n3.

made from their pelts. Studies done on other large American ungulates indicate that removal of breeding females at a level that exceeds 7 percent of the total herd will initiate population decline. With 8.2 million bison on the Southern Plains, the critical upper figure for cow selectivity would have been about 574,000 animals. Reduce the total bison number to 6 million and the yearly calf crop to 1.08 million, probably more realistic median figures for the first half of the nineteenth century, and the critical mortality for breeding cows would still have been 420,000 animals. As mentioned, a horse-mounted, bison-hunting population of 30,000 would have harvested bison at a yearly rate of less than 200,000. Hence I would argue that, theoretically, on the Southern Plains the huge biomass of bison left from the Pleistocene extinctions would have supported the subsistence needs of more than 60,000 Plains hunters.[38]

All of this raises some serious questions when we look at the historical evidence from the first half of the nineteenth century. By the end of that period, despite an effort at population growth by many Plains tribes, the population estimates for most of the Southern Plains tribes were down. And many of the bands seemed to be starving. Thomas Fitzpatrick, the Cheyennes' and Arapahoes' first agent, reported in 1853 that the tribes in his district spent half the year in a state of starvation. The Comanches were reported to be eating their horses in great numbers by 1850, and their raids into Mexico increased all through the 1840s, as if a resource depletion in their home range was driving them to compensate with stolen stock.[39] In the painted robe calendars of the Kiowas, the notation for "few or no bison" appears for four years in a row between 1849 and 1852.[40] Bison were becoming less reliable, and the evolution toward an economy based on raiding and true horse pastoralism was well under way. Clearly, by 1850 something had altered the situation on the Southern Plains.

The "something" was, in fact, a whole host of ecological alterations that historians

[38] Because of the tough meat and the thick hides that made soft tanning difficult, Indians (and whites hunting for meat) rarely killed bison bulls. See Roe, *North American Buffalo*, 650–70; and Larry Barsness, *Heads, Hides, and Horns: The Compleat Buffalo Book* (Fort Worth, 1974), 69–72, 96–98. Dean E. Medin and Allen E. Anderson, *Modeling the Dynamics of a Colorado Mule Deer Population* (Fort Collins, 1979). Whether 60,000 hunters ever worked the Southern Plains in precontact times is now unknowable, but Coronado's chronicler, Castaneda, wrote that there were more people on the Plains in 1542 than in the Rio Grande pueblos. Pedro de Castaneda, "The Narrative of the Expedition of Coronado," in *Spanish Explorers in the Southern United States, 1528–1543*, ed. Frederick W. Hodge and Theodore H. Lewis (Austin, 1984), 362. For a seventeenth-century population estimate in the Rio Grande pueblos of about 30,500, see Marc Simmons, "History of Pueblo-Spanish Relations to 1821," in *Handbook of North American Indians*, ed. Sturtevant, IX, 185 (table 1).

[39] For Thomas Fitzpatrick's report, see Berthrong, *Southern Cheyennes*, 124. The English traveler William Bollaert mentions that the Texas Comanches supposedly ate 20,000 mustangs in the late 1840s. See Hollon, ed., *William Bollaert's Texas*, 361. On the escalating stock raids and trade to New Mexico beginning in the 1840s, see J. Evetts Haley, "The Comanchero Trade," *Southwestern Historical Quarterly*, 38 (Jan. 1935), 157–76. Haley generally ascribes the situation to Comanche barbarity and Hispanic lack of respect for Lockean private property rights. See also Kenner, *History of New Mexican–Plains Indian Relations*, 78–97, 155–200.

[40] See James Mooney, *Calendar History of the Kiowa Indians* (Washington, 1979), 287–95; Levy, "Ecology of the South Plains," 19. The decline in the number of bison was becoming noticeable as early as 1844, two years before the 1846–1857 drought. See Solomon Sublette to William Sublette, Feb. 2, May 5, 1844, William Sublette Papers (Archives, Missouri Historical Society, St. Louis, Mo.). In 1845 the trader James Webb and his party traveled from Bent's Fort to Missouri without killing a bison. "Memoirs of James J. Webb, Merchant in Santa Fe, N.M., 1844," typescript, p. 69, James Webb Papers, *ibid.*

with a wide range of data at their disposal are only now, more than a century later, beginning to understand.

As early as 1850 the bison herds had been weakened in a number of ways. The effect of the horse on Indian culture has been much studied, but in working out a Southern Plains ecological model, it is important to note that horses also had a direct effect on bison numbers. By the second quarter of the nineteenth century the domesticated horse herds of the Southern Plains tribes must have ranged between .25 and .50 million animals (at an average of 10 to 15 horses per person).[41] In addition, an estimated 2 million wild mustangs overspread the country between south Texas and the Arkansas River. That many animals of a species with an 80 percent dietary overlap with bovines and, perhaps more critically, with similar water requirements, must have had an adverse impact on bison carrying capacity, especially since Indian horse herds concentrated the tribes in the moist canyons and river valleys that bison also used for watering.[42] Judging from the 1910 agricultural census, 2 million or more horses would have reduced the median grassland carrying capacity for the southern bison herd to under 6 million animals.

Another factor that may have started to diminish overall bison numbers was the effect of exotic bovine diseases. Anthrax, introduced into the herds from Louisiana around 1800, tuberculosis, and brucellosis, the latter brought to the Plains by feral and stolen Texas cattle and by stock on the overland trails, probably had considerable impact on the bison herds. All the bison that were saved in the late nineteenth century had high rates of infection with these diseases. Brucellosis plays havoc with reproduction in domestic cattle, causing cows to abort; it may have done so in wild bison, and butchering them probably infected Indian women with the disease.[43]

Earlier I mentioned modern natural mortality figures for bison of 3 percent to 9 percent of herd totals. On the wilderness Plains, fires, floods, drownings, droughts, and strange die-offs may have upped this percentage considerably. But if we hold to the higher figure, then mortality might have taken an average of 50 percent of the annual bison increase of 18 percent. Thirty thousand subsistence hunters would have killed off only 18 percent of the bison's yearly increase (if the herd was 6 million). The long-wondered-at wolf predation was perhaps the most important of all the factors regulating bison populations, with a predation percentage of around 32% of the annual bison increase. (Interestingly, this dovetails

[41] John C. Ewers, *The Horse in Blackfoot Culture* (Washington, 1955). For systematic assessment of the effects of horses on seasonal band size, camps, and resources, see John Moore, *The Cheyenne Nation: A Social and Demographic History* (Lincoln, 1987), 127–75. For a dynamic rather than a static horse ecology, see James Sherow, "Pieces to a Puzzle: High Plains Indians and Their Horses in the Region of the Arkansas River Valley 1800–1860," paper presented at the Ethnohistory Conference, Chicago, 1989 (in Flores's possesion). Clyde Wilson, "An Inquiry into the Nature of Plains Indian Cultural Development," *American Anthropologist*, 65 (April 1963), 355–69.

[42] J. Frank Dobie, *The Mustangs* (New York, 1934), 108–9. Dobie's estimate, as he pointed out, was a guess, but my work in the agricultural censuses indicates that it was a good guess. On horse/bovine dietary overlap see L. J. Krysl et al., "Horses and Cattle Grazing in the Wyoming Red Desert, I. Food Habits and Dietary Overlap," *Journal of Range Management*, 37 (Jan. 1984), 72–76. On the drier climate on the Plains between 1848 and 1874, see Weakly, "Tree-Ring Record of Precipitation in Western Nebraska," 817, 819; and Levy, "Ecology of the South Plains," 19.

[43] I follow Chapman and Feldhamer, eds., *Wild Mammals of North America*, 991–94.

closely with the Pawnee estimate that wolves got 3 to 4 of every 10 calves born.) Wolves and other canids are able to adjust their litter sizes to factors like mortality and resource abundance. Thus, mountain men and traders who poisoned wolves for their pelts may not have significantly reduced wolf populations. They may have inadvertently killed thousands of bison, however, for poisoned wolves drooled and vomited strychnine over the grass in their convulsions. Many Indians lost horses that ate such poisoned grass.[44]

The climate cycle, strongly correlated with bison populations in the archaeological data for earlier periods, must have interacted with these other factors to produce a decline in bison numbers between 1840 and 1850. Except for a dry period in the mid- to late 1820s, the first four decades of the nineteenth century had been a time of above-normal rainfall on the Southern Plains. With the carrying capacity for bison and horses high, the country south of the Arkansas sucked tribes to it as into a vortex. But beginning in 1846, rainfall plunged as much as 30 percent below the median for nine of the next ten years. On the Central Plains, six years of that decade were dry.[45] The growth of human populations and settlements in Texas, New Mexico, and the Indian Territory blocked the bison herds from migrating to their traditional drought refugia on the periphery of their range. Thus, a normal climate swing combined with unprecedented external pressures to produce an effect unusual in bison history—a core population, significantly reduced by competition with horses and by drought, that was quite susceptible to human hunting pressure.

Finally, alterations in the historical circumstances of the Southern Plains tribes from 1825 to 1850 had serious repercussions for Plains ecology. Some of those circumstances were indirect and beyond the tribes' ability to influence. Traders along the Santa Fe Trail shot into, chased, and disturbed the southern herds. New Mexican *Ciboleros* (bison hunters) continued to take fifteen to twenty-five thousand bison a year from the Llano Estacado. And the United States government's removal of almost fifty thousand eastern Indians into Oklahoma increased the pressure on the bison herds to a level impossible to estimate. The Southern Plains tribes evidently considered it a threat and refused to abide by the Treaty of Fort Holmes (1835) when they discovered it gave the eastern tribes hunting rights on the prairies.[46]

Insofar as the Southern Plains tribes had an environmental policy, then, it was to protect the bison herds from being hunted by outsiders. The Comanches could not afford to emulate their Shoshonean ancestors and limit their own population.

[44] McHugh, *Time of the Buffalo*, 226–27. For scientific discussion of predation by wolves on large ungulates, see David Mech, *The Wolf* (New York, 1970); see also Chapman and Feldhamer, eds., *Wild Mammals of North America*, 994–96. For estimates that the reintroduction of wolves to Yellowstone would reduce the bison herd there between 5% and 20%, see Barbara Koth, David Lime, and Jonathan Vlaming, "Effects of Restoring Wolves on Yellowstone Area Big Game and Grizzly Bears: Opinions of Fifteen North American Experts," in Yellowstone National Park, *Wolves for Yellowstone?* (n.p., 1990), 4–71, 4–72, and the computer simulation, 3–31. Young and Goldman, *Wolves of North America*, II, 327–33.

[45] Schulman, *Dendroclimatic Data from Arid America*, fig. 22; Weakly, "Tree-Ring Record of Precipitation in Western Nebraska," 817, 819.

[46] For emphasis on disruption by whites, see Douglas Bamforth, "Historical Documents and Bison Ecology on the Great Plains," *Plains Anthropologist*, 32 (Feb. 1987), 1–16. On the *Ciboleros*, see Kenner, *History of New Mexican–Plains Indian Relations*, 115–17. Jablow, *Cheyenne in Plains Indian Trade Relations*, 72.

Beset by enemies and disease, they had to try to keep their numbers high, even as their resource base diminished. For the historic Plains tribes, warfare and stock raids addressed ecological needs created by diminishing resources as well as the cultural impulse to enhance men's status, and they must have seemed far more logical solutions than consciously reducing their own populations as the bison herds became less reliable.

For those very reasons, after more than a decade of warfare among the buffalo tribes, in 1840 the Comanches and Kiowas adopted a strategy of seeking peace and an alliance with the Cheyennes, Arapahoes, and Kiowa-Apaches. From the Comanches' point of view, it brought them allies against Texans and eastern Indians who were trespassing on the Plains. The Cheyennes and Arapahoes got what they most wanted: the chance to hunt the grass- and bison-rich Southern Plains, horses and mules for trading, and access to the Spanish settlements via Comanche lands. But the peace meant something else in ecological terms. Now all the tribes could freely exploit the Arkansas Valley bison herds. This new exploitation of a large, prime bison habitat that had been a boundary zone skirted by Indian hunters may have been critical. In the Kiowa Calendar the notation for "many bison" appears in 1841, the year following the peace. The notation appears only once more during the next thirty-five years.[47]

One other advantage the Comanches and Kiowas derived from the peace of 1840 was freedom to trade at Bent's Fort. Although the data to prove it are fragmentary, this conversion of the largest body of Indians on the Southern Plains from subsistence/ecosystem hunters to a people intertwined in the European market system probably added critical stress to a bison herd already being eaten away. How serious the market incentive could be is indicated by John Whitfield, agent at William Bent's second Arkansas River fort in 1855, who wrote that 3,150 Cheyennes were killing 40,000 bison a year.[48] That is about twice the number the Cheyennes would have harvested through subsistence hunting alone. (It also means that on the average every Cheyenne warrior was killing 44 bison a year and every Cheyenne woman was processing robes at the rate of almost one a week.) With the core bison population seriously affected by the drought of the late 1840s, the additional, growing robe trade of the Comanches probably brought the Southern Plains tribes to a critical level in their utilization of bison. Drought, Indian market hunting, and cow selectivity must stand as the critical elements—albeit augmented by minor factors such as white disturbance, new bovine diseases, and increasing grazing competition from horses—that brought on the bison crisis of the midcentury Southern Plains. That

[47] My interpretation of the great 1840 alliance of the Southern Plains tribes has been much influenced by Jablow, *Cheyenne in Plains Indian Trade Relations,* 72–73; Levy, "Ecology of the South Plains," 19; Mooney, *Calendar History of the Kiowa Indians,* 276–346.

[48] John Whitfield, "Census of the Cheyenne, Comanche, Arapaho, Plains Apache, and the Kiowa of the Upper Arkansas Agency," Aug. 15, 1855, Letters Received, Records of the Office of Indian Affairs, RG 75, microfilm M234, reel 878 (National Archives). Letters between the principals at Bent's Fort make it clear that the Comanche trade in robes was Bent and St. Vrain's chief hope for economic solvency in the early 1840s. See W. D. Hodgkiss to Andrew Drips, March 25, 1843, Andrew Drips Papers (Archives, Missouri Historical Society).

explanation may also illuminate the experience of the Canadian Plains, where bison disappeared without the advent of white hide hunting.[49]

Perhaps that would have happened on the American Plains if the tribes had held or continued to augment their populations. But the Comanches and other tribes fought a losing battle against their own attrition. While new institutions such as male polygamy and adoption of captured children worked to build up the Comanches' numbers, the disease epidemics of the nineteenth century repeatedly decimated them. In the 1820s, the Comanches were rebuilding their population after the smallpox epidemic of 1816 had carried away a fourth of them. But smallpox ran like a brush fire through the Plains villages again in 1837–1838, wiping whole peoples off the continent. And the forty-niners brought cholera, which so devastated the Arkansas Valley Indians that William Bent burned his fort and temporarily left the trade that year. John C. Ewers, in fact, has estimated that the nineteenth-century Comanches lost 75 percent of their population to disease.[50]

Did the Southern Plains Indians successfully work out a dynamic, ecological equilibrium with the bison herds? I would argue that the answer remains ultimately elusive because the relationship was never allowed to play itself out. The trends, however, suggest that a satisfactory solution was improbable. One factor that worked against the horse tribes was their short tenure. It may be that two centuries provided too brief a time for them to create a workable system around horses, the swelling demand for bison robes generated by the Euro-American market, and the expansion of their own populations to hold their territories. Some of those forces, such as the tribes' need to expand their numbers and the advantages of participating in the robe trade, worked against their need to conserve the bison herds. Too, many of the forces that shaped their world were beyond the power of the Plains tribes to influence. And it is very clear that the ecology of the Southern Plains had become so complicated by the mid-nineteenth century that neither the Indians nor the Euro-Americans of those years could have grasped how it all worked.

Finally and ironically, it seems that the Indian religions, so effective at calling forth awe and reverence for the natural world, may have inhibited the Plains Indians' understanding of bison ecology and their role in it. True, native leaders such as Yellow Wolf, the Cheyenne whom James W. Abert interviewed and sketched at

[49] Records on the robe trade are fragmentary and frequently at odds with one another; see T. Lindsay Baker, "The Buffalo Robe Trade in the 19th-Century West," paper presented at the Center of the American Indian, Oklahoma City, April 1989 (in T. Lindsay Baker's possession). John Jacob Astor's American Fur Company was taking in 25,000–30,000 robes a year from the Missouri River from 1828 to 1830, and St. Louis was receiving 85,000 to 100,000 cow robes a year by the end of the 1840s. Baker cannot yet estimate how many of those came from the Southern Plains, but the trend toward a larger harvest seems apparent. On the Canadian experience, see Ray, *Fur Trade and the Indian,* 228.

[50] Berlandier, *Indians of Texas in 1830,* ed. Ewers, 84–85. Thirteen epidemics and pandemics would have affected the Comanches between 1750 and 1864; see Dobyns, *Their Number Become Thinned,* 15–20. On the abandonment of Bent's first Arkansas River post, see Lavender, *Bent's Fort,* 338–39. John C. Ewers, "The Influence of Epidemics on the Indian Populations and Cultures of Texas," *Plains Anthropologist,* 18 (May 1973), 106. Ewers bases his decline on an estimated early nineteenth-century Comanche population of only 7,000. If my larger estimate is accepted, the Comanche population decline was more than 90%. A falloff in the birthrate as Indian women contracted Bang's disease from brucellosis-infected bison may have contributed importantly to Indian population decline.

Bent's Fort in 1845–1846, surmised the implications of market hunting. As he watched the bison disappearing from the Arkansas Valley, Yellow Wolf asked the whites to teach the Cheyenne hunters how to farm, never realizing that he was reprising a Plains Indian/Euro-American conversation that had taken place sixty years earlier in that same country.[51] But Yellow Wolf was marching to his own drummer, for it remained a widespread tenet of faith among most Plains Indians through the 1880s that bison were supernatural in origin. A firsthand observer and close student of the nineteenth-century Plains reported,

> Every Plains Indian firmly believed that the buffalo were produced in countless numbers in a country under the ground, that every spring the surplus swarmed like bees from a hive, out of great cave-like openings to this country, which were situated somewhere in the great 'Llano Estacado' or Staked Plain of Texas.

This religious conception of the infinity of nature's abundance was poetic. On one level it was also empirical: Bison overwintered in large numbers in the protected canyons scored into the eastern escarpment of the Llano Estacado, and Indians had no doubt many times witnessed the herds emerging to overspread the high Plains in springtime. But such a conception did not aid the tribes in their efforts to work out an ecological balance amid the complexities of the nineteenth-century Plains.[52]

In a real sense, then, the more familiar events of the 1870s only delivered the *coup de grace* to the free Indian life on the Great Plains. The slaughterhouse effects of European diseases and wars with the encroaching whites caused Indian numbers to dwindle after 1850 (no more than fourteen hundred Comanches were enrolled to receive federal benefits at Fort Sill, in present-day Oklahoma, in the 1880s). This combined with bison resiliency to preserve a good core of animals until the arrival of the white hide hunters, who nonetheless can be documented as taking only about 3.5 million animals from the Southern Plains.[53]

But the great days of the Plains Indians, the primal poetry of humans and horses, bison and grass, sunlight and blue skies, and the sensuous satisfactions of a hunting life on the sweeping grasslands defined a meteoric time indeed. And the meteor was already fading in the sky a quarter century before the Big Fifties began to boom.

[51] James W. Abert, *The Journal of Lieutenant J. W. Abert from Bent's Fort to St. Louis in 1845*, ed. H. Bailey Carroll (Canyon, 1941), 15–16; U.S. Congress, Senate, *Report of the Secretary of War, Communicating in Answer to a Resolution of the Senate, a Report and Map of the Examination of New Mexico, Made by Lieutenant J. W. Abert, of the Topographical Corps*, 30 cong., 1 sess., Feb. 10, 1848.

[52] Richard I. Dodge, *Our Wild Indians* (Hartford, 1882), 286. The idea has lingered in the preserved mythologies of the Southern Plains tribes. In 1881 representatives of many of those tribes assembled on the North Fork of the Red for the Kiowa Sun Dance, where a Kiowa shaman, Buffalo Coming Out, vowed to call on the herds to reemerge from the ground. The Kiowas believed the bison had gone into hiding in the earth, and they still call a peak in the Wichita Mountains Hiding Mountain. Alice Mariott and Carol Rechlin, *Plains Indian Mythology* (New York, 1975), 140; Peter Powell, *Sweet Medicine* (2 vols., Norman, 1969), I, 281–82. There is no mention of this idea in the major work on Comanche mythology, but it is far from a complete compilation: Elliott Canonge, *Comanche Texts* (Norman, 1958). On the bison's wintering in protected canyons, see Randolph Marcy, *A Report on the Exploration of the Red River, in Louisiana* (Washington, 1854), 125–31.

[53] Wallace and Hoebel, *Comanches*, 32; Richard I. Dodge in *The Plains of North America and Their Inhabitants*, ed. Wayne Kime (Newark, 1989), 155–57. On these figures for bison, see Roe, *North American Buffalo*, 440–41.

5

Environmental Change and Social Change in the Valle del Mezquital, Mexico, 1521–1600

Elinor G.K. Melville

The often disastrous consequences of the introduction of exotic animals into a New World environment are very clearly demonstrated by the sixteenth-century history of the Valle del Mezquital, Mexico. A rapid and profound process of environmental degradation, caused by overstocking and indiscriminate grazing of sheep in the post-conquest era, leads us to ask whether the Spanish always acted in their own long-term interests in the New World.[1]

Recent analyses of the growth and development of colonial societies accord a crucial role to the natural resources: It is argued that the conquerors were forced to adapt their expectations, and their institutions, to New World realities. Differences in local resources are used to explain evidence of marked regional differentiation in social and economic growth; the strengths and deficiencies of the various regions at the time of the conquest, as defined by their relation to the growing world market, are thought to have shaped future development (or underdevelopment). A basic premise of these studies is that, given the available resources and the technology of the early modern era, the Spanish developed the best possible means to exploit the riches of the Americas.[2]

A preliminary version of this paper was read at the annual meeting of the American Historical Association, December 30, 1985. I would like to acknowledge the help of all the people who helped with earlier drafts (even though I did not always take heed of their advice): Christon Archer, Robert Claxton, Bruce Drewitt, Bonnie Kettel, Barbara Kilbourn, Herman Konrad, Michael Levin, Rebecca Scott, Gavin Smith, Jacque Solway, Shelia Van Wyck, Lorna Woods, and Eric Van Young. I am very grateful to the Social Sciences and Humanities Research Council of Canada for support during research and writing.

[1] The ecological consequences of the European diaspora have been clearly and comprehensively set out in Alfred Crosby's works, see especially *The Columbian Exchange. Biological and Cultural Consequences of 1492* (Westport, 1972) and *Ecological Imperialism. The Biological Expansion of Europe, 900–1900* (London, 1986).

[2] Charles Gibson's classic study, *The Aztecs under Spanish Rule* (Stanford, 1967), which was aimed at uncovering the details of Indian adaptation to Spanish rule, also demonstrated how Spanish institutions were modified and adapted to colonial realities and is the departure point for the modern histories. As others followed Gibson's lead and carried out detailed regional studies and investigated the internal structure and external relations of the hacienda, his ideas about the entrepreneurial spirit of the landowners in the colonial era were confirmed (*Aztecs*, 326 *ff.*), but striking differences between the regions were uncovered. See Eric Van Young's review article for

The idea that the settlers were sufficiently in control of events to implement desired changes in land use, let alone develop the best systems of production, is called into question by the case of the Valle del Mezquital. In this paper I will demonstrate that the system of land use which dominated regional production in the Valle del Mezquital after the conquest was not only not the best choice—it was not even planned. Moreover, the system of production which characterised the mature colony was neither based on the initial status of the natural resources, nor was it the result of conscious planning. My research indicates that extensively exploited latifundia were formed during the last two decades of the sixteenth century, when small-scale intensive pastoralism collapsed as a result of the deterioration of the range from overgrazing.[3]

By demonstrating a close relationship between environmental change and economic growth and development, this study challenges the idea that the natural resource base acted simply as a passive constraint on development and that external developments, such as the expansion of the international market, are sufficient to explain growth in the colonial economies.[4] Change in the natural resource base, not the initial resources, constrained regional development and made the Valle del Mezquital into one of the most economically depressed regions of Mexico; and local events assumed critical importance in shaping local developments.

SIXTEENTH-CENTURY CHANGES IN LAND USE

The Valle del Mezquital is the archetype of the barren, eroded regions of Mexico.[5] The high treeless hills, which separate the wide flat valleys and plains typical of the neo-volcanic plateau, are bare rock; the lower slopes are eroded down to a rock-like layer of hard pan (caliche) in which only scattered cacti and thorn scrub grow. Strip mining for lime has chewed away whole hills, and cement factories pollute the land with dust. Since the 1940s, ever-expanding

---

a discussion of the current understanding of reasons for regional differences in Mexico: "Mexican Rural History Since Chevalier: The Historiography of the Colonial Hacienda," *Latin American Research Review*, 18:3 (1983), 5–61. See also Magnus Mörner, "The Spanish American Hacienda: A Survey of Recent Research and Debate," *Hispanic American Historical Review*, 53:2 (1973), 183–216.

3 This paper is based on research carried out for my doctoral dissertation: "The Pastoral Economy and Environmental Degradation in Highland Central Mexico, 1530–1600" (Department of Anthropology, University of Michigan, 1983).

4 The most explicit statement of the importance of the international market for local development in Latin America was made by André Gunder Frank, *Capitalism and Underdevelopment in Latin America. Historical Studies of Chile and Brazil* (New York, 1967). Immanuel Wallerstein, *The Modern World-System*, 2 vols. (New York, 1974 and 1980), has followed with a detailed analysis of world history according to this thesis.

5 I have followed Miguel Orthón de Mendizabal's definition (in *Obras Completas*, 6 vols. [Mexico, 1946–47], VI) of the region: *south*, the northern end of the Valley of Mexico and the high mountains separating the region from the Toluca Valley; *east*, up to but not including the Sierra de Pachuca; *north*, the southern slopes of the Sierra de Juárez; and *west*, the San Francisco and Moctezuma Rivers.

irrigation systems using the waste water from Mexico City have transformed the plains and valleys of much of the region, and permanent crops are grown here for the market in Mexico City; but the rest remains the dry and barren wasteland traditionally associated with the Valle del Mezquital.

This landscape is in striking contrast with the picture obtained from the early post-conquest documents.[6] Up to the middle of the sixteenth century, the Valle del Mezquital was a densely populated and complex agricultural mosaic of planted and fallow fields, villages, quarries, lakes, dams, woods, and native grasslands. Maize production predominated in the south and west, and the maguey and nopal cactus in the drier northeast.[7] Hills and mountains were forested; willows lined the banks of rivers and streams; and few villages were without groves of trees such as cedar, native cherries, and mesquite.[8] Limestone was extracted from quarries in the hills of the southeastern quarter, which had ample woodlands to supply the lime kilns.[9] Springs, rivers, streams, dams, and *jagueyes* (depressions which catch runoff) supplied extensive irrigation systems present in most of the region; and humid bottom lands and swamps provided rich moist soils, even in areas lacking surface water for irrigation.[10]

The vegetation, both wild and domesticated, was far more varied than at the present and desert species more limited in extent. As one moved from south to north across the region, the species comprising the savannahs, woodlands, and edge habitats shifted with decreasing rainfall and increasing temperatures from grasses, oaks, pines, native cherries, willows, and the occasional mesquite (as isolated trees, not chaparral) to semidesert species of wild maguey, wild cactus, yucca, mesquite, and thorns.[11] The proportions of

6 Documentation for this study was taken from the *Archivo General de Indias* (specific sections: Audiencia de México, Contaduría, Escribanía de cámara, Indiferente Justicia, and Patronato) in Seville, Spain, and the *Archivo General de la Nación* (specific sections: Civil, General de Parte, Historia, Indios, Libro de Congregaciones, Mercedes, and Tierras) in Mexico City [hereafter referred to as *AGI* and *AGN* respectively]. The major published source was *Papeles de Nueva Espanña*, Francisco del Paso y Troncoso, ed. (Madrid and Mexico, 1905–48), Vols. I, III and VI [hereafter cited as *PNE* I, III or VI]. Where the documentary citations are extensive, reference will be made to my doctoral dissertation.

7 For a discussion of the production of maize in the southeast quarter of the region, see Sherburne F. Cook, *The Historical Demography and Ecology of the Teotlalpan* (Berkeley, 1949), 33–41. For evidence of the importance of the maguey and nopal in the north, see PNE I, 219, 220; and *PNE* VI, 22.

8 Melville, "The Pastoral Economy," ch. 2, n.s 25–30. *AGN*, "Mercedes," supplied the bulk of the information about forests and woodlands; *AGN*, "Tierras;" and *PNE* I also provided a great quantity of information; and occasional references were found in *PNE* III and VI; *AGN*, "Indios;" and *AGI*, "Escribanía de cámara," "Indiferente," and "Justicia."

9 Melville, "The Pastoral Economy," ch. 2, n. 31.

10 Melville, "The Pastoral Economy," ch. 2, n. 24. The major source for information about the irrigation systems was *PNE* I; *AGN*, "Tierras" and "Mercedes" were next in importance; followed by *AGN*, "Historia" and "Indios;" and *AGI*, "Justicia" and "México." References were mostly made to irrigation (*rregadio* or *irrigación*) but occasionally irrigation ditches or canals were mentioned (*acequias* or *canales*).

11 Melville, "The Pastoral Economy," ch. 2, n.s 25–30, 37. The names given in the docu-

cultigens such as maize, beans, squash, chia, chilies, cotton, and the domesticated maguey and nopal cactus also shifted in a roughly south to north direction with the availability of water for irrigation and increasing average temperatures.[12]

The Spaniards who surveyed this region twenty-seven years after the conquest (*circa* 1548) noted that excellent soils promised good wheat crops, good but limited native grasslands, the possibility of an expanded lime industry, and extensive woodlands necessary for mining ventures. There were, however, several obstacles to Spanish expansion at this date: the still dense indigenous populations (which took up space and used the available water resources to irrigate their fields); the undependable rainfall; and the frequent frosts.[13]

By the time of the conquest in 1521, this region had been exploited by agriculturalists for centuries. They had modified the landscape by clearing land and growing crops, building dams and irrigation systems, terraces, roads, and villages; and the population in the southeastern quarter was so dense that houses were practically contiguous.[14] There is evidence, however, that the human population was in reasonable ecological balance with its physical environment. For example, forests were still extensive; and the extent of the irrigation systems (only eight out of a total of thirty-seven head towns were without irrigation) not only indicates the essential aridity of the region and the need to use irrigation to secure crops,[15] but also the good health of the catchment area and the fact that sufficient water was internally generated for these systems. Finally, there is an almost total lack of evidence of environmental degradation before the last three decades of the sixteenth century.[16]

---

ments for these plants were: *pasto, zacate* (grasses); *rrobles, encinos* (oaks, live oaks); *pinos, oyamel* (pines); *capulies* (native cherries); *sauces* (willows); *ahuehuetes* (cedars); *mesquite* (mesquite); *lechuguilla* (wild maguey); *tunal, nopal* (wild cactus); *palmas sylvestres* (yuccas); *espinos* (thorns). The records of the land grants found in *AGN*, "Mercedes," were the most fruitful source for this type of information and were used together with *AGN*, "Tierras;" *PNE* I; and *AGI*, "Justicia."

[12] *PNE* I, 60, 125, 159–60, 193, 217–8, 220. *PNE* III, 69, 72. *AGN*, "Mercedes," vol. 2, fols. 48, 95–6; vol. 5, fol. 260; vol. 6, fol. 515; vol. 7, fol. 349; vol. 9, fols. 132–3; vol. 12, fol. 485. *AGN*, "Tierras," vol. 3, exp. 1; vol. 1529, exp. 1. *AGI*, "Justicia," leg. 124 #1, fol. 19.

[13] *PNE* I, 2–3, 21, 22, 57, 59, 60, 110, 125, 159–60, 166, 193–4, 207, 208, 209–10, 217–20, 223–4, 289, 292, 310.

[14] Archaeological surveys of the Tula River basin and headwaters demonstrate very dense conquest population, see A. B. Mastache de E. and Ana Maria Crespo O., "La Ocupación Prehispánica en el Area de Tula, Hgo.," in *Proyecto Tula*, Eduardo Matos Moctezuma, ed. (Mexico, 1974–76), no. 33; and W. T. Sanders, J. R. Parsons, and R. Santley, *The Basin of Mexico* (New York, 1979), 179, 213–6. See also Peter A. Gerhard, *A Guide to the Historical Geography of New Spain* (Cambridge, 1972), 295.

[15] *AGN*, "Tierras," vol. 3, exp. 1; vol. 64, exp. 1; vol. 79, exp. 6; vol. 1486, exp. 8; vol. 1487, exp. 1; *AGN*, "Mercedes," vol. 5, fol. 122; *AGI*, "Justicia," vol. 207:2, ramo 3. *AGI*, "Escribanía de cámará, leg. 161–C, fol. 250.

[16] I found evidence of only two instances of environmental degradation prior to 1560, both pertaining to the Jilotepec sub-area, *AGI*, México, leg. 96, ramo 1, and *AGN*, "Tierras," vol. 1872, exp. 10.

By the end of the sixteenth century the picture had changed. The Valle del Mezquital had been transformed into a sparsely settled mesquite desert considered to be fit only for sheep, and the decimated Indian communities were congregated in villages separated by an homogenous vista of mesquite-dominated desert scrub. Hills and mountainsides were deforested, and Spaniards and Indians fought over the trees now confined to *quebradas* (gorges) in areas which had been forested only thirty years before.[17] The heavy summer rains washed the stones off the now denuded hills onto formerly fertile slopes; sheet erosion removed the fertile topsoils and exposed the *tepetate* (hard pan); and gullies cut down through all the layers. Springs all over the region were failing, and both Indian and Spanish landowners had problems with their water supply. Instead of the intensive irrigation agriculture practised by the Indians at mid-century, regional production was now dominated by grazing on extensively exploited latifundia.

Three major processes underlay the transformation of the landscape of the Valle del Mezquital: the conversion of land use to grazing; the demographic collapse of the indigenous populations; and the ecological changes that accompanied the expansion of intensive sheep grazing. I argue that these processes also transformed the early colonial systems of production associated with intensive sheep grazing and initiated latifundia development.

Before discussing these processes, I will comment on the documents used for this study, especially those used to document the process of environmental change, because the evidence is scattered through a wide variety of sources which all pose some problems in their use. The best known and most obvious sources for sixteenth-century Mexican ecological history are the *Relaciones Geográficas*—regional descriptions that were drawn up circa 1548, 1569–70, and 1579–81 to supply the crown and the church with information about the Spanish possessions in the New World. While these documents are a rich source of information about populations and settlement patterns, the subsistance base and modes of land use of the Indian communities, as well as the natural resource base, they are clearly subject to observer bias.[18] The interests and capabilities of the local officials (clerics and representatives of the crown) obviously influenced the completeness and accuracy of these descriptions. Some officials were clearly fascinated by the history and the culture of the indigenous peoples of their regions and have left invaluable records, while others filled out the questionares with the minimum of information. While these reports were ostensibly made to record the Indians' use of the natural

[17] A more complete discussion of ecological changes is given below. References relating to deforestation and fighting over trees is to be found in: *AGN*, "Mercedes," vol. 6, fol. 456; vol. 7, fol. 87, *AGN*, "Tierras," vol. 2697, exp. 11; *PNE* I, 217–8. *PNE* VI, 33.

[18] For a discussion of the distorting effect of the "preconceptions and expectations" of settlers in descriptions of early New England, see William Cronon, *Changes in the Land. Indians, Colonists and the Ecology of New England* (New York, 1983), 20. See also ch. 1, in which Cronon discusses the types of sources available for the study of ecological changes in New England.

resources, they were actually made to assess the potential for Spanish expansion and this requirement shaped the perception as to which natural resources were worth mentioning. Carefully used, however, these sources not only supply general descriptions of the landscape at specific points in time but also—and perhaps more importantly for the purposes of this study—indicate broad changes over time.

The most fruitful sources for the details on ecological change are those surveys carried out for land grants, land suits, and *congregaciones* (concentration of Indian populations in selected centers). Unfortunately the most common type of survey, the boundary survey, is subject to statistical error because the information is limited to a strip of land around the edge of a property and can be regarded at best as a random sample. Very often notes were recorded, however, about the condition of the vegetation and soils of a hillside or a valley, or the productivity and history of use of the land; and these give us a somewhat broader picture. The surveys of village resources made in preparation for *congregaciones*, while providing a more complete idea of resources, are subject to the same type of observer bias noted in geographic relations since the Spanish had an obvious interest in the lands vacated by the Indian communities (although the takeover of these pueblo lands was prohibited).[19]

The problem of nomenclature was encountered in all sources: Generic terms were most often used to describe the vegetation (oaks, pines, thorns), and there is the possibility that a plant was assigned to an incorrect genera because it looked like an Old-World species. It proved possible to correct for this to some extent by collecting a lot of data from different sources and comparing the results with evidence of known plant associations for this and other semiarid regions of Mexico. In spite of these problems, surprisingly detailed evidence of the character of the landscape at specific points in time, as well as the processes by which the landscape changed, was obtained by using several different types of sources together.

The Spanish introduced cattle, horses, sheep, and pigs into the Valle del Mezquital probably as early as the 1520s but definitely in the 1530s. They also accelerated the production of lime in the southeastern quarter of the region for the rebuilding of Mexico–Tenochtitlan and founded silver and lead mines in Ixmiquilpan in the 1540s and in the Sierra de Pachuca in 1552. European crops and fruit trees were grown in a limited way as tribute crops for *encomenderos* (recipient of a grant of Indian tribute and labor) and the crown.

The initial expansion of pastoralism developed according to an alien perception of the natural resource base and rights to its exploitation in a process that combined legal resource exploitation, illegal land grabbing, and force. In

[19] See Charles Gibson, *Aztecs*, 281, for a discussion of Spanish invasion of vacated pueblo lands.

the early years of the expansion of pastoralism, livestock were introduced directly into the densely populated agricultural lands of the Indians in which they grazed on native grasslands and fallow fields, and browsed in edge habitats and crops. Spanish custom, developed to promote the needs of small pastoralists, recognised grass as a common resource wherever it grew; and crop lands were open to grazing after the harvest.[20] This meant that the expansion of grazing animals into the Indian crop lands could be viewed as an entirely legal and necessary use of an unexploited natural resource—especially given the Spaniards' consumption of animal products. After being a habitat for the wild animals that formed part of the subsistance base of the Indians, grass became a common resource of the Spanish pastoralists. Land for stations (corrals and shepherds' huts), on the other hand, was acquired illegally: The Spaniards simply took village land for the stations, often placing them next to the houses of the Indians. The practise was so extensive that when Viceroy Mendoza began to give out land grants for station sites in the 1530s, most of the early grants legalised the squatters' rights.[21]

The pastoralists maintained access to the pasture by force. Extremely violent relationships quickly developed between the herdsmen (mostly black slaves) and the villagers, and were, I believe, an important element in the process by which access to grazing was maintained in an era of common grazing in a predominantly agricultural region. For example, in 1556 the Indians of Jilotepec accused the local Spanish livestock owners of inciting their slaves to steal the women, food, and belongings of the Indians. The slaves, they said, came in large bands by order of their owners to rob the Indians, who were beaten with staffs when they tried to defend themselves. On one occasion an Indian who tried to rescue his wife from the station in which she was being held was tied to the tail of a horse and dragged until he died—naturally, said the Indians, the slave owners pretended this had not happened.[22] Undoubtedly such stories were improved in the telling to support the Indians' claims of bad treatment, but they do indicate unequal power relations; and it is clear that slaves were one of the instruments used by the Spanish to implement their rule.

At first the herds were small. Although the animals were at a relatively low density in the immediate post-conquest decades, the fact that they were maintained within lands used for agricultural production meant that they had

[20] See François Chevalier, *La formación de los latifundios en México* (Mexico, 1975), 12, 119–20; and also David E. Vassberg, *Land and Society in Golden Age Castile* (Cambridge, 1984), especially 13–18 for a discussion of stubble grazing and 19–56 for a discussion of municipal property.

[21] Simpson, Gibson, and Chevalier also noted that grants were very often issued for lands in the possession of the grantees; see Lesley B. Simpson, *Exploitation of Land in Central Mexico in the Sixteenth Century* (Berkeley, 1952), 6; Gibson, *Aztecs*, 275; Chevalier, *La Formación*, 131.

[22] *AGN*, "Mercedes," vol. 4, fols. 330–2. Herman W. Konrad has come to the same conclusion regarding the slaves' relations with the villagers and thinks that perhaps the station owners ignored orders to control their slaves' actions (verbal communication, 1985).

caused more damage than the small numbers imply.[23] As part of his efforts to relieve the destruction of the Indian crop lands by Spanish livestock, Viceroy Velasco expelled cattle from the Valle dei Mezquital and other densely populated central regions in the mid-1550s—sheep, being less of an obvious menace, were allowed to stay.[24]

During the 1550s the flocks of sheep quadrupled in size from approximately 1,000 to 3,900 head. This sudden increase in the sheep population in a region that was still densely populated by Indian agriculturalists[25] was due to a combination of factors: The animals were by this time acclimatised to the highlands; the pressure on grazing space had been lessened by the expulsion of cattle; and the domesticated grazing animals were moving into underused ecological niches, such as grasslands, edge habitats, and forests.

The expansion of the grazing animals into the nondomesticated niches helps to explain both the underdevelopment of Spanish agriculture and the incredible speed with which pastoralism expanded. The slow development of Spanish agriculture can be partly explained by fact that the Indian agriculturalists supplied the Spaniards' needs for more than fifty years after the conquest. Spaniards here, as elsewhere in New Spain, really only began to take a serious interest in agriculture after the severe 1576–81 epidemic of unknown origin (the Great *Cocoliztle*) when Indian production declined abruptly. The slow development of Spanish agriculture does not in itself explain the rapid expansion of the grazing animals since, as noted above, the region was still densely populated by Indian agriculturalists when the sheep population began to expand in the 1550s. It also does not explain why the region did not develop into an agricultural region later in the century, when the early descriptions reported excellent natural resources for grain production; irrigation systems were already in place; and there were growing markets for agricultural products in the nearby mines, Mexico City, and the mule trains going north to Zacatecas. I suggest that the explanation lies in the contrasting ways by which the Old World agricultural species and grazing animals were introduced into this region (and indeed all of New Spain).

European agricultural species were introduced by way of the *encomienda*–village system of production. Following the conquest, the *conquistadores* received grants of Indian tribute and labor (the *encomienda*) as booty. The encomenderos could specify the type of tribute they required and use their tributaries for a wide range of entrepreneurial activities, but they had no legal

23 See Melville, "The Pastoral Economy," ch. 2, n.s 35 and 40, for references to problems with animals in the Valle del Mezquital. The problems caused in Indian lands by grazing animals is well documented, see for example Gibson, *Aztecs*, 280, and Simpson, *Exploitation*, 4–6.

24 Chevalier, *La Formación*, 133–5. Evidently the order to remove cattle was enforced. For example, Rodrigo de Castañeda was compelled to remove his cattle from Jilotepec in 1557, see *AGI*, "México," leg. 1841, fols. 1r–8r.

25 An investigation carried out in 1564 of the towns to be granted in encomienda to a son of Moctzuma shows that there was still little room between the towns in the Tula sub-area at this date. See *AGI*, "Justicia," leg. 207, #2, ramo 3.

rights to the land within their encomiendas. They thus had direct control over *what* was produced but were ultimately dependent on Indian land and labor: The encomienda was embedded in the village systems of production.

Pastoralism, by contrast, was not dependent on Indian lands and labor. As noted above, grass was a common resource wherever it grew, making the ownership of land not strictly necessary; and, at least in the early years, black slaves were used as herdsmen, thus bypassing the need for Indian labor. Grazing developed outside the constraints of the encomienda–village system of production as a parallel system of resource exploitation. Thus the grazing animals and their owners had all the advantages of untrammeled exploitation of new ecological and social niches; and this, combined with the use or threat of force, meant that the grazing animals were able to expand extremely rapidly in spite of the density of the Indian agricultural populations.

By the time Spaniards began to take an interest in agriculture in the 1580s, sheep grazing not only dominated regional production but had so degraded the environment of large areas of the Valle del Mezquital that the land was fit only for sheep. Opportunities were lost in the early post-conquest period, and the degradation of the fragile ecology of this semiarid region prevented the exploitation of the growing markets for agricultural products later in the century.

During the 1560s and 1570s, the flocks continued their spectacular increase. Although the legal stocking rate (number of head per holding) was 2,000 head of *ganado menor* for a station of 7.8 square kilometers, 10,000 to 15,000 head was considered normal, even desirable during these decades.[26] By the 1570s regional production had shifted from intensive irrigation agriculture and was dominated by intensive sheep grazing.

It became clear in the course of the analysis of documentation of land use and land tenure that the domination of regional production by intensive grazing involved the displacement of the Indian agriculturalists. Once the sheep had reached a critical mass in their population increase, they simply took off and increased so rapidly and to such densities that they converted extensive areas of land to pastoralism by use—before they were transferred to the Spanish system of land tenure.

The Indians were displaced not only from their crop lands but also from such resource areas as forests, and traditional systems of land use were modified as the natural resource base of the Indian communities was diminished and in some cases destroyed. For example, the Indians complained bitterly that sheep trampled the stands of the shrub *tlacotl,* which was used as a wood substitute in the production of lime; ate the leaves of edible roots so that they

[26] See footnote 39. In sixteenth-century Mexico, the term *ganado menor* included sheep, goats and pigs. In the Valle del Mezquital, sheep appear to have constituted the bulk of the flocks, although by the end of the century the proportion of goats in the flocks increased.

could not be harvested; and left the soils bare.[27] Those Indian villages which had managed to retain some of the best agricultural lands often found the usefulness of these lands reduced by lack of water and heavy secondary growth.[28]

In a classic study of land exploitation in sixteenth-century Mexico, Lesley Byrd Simpson proposed that the post-conquest demographic collapse of the indigenous populations freed land for the expansion of the domestic grazing animals introduced by the Spanish. Simpson inversely correlated the human population decline with a steady increase in the animal population up to circa 1620, when the Indian population began to stabilise.[29]

In the case of the Valle del Mezquital, however, the growth of the sheep population was not related to the human population decline in a straightforward way. The human population of the Valle del Mezquital declined from an estimated tributary population at the conquest of between 452,623 and 226,311, to a documented tributary population in 1600 of 20,447.5 (that is, an overall decline of between 90.9 percent and 95.4 percent).[30] The sheep population began to increase rapidly in the 1550s, and it simply took off—after reaching a critical point sometime in the 1560s at which it outstripped the human population decline—and reached an estimated total of nearly four million head by 1579. In the late 1570s and early 1580s, there was, however, a sudden drop in flock size of up to 50 percent with a slowing up of the rate of increase; and by 1599 the estimated sheep population in the region had dropped to 72 percent of the total in the 1570s. The demographic collapse of the Indian population undoubtedly facilitated the formal takeover of land, but the sheep population increased and then decreased independently of the human population decline.

Even if the Indians of the Valle del Mezquital did not wholeheartedly

[27] *AGN*, "Tierras," vol. 1525, exp. 1; vol. 2697, exp. 11; *AGN*, "Mercedes," vol. 4, fols. 330–2; vol. 7, fol. 87.

[28] *AGN*, "Tierras," vol. 64, exp. 1; vol. 83, exp. 10; vol. 2717, exp. 10; vol. 2766, exp. 3.

[29] Simpson, *Exploitation*, ii.

[30] See Melville, "The Pastoral Economy," Appendix A, for sources used in estimating tributary population totals. Evidence for the tributary population was based almost entirely on tribute records from the Contaduría of the *AGI*. Ecclesiastical records were used to check doubtful totals or for the few cases in which totals were not available in the Contaduría. My totals agree fairly well with those estimated for this region by Sherburne F. Cook and Woodrow Borah, *The Indian Population of Central Mexico, 1531–1610* (Berkeley, 1960) but have the advantage of being based almost entirely on a single source.

According to contemporary witnesses the Indian population of New Spain had declined by as much as two-thirds and possibly five-sixths by the mid-1560s (Gibson, *The Aztecs*, 138). The tributary population of the Valle del Mezquital in 1570 was 76,946. If a 66 percent decline is applied to the 1570 total, we get an estimated tributary population in the Valle del Mezquital of 226,311 at the time of the conquest; while an 83 percent decline gives 452,623. By 1600 the tributary population had declined to 20,447.5. The total population decline for the Valle del Mezquital over the period 1519–1600 is therefore somewhere between 90.9 percent and 95.4 percent.

TABLE I

"Spanish" Landowning: Sheep Stations Taken up Each Decade

	Grantees[a]		*Squatters*[b]		
Decade Ending	*Spaniards*	*Others*[c]	*Spaniards*	*Others*[c]	*Totals*
1539	5	—	29	—	34
1549	5	—	36	—	41
1559	13	—	16	4	33
1565	24	90	49	5	168
1569	6	5	7	—	18
1579	18	26.5	54	7	105.5
1589	55	15	100	14	184
1599	49.5	95	95	39	278.5
Totals:	175.5	231.5	386	69	862

[a]*Grantees* were recipients of land grants.
[b]*Squatters* were undocumented landowners.
[c]*Others* refers to non-Spanish landowners: Indians; Mestizos (two grantees in the 1580s and one in the 1590s); and one Mulatto squatter (1590s).
SOURCES: Melville "The Pastoral Economy," Appendix E, 272–95. See also n.31 of this article.

accept sheep herding as a way of life, they recognised the profitability of pastoralism and the need to retain control of the land. Indian nobles, and some communities, asked for and received 56 percent of all the grants for sheep stations, and acquired 34 percent of all holdings in the Valle del Mezquital in the sixteenth century[31] (see Table 1). However, while they retained formal control of the land within the Spanish system of land tenure, it was lost for all intents and purposes to the traditional Indian systems of land use.

By the end of the 1570s, large numbers of pastoralists who based their operations in small holdings dominated regional production and exploitation of the natural resources. Mining and lime manufacture were, in terms of overall regional production, limited in importance and extent. Even as the region came to be dominated by grazing, however, the pasture failed; the size of the flocks dropped markedly; the animals became underweight and re-

[31] Sources of land holding are listed by *cabecera* and in chronological order. The two primary sources for land use and land tenure were: *AGN*, "Tierras," which contains documentation of land suits concerning ownership, boundaries, land use and inheritance; and *AGN*, "Mercedes," the records of land grants, grazing rights and licences of various sorts. Other sources in order of importance were: *AGI*, "México," "Justicia," "Escribanía de cámara;" *AGN*, "General de parte," "Indios"; and *PNE* I.

Of the 862 sheep stations documented for the Valle del Mezquital for the sixteenth century, only 407 appeared in the records as formal land grants. I have designated the remaining 455 stations as "squatters' holdings" because I found no evidence of formal title for these stations; they were identified by reference to holdings which had either formed the boundaries of new grants, been mentioned as being located in their vicinity, or been the subject of court cases, wills, *informes*, *diligencias*, or *relaciones*.

produced more slowly; and the production of wool, meat, and tallow began to decline. As the carrying capacity of the range declined, the intensive pastoralism of the 1560s and 1570s was no longer possible and was replaced by extensive pastoralism by the end of of the 1580s.[32]

OVERGRAZING AND ECOLOGICAL CHANGE

The argument that the rapid expansion of intensive sheep raising caused environmental degradation in the Valle del Mezquital brings us to the issue of the relationship between grazing and ecological change.

The ecological changes documented for the Valle del Mezquital during the last half of the sixteenth century have been reported for many other semiarid regions in which domestic grazing animals have been introduced in large numbers. The sequence of events in all these regions is similar: reduction of the vegetative cover, diminution of the number of grass species, and invasion by unpalatable arid-zone species; followed by gullying, sheet erosion, flooding, and a drop in the water table. The relationship between grazing and environmental change, especially vegetative change, is not a straightforward one; and several variables have been proposed, singly or in combination, to account for the ecological changes. The more frequently discussed for all semiarid regions include: overgrazing and trampling by large numbers of introduced grazing animals; climatic change; and the suppression by Europeans of regular burns in woodlands and grasslands. There is, however, a general consensus of opinion that removal of the ground cover results in the reduction of the infiltration rate of rainfall (with a consequent drop in the water table and loss of spring flow) and increased overland flow of surface water, which results in increased flooding, erosion, and gullying.[33]

Most historical studies which investigate the relationship of domestic livestock to environmental degradation, especially the increase in density of woody species such as the mesquite, stress the increase in numbers and density of grazing animals. Nevertheless, while I agree that the reduction and weakening of the native grasses under high densities of grazing animals produced conditions favourable to the invasion by arid-zone species, I argue that the abrupt *drop* in numbers and density of grazing animals in the late 1570s and 1580s, together with changes in the fire regime, is the key to understanding the rapid increase of mesquite-dominated desert vegetation in the Valle del Mezquital during the last quarter of the sixteenth century.

[32] Carl L. Johanessen writes that the "carrying capacity of the range may be described as the number of animals it can support in health, during the period when grass is palatable and nutritious, without reducing forage production in subsequent years. . . . Overgrazing occurs when the number of stock exceeds the carrying capacity of the range" (*Savannas of Interior Honduras* [Berkeley, 1963], 106).

[33] For a discussion of the variables and the supporting arguments, see James R. Hastings and Raymond M. Turner, *The Changing Mile. An Ecological Study of Vegetation Change With Time in the Lower Mile of an Arid and Semi Arid Region* (Tucson, 1965), 3–6, 275–83.

Very briefly, the timing and sequence of changes in the density of the sheep population and in the environment of the Valle del Mezquital during the last half of the sixteenth century were as follows. Large numbers of sheep grazed the native grasses, herbs, and shrubs during the 1560s and 1570s until the ground was left bare. Arid-zone species began to invade these grazed areas in the late 1560s but were kept in check by the density of the grazing animals and the regular burns carried out by herdsmen to stimulate pasture growth. In the 1570s the grasses began to fail—probably as a result of heavy grazing and repeated burns. As the pasture failed, the flocks declined in size and quality. It is most likely that fires no longer spread for lack of fine fuel (dry grasses). At the same time, arid-zone species began to expand in extent and density. I argue that the lessening of the controlling factors (the high density of grazing animals, frequent burns, and competing grasses) allowed a rapid increase in the spread of the arid-zone species in the last quarter of the sixteenth century. The formation of a stony pavement and exposure of the impervious hard-pan layer produces a micro-environment hostile to grasses. Such soil conditions, which were reported for the Valle del Mezquital during the 1580s, would have reinforced the shift to arid-zone species.[34]

Climatic change cannot be ruled out as a complicating factor in the process of environmental change in the Valle del Mezquital, especially since the second half of the sixteenth century falls within the peak years of the Little Ice Age (1550–1700); and a shift to drier climatic conditions, such as occurred in some areas during the Little Ice Age, would favour arid-zone species over grasses. However, the meagre amount of climatic data for this period makes it difficult to see a trend in any direction, and the climate appears to have been consistently unpredictable throughout the sixteenth century. The major argument against climatic change as the primary variable in the process of environmental degradation in the Valle del Mezquital is that environmental deterioration did not occur in all parts of the region to the same degree. As I will demonstrate below, there is a clear correlation between high densities of sheep and environmental degradation in certain sub-areas, which indicates that overgrazing was the primary variable in the process of ecological change in this region during the sixteenth century.

Regular burns in grasslands and woodlands were carried out by the indigenous populations in many parts of the New World to promote optimum conditions for game animals. The suppression of these burns by Europeans allowed for the spread of a dense undergrowth of woody species; however, I have no evidence that the indigenous populations of the Valle del Mezquital used fire—either as a means to clear agricultural land or to maintain the forest floor clear of undergrowth. Burning the grasslands was evidently instituted soon after the conquest, and permission to burn-off had to be obtained from

[34] See Johanessen, *Savannas*, 78, for a similar interpretation of the relationship between vegetative changes and livestock densities.

the *guarda mayor* of the *Mesta*.[35] Fire is therefore a new element in the ecological equation, but one about which there is surprisingly little data.

In order to examine the relationship between grazing and environmental change in the Valle del Mezquital, I grouped the sub-areas[36] of the region into five types according to differences in (1) the extent of land formally converted to pastoralism each decade[37] and (2) the grazing rates.[38] This typology presents the changing levels of exploitation of the sub-areas between 1530–1600 in schematic form (see Table 2). It can be seen that exploitation varied from almost complete conversion of lands to pastoralism and high grazing rates (Type 1) to limited conversion and relatively low grazing rates (Type 5); and from intensive exploitation at an early date to extensive exploitation throughout the century.

The number of head present in each sub-area forms the basis for estimates of grazing rates and is dependent on the stocking rate (the average number of head per holding), a figure where was determined for the region as a whole. The stocking rate, in turn, depended on the state of the pasture, human population densities, markets, and changed throughout the century according to shifts in these variables. In the early decades the flocks were restricted by the extent of the Indian agricultural lands, but they increased from 1,000 to 3,900 head per station when the amount of available grazing land expanded in in the 1550s. Although the legal stocking rate for a sheep station of 7.8 square kilometers was 2,000 head, there are enough examples in the 1550s to show

[35] *AGN*, "Mercedes," vol. 3, fols. 95–96, 113. Although the regulations did not allow burning in the forests, accidents sometimes did happen, and forests were burnt; see *AGN*, "Mercedes," vol. 3, exp. 249, fols. 95–96. Jeffrey R. Parsons has not found archaeological evidence of regular burns in the Valley of Mexico for the late pre-conquest period (verbal communication, 1983).

[36] The division of the region into ten sub-areas was based on geopolitical criteria. The Valle del Mezquital consists of eight wide flat plains and valleys, an area of low rolling hills forming the headwaters of the Tula River, and the high mountain valleys of the northern end of the Sierra de las Cruces. The broad division of the region is based on these ten geographic areas, and the final boundaries have been taken to be coterminous with the land under the jurisdiction of the *cabeceras* (head towns) located within their borders.

[37] The area of land converted to pastoralism was obtained by multiplying the number of stations present in each sub-area at the end of each decade by 7.8 (the area in square kilometers of a grant for a sheep station) and was expressed as a percentage of the total sub-area in order to compare sub-areas of widely different surface extension.

Land use was specified in the land grants, and these documents have been used in this study as the major source of evidence for land use practises. Although compliance with these orders is somewhat problematic, I have found that only in the 1550s, when cattle were expelled from this region, was the order disobeyed and in this case the owner was forced to comply (see n. 24). In his study Simpson asked: "What assurance do we have that land was actually used for the purposes stipulated in the grants?" and replied that this question "may . . . be safely answered in the affirmative" (*Exploitation*, 20).

[38] The grazing rates for each sub-area were calculated by dividing the total number of head within a sub-area at the end of each decade by the square kilometers of grazing land: $G = Sn/a$, where G is the grazing rate in head per square kilometers, S is the stocking rate in head per station, n is the number of stations, and a is the surface area of the sub-area in square kilometers.

TABLE 2A

Land Converted to Pastoralism[a]

	Decade Ending							
	1539	*1549*	*1559*	*1565*	*1569*	*1579*	*1589*	*1599*[b]
TYPE 1:								
Tula	.6	2.5	9.5	44.0	45.9	61.2	72.0	93.6
Southern Plain	6.4	9.6	11.3	24.2	32.2	47.2	73.0	81.6
TYPE 2:								
Central Plain	2.5	3.8	3.8	29.7	34.9	45.5	54.6	80.3
North-South Plain	—	1.0	2.0	14.5	16.5	33.1	63.1	76.7
TYPE 3:								
Alfaxayuca	4.9	6.1	12.3	31.9	31.9	45.5	59.0	61.1
Jilotepec	9.4	18.9	22.1	28.7	30.0	34.1	43.9	68.4
Chiapa de Mota	—	7.8	10.1	25.8	25.8	26.9	33.7	63.3
TYPE 4:								
Huichiapan	—	—	1.3	11.0	11.0	21.2	45.7	66.1
TYPE 5:								
Northern Valley	—	.7	.7	1.5	2.3	5.3	16.8	18.4
Ixmiquilpan	—	1.5	3.0	7.5	7.5	7.5	9.8	9.8
REGION:	2.6	5.8	8.4	21.5	22.9	31.1	45.5	60.0

[a]Land converted to pastoralism as expressed as a percentage of the total surface area of each sub-area, by decade.
[b]Estimates for the 1590s are based on totals adjusted downwards by 30 percent to account for undocumented sales.
SOURCES: Melville "The Pastoral Economy," ch. 5, n.s 8–16; Appendix E, 272–95.

that the injunction to carry 2,000 head was taken to mean 2,000 ewes, with a total flock consisting of 3,900 ewes and lambs. Rates varied greatly in the first half of the 1560s, ranging from 4,000 to 10,000 head; and an average rate of 7,500 head has been used for the period 1560–65. During the period from the mid-1560s to the late 1570s stocking rates were very high: The highest stocking rate recorded for the Valle del Mezquital was for the 20,000 sheep corralled on one station in 1576; however, the rather conservative figure of 10,000 head has been used to calculate the grazing rates in the period 1566–79.[39]

[39] See Melville, "The Pastoral Economy," ch. 5, n.s 8–16 for references to changes in the stocking rates up to 1579. Evidence for the stocking rates was taken from documentation of court cases, wills, complaints lodged by Indians, and censuses. The information contained in these sources has obvious biases. For instance, in order to make their case Indians undoubtedly reported larger numbers of animals than in fact existed on Spanish lands. The numbers admitted to by the Spanish pastoralists depended on the case they were arguing. For example, if they needed more Indian laborers assigned to them, they inflated the numbers of stock; but if they were responding to the complaints of the Indians about overstocking, they played down the numbers.

TABLE 2B

Grazing Rates on Common Pasture

	Decade ending							
	1539	*1549*	*1559*	*1565*	*1569*	*1579*	*1589*	*1599*
TYPE 1:								
Tula	.8	3	48	423	589	785	712	440
Southern Plain	8	12	56	233	414	605	703	384
TYPE 2:								
Central Plain	3	5	19	286	448	584	525	378
North-South Plain	—	1	10	139	212	425	607	361
TYPE 3:								
Alfaxayuca	6	8	61	307	410	583	567	288
Jilotepec	12	24	111	276	385	437	423	322
Chiapa de Mota	—	10	50	248	331	346	324	298
TYPE 4:								
Huichiapan	—	—	7	107	142	273	440	310
TYPE 5:								
Northern Valley	—	.9	4	15	29	69	162	86
Ixmiquilpan	—	2	15	73	97	97	95	47
REGION:	3	7	42	207	294	399	438	288

SOURCES: Melville "The Pastoral Economy," ch. 5, n.s 8–16; Appendix E, 272–95.

The average rate of 7,500 head per station used to calculate the grazing rates of the 1580s reflects a decline in the stocking rates reported for the region as whole and is based on a 1588 census of Spanish holdings in the Alfaxayuca and Huichiapan sub-areas.[40] During the 1590s the flocks further declined in numbers and in quality. Ewes were so underweight that they failed to reproduce before four years of age, whereas they had begun bearing lambs at fourteen months to two years in the era of expansion; and the animals slaughtered for meat and tallow were smaller and weighed less. The stocking rate for the 1590s is estimated at 3,700.[41]

Although there was no official grazing rate, the legal stocking rate of 2,000 head on 7.8 square kilometers may indicate an official estimate of the carrying

Censuses carried out by royal officials are probably the most reliable source, although officials can always be bribed. The best way to deal with these difficulties is to collect many samples from different sources, so that the estimation of the average stocking rate is predicated on as broad a sample as possible.

[40] *AGI*, México, leg. 111, ramo 2, doc. 12.

[41] The principal source for the stocking rate of the 1590s is de Mendizábal, *Obras Completas*, 114–7.

capacity of the pastures of New Spain at conquest, of 256 per square kilometer. It is likely, however, that if estimates were made of grazing rates, they were predicated on areas far larger than the legal area of a land grant, since common grazing was the custom and land grants were viewed as operational bases in the immediate post-conquest decades. The problem of estimating the amount of land actually used for grazing complicates the calculation of grazing rates for the sixteenth century. Given the customs of *agostadero* (grazing harvest stubble) and grazing in common, together with the fact that Indians could graze flocks within village lands (which were prohibited to Spaniards after the 1550s),[42] it can be argued that all lands provided grazing at some point in the year. For this reason the simple estimate of grazing rates predicted on the total area of each sub-area was accepted as providing an adequate indication of the changing densities of animals. Even when the total surface area of a sub-area is used to estimate grazing rates, grazing rates were excessively high by twentieth-century standards.[43]

Having constructed the typology, I then compared it to the changes in the environment documented for each sub-area. The results of this comparison indicated that those areas subjected to high densities of sheep and heavy grazing between 1560–80 exhibited environmental deterioration by the end of the 1560s and environmental degradation by the 1580s. In areas subjected to lower densities of sheep and lighter grazing, the deterioration of the environment was correspondingly less and later in appearance.

TYPE 1: TULA, THE SOUTHERN PLAIN

These southern sub-areas, which had the best initial resources for both pastoralism and agriculture, were the first to be intensively grazed and the most markedly affected by grazing. As can be seen from the typology (Table 1), both of these sub-areas were under very heavy grazing after the 1560s; and by 1580 they were over-exploited. By 1600 between 81.6 percent and 93.6 percent of the total surface area of both sub-areas was formally converted to pastoralism. The increased rate in land takeover during the last two decades of the century occurred even though there was a sharp drop in the carrying capacity, as indicated by the grazing rates: from 711.9 head per square kilo-

[42] Chevalier, *La Formación*, 121–2, 131, 141 *ff*.. Gibson, *Aztecs*, 276–7, 280. Simpson, *Exploitation*, 21.

[43] Grazing rates reported for other regions far exceeded those estimated for the Valle del Mezquital in the sixteenth century. For example, Simpson reports a grazing rate of 2,857 per square kilometer in Tlaxcala in 1542 (*Exploitation*, 13). In the Bajío in 1582, 200,000 sheep, along with 100,000 cows and 10,000 horses, grazed an area of nine leagues square (1,417.5 square kilometers): Richard J. Morrisey, "Colonial Agriculture in New Spain," *Agricultural History* 31 (July, 1957), 24–29. Using sixteenth-century estimates of adequate stocking rates for sheep (2000 head/7.8 square kilometers equal 256 per square kilometer) and cattle and horses (500 head divided by 17.5 square kilometers equals 28.5 per square kilometer), it can be seen that cattle and horses needed nine times as much grazing land; if we convert cattle and horses into sheep, we get a total of 1,190,000 head and a grazing rate of 839 per square kilometer.

meter in the 1580s to 440 head per square kilometer in the 1590s in Tula; and from 703 per square kilometer to 384 per square kilometer in the Southern Plain. The accelerated exploitation of limestone after the conquest in the northern hills of the Southern Plain and in the hills separating the two sub-areas represents another significant drain on the natural resources of these two sub-areas during the sixteenth century.[44]

Environmental deterioration was apparent by the 1560s and consisted of deforestation, denudation of the soils, conversion of abandoned crop lands and hillsides to grasslands, and the subsequent invasion of these grasslands by a secondary growth of arid-zone species: wild maguey (*lechuguilla*: Agave lechuguilla), yucca (*palmas sylvestres*: Yucca species), cacti (*nopal, tunal sylvestre*: Opuntia species), thorn bushes (*espinos*: possibly ocotillo: Fouqueria species), mesquite (*mesquite*: Prosopis species), and a type of thistle or wild artichoke (*cardon*: possibly the introduced Cynara carnunculus). Environmental degradation was evident by the 1580s: Hills were eroded and gullied, and the remaining soils in many areas were thin and stony. In 1595 deterioration of the catchment value of the Southern Plain was evident in the failure of the springs that fed the north-flowing streams. By 1601 sheet erosion of slopes had exposed extensive areas of *tepetate* (hard-pan). The deep soils in the center of the valleys were still very fertile; they had not been eroded to tepetate and were not covered by stony slope-wash debris, but because they were covered with desert scrub and lacked water for irrigation they were not considered to be suitable for agriculture.[45]

A few examples taken from different parts of these sub-areas will serve to illustrate the changes in the landscape and land use in the last quarter of the sixteenth century. The lands around Hueypustla in the Southern Plain were noted for their good soils and suitability for growing wheat circa 1548, but there were increasing references to stony soils during the last two decades of the century; and by 1606 these lands were characterised by thin, ruined soils and slopes eroded down to *tepetate*.[46]

A Spaniard who asked for permission in 1590 to convert arable land on the north bank of the Tepexi River (a tributary of the Tula River) to sheep grazing said that the soils were no longer suitable for cultivation because they were stony and eroded tepetate—they were suitable only for sheep. Sheep stations had been granted in the hills surrounding this holding in the 1560s, but by 1590 they were described as being gullied and eroded tepetate.[47] It was said

[44] For Spanish lime manufacture, see *AGN*, "Mercedes," vol. 6, fols. 455–6; vol. 7, fol. 87; vol. 8, fols. 227–8; vol. 13, fols. 71, 176; vol. 14, fol. 292; vol. 16, fols. 201–2. *AGN*, "Indios," vol. 6-2, exp. 998. *AGN*, "Tierras," vol. 2697, exp. 10; exp. 11.

[45] Melville, "The Pastoral Economy," ch. 5, n.s 27–68.

[46] "Son tierras ruinas y lomas en tepetate y tierras delgadas," *AGN*, "Tierras," vol. 2812, exp. 12.

[47] "Por ser tierra pedregosa tepetate barrancas no es por sembrarlas . . . esta rrodeada de cerros y barrancas y todo pedregal calichal y tepetate . . . no servir sino para traer en ellas ganado menor," *AGN*, "Tierras," vol. 2735, 2ª pte., exp. 9.

that only sheep were to be found in Sayula, a town in the Tula Basin, in 1580. By the early 1600s the banks of the Tula River were covered with mesquite, and sheep grazing predominated over agriculture—even though there was water for irrigation, and the soils had not been affected by erosion.[48] Just to the south of Sayula, a village was surveyed in 1561 as part of the encomienda assigned to Pedro de Moctezuma. At this time there was no mention of erosion, but in a report made in 1603 this village was described as being unsuitable as a center for congregation because it was situated on lands that had been eroded down to tepetate and was surrounded by mesquite. It was, however, "suitable for sheep."[49]

TYPE 2: NORTH–SOUTH PLAIN, CENTRAL VALLEY

Less land was converted to pastoralism at the height of the era of high stocking rates (1570s) and the densities of animals were therefore somewhat lower in these sub-areas than in Type 1, but by the end of the century there was almost complete conversion of land to pastoralism with 76.7–80.3 percent of the land formally converted. Although grazing rates dropped from very high in the 1580s to high in the 1590s (from 525 per square kilometer in the Central Valley and 607 per square kilometer in the North–South Plain, to 377 per square kilometer and 361 per square kilometer) they still represented a continuous pressure on the land.

The sequence of environmental change recorded for these sub-areas is similar to that reported for Type 1, although the appearance of actual degradation is somewhat later. By 1580 the forests in the foothills of the Sierra de Pachuca had been heavily cut by the Indian communities for sale to the miners in Pachuca, and deforestation was noticeable in the limestone hills of the southwestern Central Valley and the western North–South Plain. The process of destruction of the ground cover was completed by pastoralists who cleared and burned the grasslands, thus providing the soil–climate suitable for desert species.[50]

The formerly densely populated flat lands of the southern half of the North–South Plain were said to have "always been grazing lands" in 1580. The Central Valley and the northern half of the North–South Plain, areas that had been considered unsuitable for grazing for lack pasture and surface water in the 1540s, were also reported to have good grazing at this date.[51] Since the primary native vegetation in the northern end of the North–South Plain consisted of desert species, which were invading the southern half and the Central Valley, it would appear that not only had grazing replaced agriculture but also

48 "No hay ni se halla en este pueblo mas de ovexas y desto hay buen multiplico," *PNE* VI, 181.

49 "Que esta en un llano sobre tepetate y entre unos mesquitales no comodo por congregacion . . . comodo el sitio para ganado menor" (*AGN*, Historia, vol. 410, exp. 5).

50 Melville, "The Pastoral Economy," ch. 5, n.s 89–92.

51 *Ibid.*, ch. 5, n.s 69–73.

that desert species had replaced grasses as the norm for adequate fodder by 1580.

By the 1590s the North–South Plain was covered by mesquite-dominated scrub so dense that it was impossible to cut a way through it, and incipient forest regeneration in the southern half of the North–South Plain was displaced in the early 1590s by desert species. The soils underlying this heavy secondary growth were thin, stony, and infertile. Erosion to *tepetate* on the lower slopes of the eastern hills of the Central Valley was apparent by 1599, and the failure of springs and loss of stream flow was reported here by that same date. In 1607 the encomendero of Tlacotlapilco in the northern end of the North–South Plain commented that the person who took up and populated a station in this area merited a prize, because that was the only way to clear out the sheep stealers and escaped slaves who found refuge in these badlands.[52]

During the last two decades of the sixteenth century, formal transference of land from the Indian to the Spanish systems of land tenure in the four sub-areas comprising Types 1 and 2 outstripped the rate of Indian population decline. Grants were made of the leftovers (*sobras y demasias*) between earlier grants; some grants were a fraction of the legal size—800 *pasos* (paces) square (1.24 square kilometers) instead of 2,000 pasos square (7.8 square kilometers); and Indian communal lands were being granted as *mercedes* (land grants).[53] The majority of these holdings were for sheep grazing; only a very small percentage of the total surface area of these sub-areas was granted for agriculture during the sixteenth century (see Table 3).

By 1600 Indian communities were left with less than the legal square league of communal lands. The Tula sub-area, in which twelve *cabeceras* (head towns) required 17 percent (210.7 square kilometers) of the total surface area, only .8 percent (9.7 square kilometers) remained unconverted to the Spanish system (see Table 3). In the Southern Plain six cabeceras theoretically had title to community lands comprising at least 105.3 square kilometers or 22 percent of the total surface area, but only 7.6 percent of the surface area (36.7 square kilometers) remained unconverted. In the North–South Plain 17.2 percent (129.5 square kilometers) remained unconverted by 1600, but the nine cabeceras in this sub-area needed 21 percent (158 square kilometers) of the surface area for communal lands. In the Central Valley 18.4 percent (110.9 square kilometers) remained—barely sufficient for the six cabeceras that should have had 104.5 square kilometers, or 17.4 percent of the surface area.[54] It must be remembered, however, that Indian *caciques*

[52] *Ibid.*, ch. 5, n.s 74–88, 93–102.

[53] *Ibid.*, ch. 5, n.s 24–26.

[54] One quote from an application for a *merced* in the Tula sub-area in 1594 makes the position very clear: "Y declara y declaro no aver lugar de dar sitio de estancia en ella a otra persona y asi lo mande poner por auto" (*AGN*, "Mercedes," vol. 18, fol. 156).

TABLE 3

Land Converted to the Spanish System of Land Tenure[a]

Sub-Area	*Percent of Surface Converted to Pastoralism*	*Percent of Surface Converted to Agriculture*	*Percent of Surface Converted to Spanish Tenure*
Tula	93.6	5.6	99.2
Southern Plain	81.6	10.8	92.4
Central Valley	80.3	1.3	81.6
North-South Plain	76.7	6.1	82.8
Alfaxayuca	61.1	1.3	62.4
Jilotepec	68.4	3.1	71.5
Chiapa de Mota	63.3	5.8	69.1
Huichiapan	66.1	7.3	73.4
Northern Valley	18.4	.5	18.9
Ixmiquilpan	9.8	.4	10.2

[a]Land converted to the Spanish system of land tenure, as expressed in a percentage of the surface area of each sub-area.
SOURCES: Melville "The Pastoral Economy," Appendix E, 272–95.

(headmen) and communities held over one-third of the sheep stations in the Valle del Mezquital by the end of the century. So that while the traditional Indian resource base and systems of land use had been drastically modified and the total area of land at the disposal of the communities reduced, the Indian caciques, and the villages to a lesser extent, benefited by a system that legally allowed them to hold land within the Spanish land tenure system.

TYPE 3: JILOTEPEC, ALFAXAYUCA, AND CHIAPA DE MOTA

This type does not exhibit the very high densities of animals nor the complete conversion of land found in the first two types; however, the sub-areas making up this type were quite intensively exploited from an early date and demonstrate environmental degradation by the end of the century.

As early as 1551 the Indian communities of Jilotepec complained that the animals grazed by the Spaniards in the savannas (30,000 sheep and many mares and cows) had laid the soils bare. After the 1550s the pressure on the densely populated eastern half of the Jilotepec plateau was relieved by the measures enforced by Velasco. Cattle were removed from the Valle del Mezquital; sheep were to be grazed 3,000 pasos from Indian villages; and grants were made in lightly populated areas of the western half of the plateau and in the mountains to the south. Following thirty to forty years of relatively light grazing, scattered deterioration of the environment—deforestation of oak woodlands, and the appearance of the yucca—was reported in the western half of Jilotepec.[55]

After forty years of grazing (1530–70), the flat lands of the northern half of the Alfaxayuca Valley were covered in a dense cover of mesquite-dominated scrub. Young live oaks reported as growing in the valley in the 1560s appear to have been superceded by desert scrub later in the century, and by 1611 the formerly fertile Indian agricultural lands in the north were covered by debris eroded off the surrounding high hills.[56] In Chiapa de Mota environmental degradation in the form of erosion first appears in the records for the 1590s. There are no reports of deterioration prior to this date, and it is probable that the heavy forests masked changes on the steep mountain slopes.[57]

TYPE 4: HUICHIAPAN

Environmental deterioration was apparent by the 1580s in areas which had been grazed from the 1550s and 1560s; environmental degradation, evident in the failure of springs, was reported in the 1590s.[58] Spanish settlement of the Huichiapan Plateau was slowed by the threat of invasion by the nomadic Indians (Chichimecs), who moved further south with the opening of the route to the northern mines and made this sub-area so unsafe for both Indian villagers and Spanish settlers that it was called *tierra de guerra* (war zone). Early grants for grazing were concentrated along the southeastern and eastern borders with Jilotepec and Alfaxayuca. Because the estimated grazing rates are predicated on total surface area, the early rates indicate much lower densities than probably obtained in areas actually grazed.

The history of Spanish agriculture in the Huichiapan sub-area provides a clear instance of the problems facing agriculturalists in semiarid regions in which the lands forming the catchment areas are intensively grazed. It was not until the 1570s that grants for both grazing and agriculture were made for all parts of the plateau, even though the Chichimecs still posed a threat and Tecozautla in the northwest sector of this sub-area was a walled town in which all men (including Indians) had to bear arms.[59] The number of Spanish-owned agricultural holdings in this sub-area in 1588 (eighty-two in all) demonstrates very clearly the interest shown by the Spanish in intensive agriculture after the 1576–81 epidemics. Of the total number of Spanish holdings at this date, 50 percent were sheep stations and the other 50 percent were *labores* (agricultural holdings approximately 85–255 hectares in size).[60] There were more

[55] Melville, "The Pastoral Economy," ch. 5, n.s 103–12.
[56] *Ibid.*, ch. 5, n.s 113–7.
[57] *Ibid.*, ch. 5, n. 118.
[58] Melville, "The Pastoral Economy," ch. 5, n.s 119–36.
[59] *AGN*, "Tierras," vol. 79, exp. 6.
[60] *Labores* consisted of two to six *caballerias de tierra* (agricultural land grants) worked as a unit.

of these small intensively worked agricultural holdings in Huichiapan by the 1590s than in any other sub-area of the Valle del Mezquital. By contrast, Alfaxayuca and other sub-areas in which pastoralism had been introduced at an early date and dominated production had few agricultural holdings that were not ancillary to a station.[61]

Intensive methods of cultivation and irrigation were used to grow wheat, barley, grape vines, and fruit trees on the labores, and they were very productive and profitable until the end of the century when many were abandoned for lack of water. By 1600 springs were failing in Tecozautla in which 500 fig trees were lost for lack of irrigation, and there was insufficient water to irrigate 4,000 grape vines belonging to Catalina Méndez and to supply the village of San José Atlán to the south of the cabecera, Huichiapan. When an agreement between Méndez and the village was first completed in 1590 to share the water, there was enough for all; but a new agreement in 1600 had to be drawn up according to which Méndez had sole use of the water from Saturday night to Monday morning, restricting the village to the rest of the week. Indian witnesses in the Méndez case made it very clear that the problem did not arise from increased use by the failure of the springs: "When the agreement was made there was twice as much water in the springs, since then some springs have dried up and others don't give as much water as they did, and if it were necessary to irrigate the hacienda and the Indian lands today according to the agreement between them, there would not be sufficient water."[62]

Although one-half of the Spanish holdings in the 1580s were for agriculture, sheep grazed an area at least five times as large as the agricultural lands—and they grazed the hills forming the catchment area. Deforestation and removal of the ground cover in the high eastern hills meant that the ground water was not replenished and spring flow was reduced along the lower western edge of the plateau, where the rich agricultural lands were situated.

TYPE 5: IXMIQUILPAN, NORTHERN VALLEY

These two sub-areas did not attract many pastoralists in the sixteenth century, and the number of stock which grazed there was never very large. Less than 20 percent of the surface area was converted to pastoralism, and grazing rates remained low to very low throughout the century. Little environmental deterioration was recorded for these arid sub-areas during the sixteenth century, and it is possible that they were considered to be so poor in natural resources that

61 Melville, "The Pastoral Economy," ch. 5, n. 135. Huichiapan had forty-one *labores;* the Southern Plain, two, Tula, two, Jilotepec, one.

62 The quote reads: "Quando se hizo el concierto avia doble mas agua en los fuentes y despues aca se an secado algunos ojos y no mana tanta cantidad como solia y si huviesen de rregar el dia de hoy la dicha hazienda y los dichos yndios por el orden y concierto que an tenido no ay bastante agua," *AGN*, "Tierras," vol. 3, exp. 1.

changes went unnoticed and unrecorded. It is very difficult to distinguish long-term trends in the environment from those associated with sixteenth-century changes in land use. For example, the appearance of a dense thorn scrub on the flat lands of the Northern Valley by the late 1570s obscured earlier specialisation in the cultivation of the maguey and nopal. Since this type of vegetation is native to the area and because the invasion coincided with the epidemic known as the Great *Cocolistle* (1576–81), it could simply have been the regrowth of primary vegetation on fallow lands. However, gully formation was accelerated in the Northern Valley in the 1590s and followed a sharp increase in the number of stations and the density of animals when grazing rates rose from 68.8 per square kilometer to 162.2 per square kilometer in the 1580s.[63]

The mines that were founded in Ixmiquilpan in the 1540s caused a heavy drain on the stands of mesquite because these trees were favoured by the miners for wheels and lanterns.[64] By the beginning of the seventeenth century there was a dense growth of thorns, yucca, and thistles on former agricultural lands in Ixmiquilpan; and while these plants represent the normal vegetation for this area, the presence of the thistle indicates disturbance by grazing animals.[65]

The sudden increase in interest in these arid regions during the last two decades of the century, especially the Northern Valley, reflected a lack of room for continued expansion in the rest of the region; it also indicated an acceptance of desert species as adequate fodder for the goats, which made up an increasing proportion of the flocks at the end of the century, and sheep.

The intensity of exploitation by grazing in the Valle del Mezquital in the sixteenth century and the severity of environmental degradation from over-grazing depended on the initial status of the natural resources. Those areas with excellent resources (water, pasture, and soils) were the first to be intensively grazed, were over-exploited as early as the 1560s, and exhibited the most advanced state of degradation by the end of the sixteenth century. As pastoralism spread, however, the original diversity of the vegetative cover was lost everywhere; and the region was considered to be fit only for sheep. Continued grazing (albeit at reduced levels of intensity) maintained pressure on the environment and the possibility of recovery was reduced. Indeed the region continued to deteriorate under grazing until the late 1940s, when irrigation systems using the waste waters of Mexico City were initiated.

This process of environmental deterioration is not the first to afflict the Valle del Mezquital. In his study of the southeastern quarter of the Valle del Mezquital (the Teotlalpan), Sherburne F. Cook demonstrated that three cycles of erosion and deposition have occurred here during the past one thousand

63 Melville, "The Pastoral Economy," ch. 5, n.s 137–46.
64 *PNE* VI, 4.
65 Melville, "The Pastoral Economy," ch. 5, n.s 147–9.

TABLE 4

Timetable of Erosion Cycles in the Teotlalpan

Cycle 1 (1000–1200 A.D.):	Removal of the A horizon; Toltec Era.
Cycle 2 (1400–1521 A.D.):	Sheet erosion and deposition of slope wash; Aztec Era.
Cycle 3 (1600–1800 A.D.):	Gullying and barranca formation; Post–Conquest.

SOURCE: Cook, *The Historical Demography*, 41-59.

years.[66] He developed a timetable linking these erosion events to increases in population densities and such destructive land use techniques as hillside agriculture and concluded that the periods of maximum deterioration "had coincided with the era of high cultures and great populations before the Conquest"[67] (see Table 4).

While he conceded that the Spanish vastly accelerated the destructive processes by increased deforestation for lime manufacture and mining and the introduction of the plough and grazing animals, Cook thought that by 1521 the processes of environmental degradation "had gone a long way towards completion, and that had the Spanish not arrived when they did the Teotlalpan, in any case, *was ecologically and demographically doomed to destruction*" (emphasis in original).[68] Evidence of accelerated environmental degradation in a period of rapid population decline (circa 1548–80) did seem, however, to cause some difficulties for Cook, who attempted to resolve these problems by postulating that: "There can be no alternative to the conclusion that hillside agriculture, as introduced into and practised extensively in the Teotlalpan in the fifteenth century, was accompanied by severe wastage of the mature top soil. This wastage was caused primarily by sheet erosion, rather than gully erosion."[69] He concluded that "the Spanish accounts . . . lead to the supposition that agriculture in the Teotlalpan had undergone a sharp decline in the sixteenth century. If it is assumed that such a decline had already been initiated under the last Aztec rulers, *or at least was on the point of becoming manifest,* then the influence of the Spaniards was simply to induce a rapid acceleration of the process" (emphasis added).[70]

I found no documentary evidence for the extensive sheet erosion and deforestation that Cook proposed as being present at conquest, rather, I found that the region was ecologically quite stable up to the middle of the sixteenth century. Like Cook, I did find evidence that the situation deteriorated between 1548 and 1580; and recent archaeological surveys of the Valley of Mexico

66 Cook, *The Historical Demography*, 41–59.
67 Sherburne F. Cook, *Soil Erosion and Population in Central Mexico* (Berkeley, 1949), 84.
68 Cook, *Historical Demography*, 54.
69 *Ibid.*, 52.
70 *Ibid.*, 54.

and the Tula River Basin, which include parts of the Teotlalpan in their survey areas, not only confirm Cook's timetable of preconquest population cycles and the move to hillside agriculture in the immediate preconquest era (Late Horizon 1350–1519 A.D.) but also demonstrate the extensive use of erosion control techniques such as terracing.[71]

It is instructive, at this point, to compare the post-conquest history of the Valle del Mezquital with the conclusions reached in a study of the relationship of the human populations of southern Greece to their environment over the last 50,000 years.[72] Tjeerd H. van Andel and Curtis Runnels concluded that two modes of land stabilization (terracing; or total abandonment) and two modes of destabilization (overly intensive use, including clearing very marginal lands; or a shift to pastoralism) alternated to produce the cycles of erosion and recovery evident in the southern Argolid. They note that while terraces require unceasing maintenance, total abandonment of terraced land does not necessarily lead to erosion since rapid growth of grasses and bushes binds the soils and walls and prevents slippage. By contrast, partial abandonment and a shift to pastoralism can lead to catastrophic erosion. When care is not taken to maintain the terrace walls by replacing stones dislodged by grazing animals or by protecting them with thorny branches, they are destroyed by the movement of animals across them; and the contained earth then slides down into the valleys. These authors correlate the process of partial abandonment and expansion of grazing onto terraced lands with economic downturns, when agriculturalists use their good bottom lands for grains and rent their terraces to shepherds.[73]

If this model of the replacement of hillside agriculture by grazing and the resulting erosion is kept in mind, it is possible to accomodate the results of my study with Cook's model of human population maximums and ecological deterioration by modifying the erosion timetable to take into account the lack of evidence of environmental degradation in the immediate post-conquest era and the documented consequences of intensive sheep grazing in the last half of the sixteenth century.

I suggest that truncation of the A horizons (Cycle 1) under the intensive indigenous systems of agriculture continued up to the late 1570s, when the 1576–81 epidemic caused a sharp reduction in Indian population and agriculture. Deforestation, the rapid expansion of grazing, and removal of the vegetative cover from overgrazing during the 1560s and 1570s, accelerated this process and led to the catastrophic erosion events of the 1580s and 1590s that left bare stretches of tepetate on the piedmont and deep deposits of slope wash on the flats (Cycle 2). I principally agree with Cook's dating of the last

[71] Mastache de E. and Crespo O., "La Ocupación Prehispánica en el Area de Tula, Hgo.," 76–77; and W. T. Sanders *et al. The Basin of Mexico*, 179, 213–6.

[72] Tjeerd van Andel and Curtis Runnels, *Beyond the Acropolis: A Rural Greek Past* (Stanford, 1987).

[73] *Ibid.*, ch. 8.

TABLE 5

Revised Timetable Events

Cycle 1 (1000–1580 A.D.):	Removal of the A horizon; indigenous agricultural systems.
Cycle 2 (1580–1600 A.D.):	Sheet erosion and slope-wash deposition; deforestation and overgrazing.
Cycle 3 (1580–Present):	Gullying; continued grazing or overgrazing.

erosion cycle, gullying and barranca formation (Cycle 3), but in some parts of the region this process started as early as the 1580s and has continued up to the present (see Table 5).

THE FORMATION OF LATIFUNDA

During the last two decades of the sixteenth century, more land was transferred to the Spanish system of land tenure for grazing than in the previous five decades.[74] The reason for the intensified interest in sheep raising in a region characterised by declining production is to be found in the market for pastoral products at the end of the century. Prices for all foodstuffs rose sharply in the 1580s and 1590s, in part because of the general inflation and dropping supplies of the late sixteenth century. It is generally agreed that the supply of agricultural products dropped because of the demographic collapse of the Indian population, especially after the 1576–81 epidemic, since the Indians up to that time were the primary producers of agricultural products. The drop in the supply of pastoral products was based on declining production (a decrease in the herds which was reported for all parts of New Spain during the last quarter of the sixteenth century). Even though the number of consumers dropped with the dwindling population of Indians, who were not replaced in equal numbers by the growing Spanish and Mestizo (people of Spanish and Indian descent) populations, the demand for meat exceeded the supply.[75] Pastoralists were assured of a market and ample profits if production could be sustained—the problem was to maintain or increase the number of stock.

The pastoralists of the Valle del Mezquital responded to the decline in the carrying capacity of the region and high prices for their products by attempting to restrict access to the pasture on their lands. The rich and powerful *latifundistas* of this early period, who had the capital needed to buy out the small land owners, acquired title to large numbers of stations, increasing their

74 The number of stations (cumulative totals) in the region by the end of each decade: 1539 (34), 1549 (75), 1559 (108), 1565 (276), 1569 (294), 1579 (399.5), 1589 (583.5), 1599 (862).

75 Chevalier, *La Formación*, 139–140. Simpson, *Exploitation*, 22 *ff.*, André Gunder Frank, *Mexican Agriculture, 1521–1630*, (Cambridge, 1979), ch. 6.

flocks at the expense of the smaller individual pastoralists and thus monopolised access to the land. For example, in 1603 when Jerónimo López entailed three estates totalling 500 square kilometers in the eastern half of the region, the annual proceeds from his urban and rural holdings totalled 51,720 *pesos de oro comun.*[76]

The most striking aspect of the haciendas that developed in the Valle del Mezquital at the end of the sixteenth century is their size. While they do not compare with the vast northern estates, which measured up to thousands of square kilometers in size, they were large in comparison with the haciendas in the Valley of Mexico. They represent a marked monopolization of land and indicate the extent to which the ownership of the means of production had shifted from many small land owners to a small number of large landowners.[77]

Contrary to current thought on the development of extensively exploited latifundia in Mexico, the emergence of a small powerful elite that monopolised land and production in the Valle del Mezquital by the early seventeenth century was neither based on inherently poor natural resources and a small initial population, nor did it emerge as a direct result of the displacement of Indian agriculturalists. Rather, it was based on the decline of the productive potential of the region. The very rapid process of environmental deterioration that accompanied the over-exploitation of the region by grazing, with mining and limeworking to a lesser extent, produced an environment in which (given the technology of the age) only large holdings were economically feasible.

The formation of haciendas in late sixteenth-century New Spain is generally associated with the monopolization of land; and both François Chevalier and André Gunder Frank note that the hacienda developed out of small holdings during a period of waning supply and high prices for both agricultural and pastoral products.[78] Although recent research has demonstrated an immense amount of variability in the size of colonial haciendas, as well as their internal structure and relation to regional and interregional trade networks, it does appear that especially large holdings were formed in pastoral regions—

[76] Mendizábal, *Obras*, vol. 6, 112. The *peso* of common gold was a low-grade gold coin of approximately 14 carats, equal in market value to eight silver *reales*. See Wilbur Meeks, *The Exchange Media of Colonial Mexico* (New York, 1948), 34–38.

[77] Examples of some haciendas in the Valle del Mezquital in the early seventeenth century by approximate size: 500 square kilometers (Mendizábal, *Obras*, VI, 112); 420 square kilometers (*AGN*, "Tierras," vol. 2711, exp. 10); 293 square kilometers (*AGN*, "Tierras," vol. 1520, exp. 5); 180 square kilometers (*AGN*, "Tierras," vol. 2692, exp. 6); 124 square kilometers (*AGN*, "Tierras," vol. 2813, exp. 13). Ixmiquilpan was monopolised by ten haciendas, each averaging 102.8 square kilometers (*AGN*, "Civil," vol. 77, exp. 11, fol. 80v). The huge Santa Lucia Hacienda of the Jesuits, which extended over the eastern half of the Northern Valley and Ixmiquilpan sub-areas by the early eighteenth century, already extended well into the Northern Valley by the early seventeenth century (Herman W. Konrad, *A Jesuit Hacienda in Colonial Mexico. Santa Lucia, 1576–1767*, (Stanford, 1980), ch. 3).

[78] Chevalier, *La Formación*, 144. Frank, *Mexican Agriculture*, ch. 6.

even in central regions such as the Toluca and Puebla Valleys—as well as in the Valle del Mezquital.[79]

I argue that the carrying capacity of many parts of New Spain deteriorated under intensive grazing in the sixteenth century and that environmental degradation, by making large areas of grazing land necessary to maintain profits, was an important variable in the development of large-scale haciendas in many pastoral regions. Similar ecological changes to those documented for the Valle del Mezquital are indicated for other pastoral regions by the generalised decline in flock size.[80]

Many early writers commented on the spectacular increase of the grazing animals introduced shortly after the conquest, and New Spain was known for cheap and plentiful meat,[81] but in the last decades of the century herds were decreasing; animals were underweight; and production of pastoral products declined. Several reasons were given by contemporaries for the decline in the herds: the extremely wasteful practise of killing huge numbers of animals for their tallow or hides alone; the high consumption of meat by the Indians; cattle rustling by nomadic Indians; and the depredations of the bands of wild dogs that roamed the land.[82] Viceroy Martín Enríquez wrote in 1574 that "livestock does not multiply as it used to: cows used to bear calves before two years of age—because the land was not trampled and pasture was fertile and extensive; but now that [the pasture] has failed the cows do not bear before three or four years of age."[83]

Most modern writers would agree with Viceroy Enríquez that the primary cause of the decline in the herds was deterioration of the natural resource base from overstocking and overgrazing. Simpson saw a direct correlation between overgrazing by sheep and goats and the decline in flock size because sheep and goats destroy their own subsistance base, but he thought that perhaps the shortage of cattle was confined to the Spanish cities and simply a factor of growing demand. However, Chevalier wrote that diminution of the herds of cattle was reported in the north as well as the center of Mexico and quite clearly resulted from deterioration of the natural resource base.[84]

Measures were taken to correct the drop in supply of pastoral products

[79] Jack A. Licate, *Creation of a Mexican Landscape. Territorial Organization and Settlement in the Eastern Puebla Basin, 1520–1605*, (Department of Geography, University of Chicago, Research Paper no. 201, 1981), 124. James Lockhart, "Españonles entre indios: Toluca a fines del siglo XVI," in *Estudios sobre la ciudad iberoamericana*, Francisco de Solano, coordinator (Madrid, 1975), 448.

[80] Several writers have correlated the spread of pastoralism and overgrazing in Mexico with deforestation and erosion. For example, see Gibson, *Aztecs*, 5–6, 305; Simpson, *Exploitation*, 23; Wolfgang Trautmann, *Las transformaciones en el paisaje cultural de Tlaxcala durante la época colonial* (Weisbaden, 1981), 177.

[81] See Simpson, *Exploitation*, 2 *ff.*, for contemporary witness to the increase of grazing animals.

[82] Chevalier, *La Formación*, 137 *ff.*

[83] Quoted in Chevalier, *La Formación*, 138.

[84] Simpson, *Exploitation*, 23; Chevalier, *La Formación*, 139 *ff.*

during the last decades of the sixteenth century. For example, Indian meat consumption was restricted, and laws were passed that forbade the killing of female stock and limited the numbers of animals butchered. Viceroy Luis de Velasco the Younger wrote to his successor in 1595 that although he had taken measures to correct the situation, he did not "see that [he had] been able to make good the damage or to restore the herds to their former size."[85]

I suggest that a process of environmental degradation associated with the expansion of intensive pastoralism underlay a structural shift in the pastoral sector of the political economy of New Spain during the last quarter of the sixteenth century. At mid-century, supply kept pace with and even exceeded demand; prices were low, although rising with the general inflation of the sixteenth century; and production was in the hands of a large number of small and medium landowners. By the end of the sixteenth century, the pastoral sector was, however, characterised by an absolute decline in productive potential as well as production levels; increasing demand (a factor of both declining supply and growing numbers of Spaniards and mestizos); high prices; and control of the means of production by a small group of large landowners. Although the pastoral sector of the emerging Mexican economy at the beginning of the seventeenth century was monopolised by the owners of the great estates—in confirmation of the classic picture of Mexican society—this situation was neither an inevitable outcome of either the conquest of Mexico by the Spaniards nor the development of mercantile capitalism; it grew out of local processes and history.

[85] Quoted in Simpson, *Exploitation*, 22.

6

Indigenous and Colonial Land-Use Systems in Indo-Oceanian Savannas: the Case of New Caledonia

Jacques Barrau

In his classic works on the origin and variation of cultivated plants, Vavilov defined an Indo-Malayan centre which included the entire Indonesian (Malay) archipelago and the whole of mainland Southeast Asia (Indo-China), a region of great floristic diversity where many crops originated (Vavilov 1951). Vavilov's Indo-Malayan centre stretches from southern China to eastern Melanesia and thus includes those parts of the Intermediate Tropical Zone that comprise the Southeast Asian and Australian savanna regions (Harris: this volume, Fig. 1). These two regions, separated by the wetter, rain-forest region of Malaysia-western Indonesia, share many cultivated plants and agricultural traditions, although they have also experienced significantly different agricultural histories. Elsewhere in this book, Condominas describes systems of resource use among some of the indigenous agriculturalists of mainland Southeast Asia. In this chapter I focus on Melanesian land-use systems and in particular on the southwestern Pacific island of New Caledonia, where the historical contrast and conflict that developed in many parts of the tropics between indigenous and colonial systems can be demonstrated with great clarity.

Melanesian Savanna Environments and Horticultural Traditions

I have described elsewhere the contrast that exists between perennially humid and seasonally dry environments in the Indo-Pacific area (Barrau 1965). It is reflected in the contrast between the evergreen rain forests that dominate areas where the dry season lasts for less than 2.5 months and the semi-evergreen and deciduous or "monsoon" forests that occupy areas of 2.5 to 5.0 months' dry season. In the Australian region these relatively humid savanna environments, where mean annual rainfall ranges from 2000 to 1000 mm, occur in the Sunda Islands, in south-central New Guinea, in parts of north-eastern Australia and in the northern two-thirds of New Caledonia (Harris: this volume, Fig. 1). Here, open-canopy savanna woodlands and grasslands coexist with the semi-evergreen and deciduous forests that appear to be the dominant type of natural vegetation. Modification of these forests by burning and clearance through many millennia of human occupation has probably resulted in the widespread extension of woodlands and grasslands at the expense of closed-canopy forests. Certainly this has occurred in recent centuries on the island of New Caledonia, as is shown later in this chapter.

Although the hunting and gathering activities of non-agricultural peoples in Melanesia have modified to some extent the natural vegetation of the region, more profound alterations have resulted from the widespread and ancient practice of swidden (shifting or long-fallow) cultivation. Although I do not accept simple unilinear models of agricultural evolution in the Indo-Pacific area (Barrau 1962), I believe that swidden cultivation has been a common practice from early horticultural times.* The use of fire to aid forest clearance is an integral part of swidden cultivation. Seasonal burning has long been, and still is, practised wherever swidden cultivation is carried out in forested landscapes, although the forests' resilience to swidden varies with the humidity of the environment. In those parts of the Indo-Pacific area where the dry season exceeds 2.5 months, fire has been a potent factor in the shift from forest to more open woodlands and grasslands. Other activities involved in swidden cultivation also tend to encourage the establishment of grassland at the expense of forest. As Conklin (1959) showed in his study of the Hanunóo swidden gardeners of southern Mindoro in the Philippine Islands, these activities include weeding, cutting grass and other herbs for thatch and fodder, grazing by domestic livestock, and burning for reasons other than forest clearance, for example to promote grass growth and thus attract game, to scare predators, to open trails, and to make fire breaks; also accidental burning occurs. Conklin's analysis of the factors that encourage a succession from forest to grassland is relevant to swidden cultivation throughout the savanna environments of the Indo-Pacific area, as well as to comparisons between indigenous and colonial systems of land use in New Caledonia and elsewhere; but before examining the case of New Caledonia it is necessary to mention some of the significant differences in horticultural traditions between Southeast Asia and Melanesia.

As first Haudricourt and Hedin (1943) and later Sauer (1952) pointed out, and as recent archaeological evidence partly confirms (Barrau 1974, Gorman 1974, Yen 1977, Jones: this volume), the Indo-Pacific area shares a common history of plant domestication and cultivation. But, whereas Melanesia has retained what seems to be a very ancient food-plant pattern characterized by the dominance of perennial, vegetatively propagated plants, mainly root crops, Southeast Asian horticulture incorporates cereals, particularly rice and *Setaria* millet, as well as the food plants still dominant in Melanesia. Moreover, domestic cattle, the water buffalo (*Bubalus bubalis*), and the zebu (*Bos indicus*), were all bred in Southeast Asia in the precolonial era as well as domesticated forms of the native banteng (*Bos javanicus*). By contrast, Melanesia is characterized more completely by what Gourou (1948) described as *La*

* For reasons given elsewhere (Barrau 1967), I prefer horticulture to agriculture as a comprehensive descriptive term for Indo-Pacific techniques of cultivation; for a discussion of the possible evolutionary relationship of swidden cultivation to garden cultivation or "fixed-plot horticulture" see Harris (1973: 399–402).

civilisation du végétal, with the pig, the dog, and the chicken being the only domesticated animals present there in pre-European times; and even the pig and the dog were absent in some Pacific islands, for example New Caledonia, at the time of European discovery. The use of fire as a hunting tool was, however, known in the Pacific, and I have also seen it used as such in the savanna grasslands of New Guinea.

If we examine the general horticultural tradition of Melanesia from a historical point of view (Barrau 1958, 1965), we find that, in the course of time, there have been long-term adjustments of cultivation techniques to changes in the environment, changes partly due to human interference with natural processes. Whereas swidden cultivation is still the usual practice today where the rain forest remains the dominant plant community, horticulture based on relatively fixed modes of land utilization involving such techniques as irrigation, drainage, terracing, and ridging are prevalent where savanna woodlands and grasslands have become the dominant features of the plant cover. Moreover, particularly in grasslands, tilling the ground requires more elaborate work than simply digging with a stick, as is done in forest swiddens; and, where rhizomatic grasses are abundant, it is necessary to break the sod with a digging stick which is used as a ploughing tool, as in the Markham valley of New Guinea and in New Caledonia where a wooden spade was also used (Barrau 1956).

The Case of New Caledonia

The Precolonial Landscape

Such changes are clearly discernible in the land-use history of the humid savanna environment of New Caledonia. There the Melanesians grew (and still grow) taro (*Colocasia esculenta*) on irrigated terraces, and yams (*Dioscorea* spp., particularly the greater yam, *D. alata*) in carefully constructed mounds. Both terraces and mounds are permanently built but periodically cultivated and subject to clearance by burning. These, and earlier swidden, methods of subsistence gardening were responsible, in precolonial days and mainly on the somewhat drier, western side of this island, for some extension of savanna woodlands and grasslands (Virot 1956). Nevertheless, the intensity of land use was sufficiently low for the system to maintain a continuing balance between forest and grassland. Moreover, the savanna landscape thus developed had some economic usefulness for New Caledonian Melanesians who, for example, used *Imperata* leaves extensively for thatching, and the bark of *Melaleuca leucadendron* for building, their houses. This tree, originally from the swampy lowlands but resistant to fire, colonized lands exposed to burning until it became the dominant tree of the local savanna woodlands (Fig. 1), just as other species of *Melaleuca* and *Eucalyptus* did in the southern lowlands of New Guinea.

The precolonial landscape of most of New Caledonia consisted of villages and gardens interspersed in savanna woodlands and grasslands, riverine forests, and stands of deciduous and evergreen forest. Elsewhere, on the remaining quarter or so of the land's surface, peridotite and serpentine geological formations gave rise to soil toxicity that almost prohibited any cultivation. It was in these areas that the main mineral resources of the island were later discovered.

Fig. 1 A cattle ranch near Purendimié, New Caledonia, in savanna woodland dominated by *Melaleuca leucadendron*; the stand of coconut palms indicates a site of former Melanesian occupation (photo by J. Barrau).

For the Melanesians, space was divided between the *séjour paisible* (the peaceful sojourn, cf. Leenhardt 1937) and the uncivilized domain of the woods, the bush, and the long abandoned fallows, which were lands of the unknown and the unpredictable. While carrying out ethnobotanical research with the Melanesian people of New Caledonia, I came to realise that, in the past, they were very careful about controlling burning within the realm of the peaceful sojourn, but that this was not so true in the wilderness outside, where fire could even be regarded as a means of conjuring evils. Indeed, it seems that fire as a clearing tool had both a technological and a magical role, particularly where forests were perceived as the haunt of fearful and undominated forces, as was, for example, the case in parts of Indonesia (Lombard 1974). Yet, in New Caledonia, although savanna woodlands and grasslands had spread, particularly on the western, leeward side of the island, a balance between these plant communities and forests was maintained, for the latter were regarded with mixed feelings of fear and respect; moreover they provided products essential both for the material life of the Melanesians, such as wild foods and materials for house and canoe building, and for their ideological life, for example sacred wood carvings. In addition, as Leenhardt (1937: 44) has rightly noted, "la nature est, pour eux, chargée d'une histoire humaine qui est aussi la leur propre", and this was reflected in a kind of gardened landscape (Fig. 2).

Fig. 2 The Melanesian *séjour paisible* or "peaceful sojourn", a present-day gardened landscape in a Melanesian Reserve near Bourail, New Caledonia (photo by J. Barrau).

The Establishment of Colonial Control

Discovered by James Cook in 1774, New Caledonia became a French colony in 1853. From 1840 onwards a few traders and missionaries established posts along its shores, but effective European settlement, both penal and free, followed the hoisting of the tricolor flag over the island. From the beginning, comparisons with the nearby Australian colonial model were frequent and largely based on similarities between landscapes, particularly those of leeward New Caledonia, where savanna woodlands and grasslands were broadly comparable to those of tropical and subtropical Eastern Australia (Delignon dit Buffon 1898: V–VI). Moreover, with their historical experience of livestock farming, Europeans could not see these grasslands without immediately thinking of cattle.

The Australian model played a decisive role in the establishment of the New Caledonian cattle industry. The techniques and vocabulary of the industry came from Australia and even today they remain basically Australian. By 1859 1000 head of cattle had been introduced into New Caledonia from Australia; in 1883 there were 87,923 cattle in the island, and today there are about 100,000. Cattle were *the* means of land alienation by Europeans and indeed the process of white settlement can be caricatured as an advancing line of cattle followed first by surveyors and gendarmes, and then, at a distance, by the owners of the cattle. Terrified by these strange, huge animals, which destroyed their gardens and broke "the peace of their sojourn", the Melanesian gardeners, particularly those of western New Caledonia, retreated to the hills.

Conflicts between the Melanesians and the invading white graziers gave rise to rebellions that were ferociously repressed by the French colonial authorities. The anthropologist Leenhardt, who came to the island in 1902 as a missionary, thus vividly described the impact of the French settlement on both the environment and the society of precolonial New Caledonia:

> Le pays devait avoir un aspect extraordinairement agréable et vivant ... Il n'y avait pas jadis de gros animaux dans l'île: les colonisateurs y ont introduit le bétail. Celui-ci est monté dans les vallées, il a piétiné les conduites d'eau, ravagé les cultures, effrité le sol que l'herbe rasée ne préservait plus contre les pluies torrentielles. Le Canaque* a reculé devant l'invasion bovine soutenue par la puissance des Blancs. Le paysage verdoyant qu'avait assuré son industrie s'est desséché. On ne devine sa grâce ancienne qu'à certains vestiges dans les réserves indigènes constituées pour sauvegarder le peuple d'un plus grand désastre (Leenhardt 1937: 17).

Cattle Grazing and the Extension of New Caledonian Savannas

The truth is that, in their new environment, the white graziers had made a big leap backward to archaic animal husbandry. Two excerpts from nineteenth-century reports quoted by Delignon dit Buffon (1898: 161) sum this up:

> On met un troupeau sur un terrain non entouré de barrières, et on lui dit comme dans l'évangile: croissez et multipliez. Il ne reçoit pas d'autres soins; *and* Après avoir épuisé les pâturages concédés ... les troupeaux devenus trop considérables se sont répandus, grâce à la tolérance de l'administration, sur les terres restant au Domaine et les ont épuisées de même (cf. also Appendix 1).

Although many attempts were made to establish a kind of white-settler peasantry in New Caledonia, ranching and (almost) open-range grazing remained prevalent in the local rural economy. In 1955, according to Le Borgne (1964: 174–6), out of a total area of 1,675,000 hectares, more than 40,000 ha were used for grazing and only 5000 ha for cropping. The majority of the livestock and the ranches were located on the leeward side of the island, where savanna woodlands and grasslands were initially present and where they were later extended considerably as a result of the colonial system of land utilization (Fig. 3). Towards the end of the nineteenth century, the graziers were forced to fence their *stations*, as the ranches are called locally (Fig. 1). This represented a slight technological improvement in management, and, if it had not been for the accidental introduction of the Australian cattle tick (*Boophilus microplus*) by the horses of the United States Army during the Second World War, New Caledonian cattle husbandry would have remained much as it was in the early days of white settlement. But, due to the havoc played by this introduced pest, the graziers were forced periodically to dip and spray their beasts, a closer form of management which amounted to a domestication process. Overgrazing led to impoverishment of the grasslands, which, together with recurrent losses of cattle due to drought and the poor productivity of the local mixed breeds, finally persuaded the graziers to adopt from the Australian livestock industry new techniques of pasture and cattle

* The name given to the New Caledonian Melanesian by the French settlers.

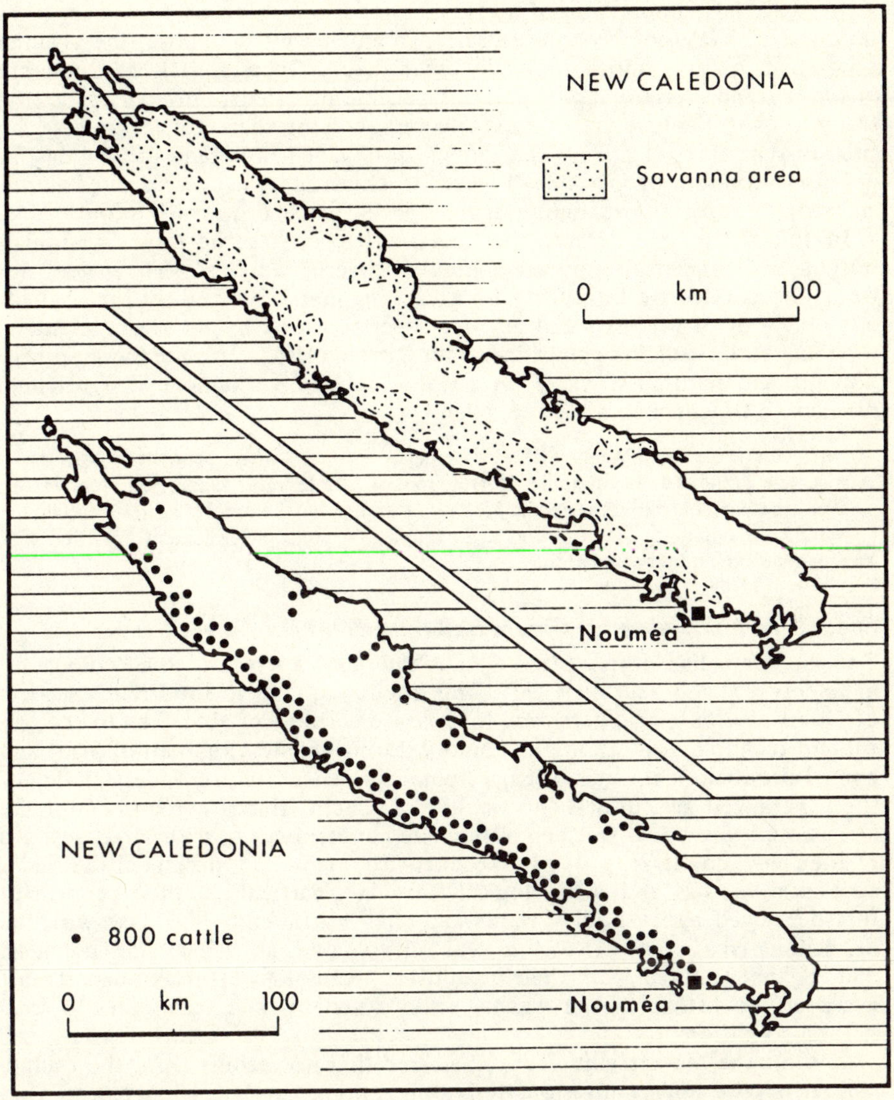

Fig. 3 New Caledonia: distribution of savanna woodlands and grasslands, and of cattle in 1955 (after Le Borgne 1964).

improvement. As a result, New Caledonian animal husbandry has improved considerably during the last twenty years.

Both the degradation of the environment and the inefficiency of cattle production made this improvement necessary. For, since its inception, the cattle industry of New Caledonia had been based on a myth: that of *natural* pasture. The gardened savanna landscape of the Melanesians gave the white settlers the false impression of a lush and nutritious resource of primary grass production. Very quickly, due to selective grazing by the cattle, the savanna ecosystem became more biotically specialized. Species such as *Imperata cylindrica* and *Heteropogon contortus* became prevalent; burning was extensively used to induce more palatable regrowth of these tough grasses; the fire-resistant tree, *Melaleuca leucadendron*, invaded more and more areas, as did also a number of introduced weeds, such as *Lantana camara* and *Psidium guajava*; soil erosion was induced and locally reached dramatic proportions.

In this story, fire deserves particular attention for the New Caledonian settlers and graziers developed a kind of "Nero complex". They were obviously fascinated by the spectacle of burning grasslands and set fire to them not only out of necessity but also for sheer pleasure. Even visiting scientists were the victims of this syndrome. Thus Sarasin (1917: 28), the Swiss anthropologist and naturalist, wrote, in a rather surprising example of pyromanic lyricism to flow from a naturalist's pen:

> Un incendie de bois de niaoulis* et d'herbages desséchés offre, pendant la nuit, un spectacle grandiose. La muraille de feu, poussée par le vent, avance, en crépitant, avec une extrême rapidité. Pareils à de gigantesques flambeaux, les arbres s'enflamment pour s'éteindre bientôt après ... Le feuillage se consume également très vite, en répandant une odeur aromatique ...

Specialized Colonial versus Generalized Indigenous Ecosystems

The late Canadian ethnobotanist Rousseau (1972) wrote of the "colons qui amènent avec eux leur écologie". This applies equally to the French settlers of New Caledonia, who perceived their new environment according to ecological and technological criteria elaborated during the development of plant and animal production in their original homeland. Their lack of understanding of the generalized agricultural (horticultural) system (Harris 1969: 4–7) of the Melanesians was equal to their obtuseness concerning the natural ecosystems of their new country. Into this exotic environment they projected their own ecological models and technological formulae, particularly their propensity for specialized agricultural systems and their cattle cult. The clash with the indigenous *civilisation du végétal* was unavoidable, as well as the ecological damage that resulted from their unadjusted technology. It took them at least a century to realize that it was necessary to adapt their techniques to local natural conditions.

It took the Melanesians about the same time to recover from the trauma caused by the bovine invasion. It is only during the last twenty years that they have started to raise cattle on the limited area left to them by the white colonists. And when they do so, they take great care to segregate within their Reserves (Fig. 2) the grazing space from the gardening space. But they have not forgotten the brutal alienation of their land and many bushfires today are lit by Melanesians on white properties which once were the peaceful sojourn of their ancestors.

In his essay on agricultural involution in Indonesia, Geertz (1963: 17) notes

* *Niaouli* is the local vernacular name of *Melaleuca leucadendron*.

that "Much of the most effective human utilization of the natural habitat consists in changing generalized communities into more specialized ones". However, as Geertz himself admits, in the tropical rain forest swidden cultivation, which represents a generalized domestic ecosystem, is a very effective way of utilizing the natural habitat. In general, it is true that the development of neotechnic or modern techniques of plant and animal management have involved changes from generalized to specialized and even hyperspecialized ecosystems, an evolution largely responsible for many present-day ecological problems. In fact, human interferences in natural ecosystems to insure plant and animal production have always taken the form of specialization processes. Such processes can only be ecologically effective when and where people have long been adjusted or accustomed to the natural habitat concerned, or when and where there has been a rational approach to the ecological understanding of such habitats. In this connection, it is interesting to note that in 1869 Marx, in a letter to Engels commenting on the work of Fraas (1848) on the history of agriculture, remarked that cultivation "dès lors qu'elle progresse naturellement sans être dominée consciemment ... laisse des déserts derrière elle" (Marx and Engels 1974: 62). And Marx quoted as an example the case of Mesopotamia which has been part of the agricultural and pastoral cradle of European civilization. In that region, contradiction and competition between plant production and animal production, between crops and cattle, certainly played an important part in the desertification of sensitive ecosystems.

Such a danger particularly threatens when people colonize new environments which they perceive according to their past experience of different natural surroundings and according to the ecological knowledge they have acquired and developed there. Writing about the history of the white man's exploitation of the grasslands of the North American plains, Carter (1964: 359) remarked that this "illustrates the dominance of men's ideas over the realities of mere physical geography". Colonial misadventures, such as the one just described in the New Caledonian savannas, have often been due to a culturaly biased perception of exotic environments. When the ecological damage thus caused becomes evident, the only solution, if it is not too late, is to resort to considerable input of energy to restore and then maintain the productivity of the new, specialized ecosystems. In New Caledonia, during the last twenty years, the extraordinary increase in importation and use of fertilizers, pesticides, and agricultural machinery, the extension of sown grass-and-legume pastures, the general adoption of grazing rotations, and the improvement of water management in grasslands, all clearly illustrate the validity of this rule. If it is not complied with, the only alternative is a severe and increasing degradation of the environment.

Melanesian Adjustment to Savanna Environments

If we look back at the precolonial situation in Melanesian New Caledonia it is evident that the elaborate horticultural techniques of the islanders already represented specialization by comparison with the generalization of the swidden cultivation system (cf. Condominas: this volume). Ecologically, swidden cultivation seldom functions satisfactorily in the drier environments beyond the humid tropical forests: the diverse plant community of a swidden garden and its complex, closed-canopy structure make such a garden a miniature of the surrounding forest. To a large extent, the swidden cultivation system functions within the flow of energy of the ambient forest ecosystem. To quote Geertz (1963: 27) again: "The delicacy of swidden equilibrium increases at equal pace with this transition toward a more subtropical environment because of the steadily diminishing power of the natural community rapidly to reconstitute itself after human interference". This statement illuminates the ecological status of savanna woodlands and grasslands in the drier environments of Melanesia. Developed and expanded by human activity, they constituted a habitat where indigenous peoples had to adjust their techniques to the ecological conditions they were generating. Hence the need to resort to cultivation techniques that involved a more complete ecological transformation of the natural habitat. In seasonally dry parts of Melanesia where savannas are present, this is reflected in the development of irrigation and drainage systems, and in the creation of permanently-built-but-periodically-used ground structures such as yam mounds, taro terraces, and rice terraces. In Indonesia too, Geertz's (1963: 15–37) exemplary description of the contrast between swidden and sawah (wet-rice) cultivation clearly illustrates this point.

In their adjustment to savanna conditions, the Melanesians of New Caledonia developed gardening and landscaping techniques which created relatively stable environments for the production of their subsistence crops. In doing so, they brought about some specialization of their domestic ecosystem. Yet, in their cultivation techniques and behaviour, they retained some of the traits of the swidden cultivators: individual treatment and careful, almost friendly, manipulation of their cultigens, as well as a constant concern to avoid conflicts with the natural forces in their domestic surroundings. They thus managed to reach a fairly harmonious balance between their subsistence activities and the sensitive savanna environments of precolonial New Caledonia.

Conclusion: Is Colonial and Neocolonial Savannization Unavoidable?

Unlike indigenous systems of land use in Melanesia and elsewhere in the tropics, colonial systems of agricultural exploitation have, as a rule, acted drastically on savanna environments, using technical formulae elaborated in completely different ecological contexts, perceiving these exotic environments through the prisms of their own cultural histories, and acting there as elsewhere in accordance with their quest for fast profits. It has taken colonial "developers" many years and many unsuccessful experiences, including some technological regressions, to realize here and there in the Intermediate Tropical Zone that eco-technological adjustments were necessary in these easily degraded savanna ecosystems. Has the lesson been learned? When one looks at some present-day "neocolonial" ventures, such as the large-scale conversion of forest to pasture that is taking place in parts of northern South America (Parsons: this volume), one wonders if "savannization" is not going to increase its pace for the worse in the tropical world.

Moreover, these colonial systems, through land alienation and the deep socioeconomic changes they brought to the colonies, made it difficult for the indigenous societies to maintain as integral wholes their traditional modes of land use. By means of education, European-derived technological models for the exploitation of savanna environments have been impressed on indigenous elites. Thus there is too often a tendency among them summarily to dismiss the old ways of dealing with nature and to consider them as archaic, unproductive, and finally uncivilized. But the many failures of colonial and neocolonial systems of plant and animal management show that the time has come to reconsider the technological and ecological basis of agricultural land use in the tropics. A fresh look at the ways indigenous systems operated in sensitive savanna ecosystems may provide a valid basis for such a re-evaluation.

In this chapter I have examined one such indigenous system and traced the conflict that developed between it and the colonial system that was imposed upon it. The history of land use in the savanna environment of New Caledonia is one of an eco-cultural confrontation between a *civilisation du végétal* and a *civilisation de l'animal*, a confrontation that in many ways parallels that between Amerindian and European systems of land use in tropical American savanna environments (Hammond, Parsons: this volume). It has also, more broadly, been a confrontation between a generalizing society and a specializing one. As such, it represents in insular microcosm a contrast and conflict that has recurred many times and on a larger scale throughout the tropics. Perhaps, before they too readily apply formulae for hyperspecialized "development", tropical agriculturalists should consider what ecological lessons might be learned from the long and laborious adjustment of Melanesian horticulturalists to their savanna environments, which, in New Caledonia, brought into being the diverse and productive landscape that was observed by the first French settlers and so soon assaulted by their cattle (Appendix 1).

Appendix 1

Paradoxically enough, many early observers of colonial New Caledonia were aware of the ecological damage done by open-range cattle grazing. Several authors stressed the ecological dangers of this form of specialized exploitation. Thus, in the first geographical monograph on the island, Bernard (1895: 364) described the situation in the following terms:

> L'histoire de l'élevage en Calédonie doit être considéré comme l'histoire d'une véritable erreur géographique: on avait ouï dire que l'élevage avait fait la fortune de l'Australie, on n'a imaginé rien de mieux que de la pratiquer à la manière australienne, sans tenir compte de la différence des deux contrées. On a servilement imité les usages des grandes exploitations australiennes, comme si la Calédonie disposait d'espaces indéfinis et indéfiniment plats; comme si ce qui convient aux vastes steppes de l'Australie et de l'Argentine pouvait réussir dans une île étroite et montagneuse. Cette erreur a ruiné les éleveurs et la colonie. Pour les éleveurs, la rassemblement annuel des troupeaux dans un parc par les stockmen présentait des difficultés très grandes, à cause du puissant relief du sol; la race bovine, abandonnée à elle-même, laissée sans soins et sans surveillance, a dégénéré, est devenue la proie des épizooties; les pâturages, dont l'étendue n'est pas indéfinie, ont été ruinés par les mauvaises herbes. Pour la colonie, les inconvénients one été beaucoup plus sérieux encore: les troupeaux, vaguant à travers l'île, ont envahi les cultures canaques, ce qui fut la cause de l'insurrection de 1878; ils ont dévasté les cultures européennes; enfin c'est en vue

de l'élevage qu'on a fait à des particuliers des concessions très étendues, créé un régime de grande propriété qui ne saurait convenir à la Calédonie à aucun point de vue.

Ten years later, Lafforgue (1905: 10, 34, 39) was even more critical in his account of the grazing industry:

Le sol de la Nouvelle Calédonie n'est pas d'une richesse remarquable ... par contre, le pays était admirablement disposé pour l'élevage. Nous parlons volontairement au passé, car, par suite d'abus que l'on en a fait, les pâturages se sont appauvris et ont été peu à peu envahis par les mauvaises herbes ... dans une brochure éditée à l'occasion de l'exposition universelle de 1889, nous trouvons les lignes suivantes:

Le bétail est à peu près abandonné à lui même; la race, par suite de la mauvaise qualité des pâturages et des sécheresses prolongées, va toujours s'affaiblissant; la nourriture mediocre et peu substantielle que consomment les animaux provoque un développement intestinal considérable au dépens du système musculaire. Mais un jour viendra où l'éleveur ne gardera que quelques bêtes de choix, où l'on se souciera d'aménager des sources et des abreuvoirs, où l'on songera à semer des fourrages de meilleure qualité, à faire en un mot de l'élevage sérieux.

Lorsque l'élevage a débuté en Nouvelle Calédonie on s'est trouvé en présence de terrains neufs que l'on considérait comme inépuisables tant pour la fertilité que pour la superficie. Cette croyance s'était si bien enracinée dans l'esprit des éleveurs que beaucoup d'entre eux commencent à peine à s'en débarrasser.

Il est assez difficile de se faire une idée exacte de ce que pouvaient être les pâturages calédoniens au début de l'élevage ... Cependant, comme une très forte proportion de terres étaient cultivées et bien entretenues par les canaques, il est permis de supposer que sur ces cultures se sont créées de bonnes prairies naturelles. De l'avis de tous les vieux Calédoniens, en effet, les pâturages actuels ne peuvent donner qu'une idée très imparfaite de leur ancienne fertilité. La cause de ce dépérissement c'est dit-on, la trop grande quantité de bétail. Les bonnes espèces consommées trop vite et exclusivement n'avaient pas le temps de se reproduire, et peu à peu elles ont cédé la place aux mauvaises herbes. Il est permis de supposer qu'au début les irrigations des indigènes ont encore fonctionné quelque temps donnant aux terres un peu de fraîcheur et rendant moins sensibles les effets des sécheresses. Mais par suite de la méthode du laisser-aller généralement adopté, tous ces travaux considérés comme inutiles et d'un entretien trop coûteux, furent abandonnés.

But colonial greediness proved too strong and both the local administrators and the settlers remained deaf to these warnings.

References

Barrau, J. 1956. *L'Agriculture Vivrière Autochtone de la Nouvelle Calédonie.* Noumea: Commission du Pacifique Sud.

Barrau, J. 1958. *Subsistence Agriculture in Melanesia.* Honolulu: Bernice P. Bishop Museum, Bulletin 219.

Barrau, J. 1962. *Les Plantes Alimentaires de l'Océanie, Origines, Distribution et Usages.* Marseille: Musée Colonial de la Faculté des Sciences de Marseille.

Barrau, J. 1965. L'humide et le sec. An essay on ethnobiological adaptation to contrastive environments in the Indo-Pacific area. *Journal of the Polynesian Society* **74**, 329–46.

Barrau, J. 1967. De l'homme cueilleur a l'homme cultivateur: l'exemple Océanien, *Cahiers d'Histoire Mondiale* **10**, 275–92.

Barrau, J. 1974. L'Asie du Sud-Est, berceau cultural. *Etudes Rurales, No. Spécial, Agriculture et Sociétés en Asie du Sud-Est* 53–56, 17–39.

Bernard, A. 1895. *L'Archipel de la Nouvelle Calédonie.* Paris: Hachette.

Carter, G. 1964. *Man and the Land: A Cultural Geography.* New York: Holt, Rhinehart and Winston.

Conklin, H.C. 1959. Shifting cultivation and the succession to grassland climax. *Proceedings of the 9th Pacific Science Congress* (Bangkok, 1957) 7, 60–2.

Delignon dit Buffon, L. 1898. *Les Aliénations de Terre et la Colonisation Libre Agricole en Nouvelle Calédonie.* Paris: Chalamel.

Fraas, C.N. 1848. *Historisch Encyclopädischer Grundriss der Landwirtschaftslehre.* Stuttgart: Franckh.

Geertz, C. 1963. *Agricultural Involution. The Process of Ecological Change in Indonesia.* Berkeley and Los Angeles: University of California Press.

Gorman, C.F. 1974. Modèles à priori et préhistoire de la Thailande: à propos des débuts de l'agriculture en Asie du Sud-Est. *Etudes Rurales, No. Spécial, Agriculture et Sociétés en Asie du Sud-Est* 53–56, 41–71.

Gourou, P. 1948. La civilisation du végétal. *Indonésie* 1, 385–96.

Harris, D.R. 1969. Agricultural systems, ecosystems and the origins of agriculture. 3–15 in *The Domestication and Exploitation of Plants and Animals*, Ucko, P.J. and Dimbleby, G.W., eds. London: Duckworth.

Harris, D.R. 1973. The prehistory of tropical agriculture: an ethno-ecological model. 391–417 in *The Explanation of Culture Change. Models in Prehistory*, Renfrew, C., ed. London: Duckworth.

Haudricourt, A.G. and Hedin, L. 1943. *L'Homme et les Plantes Cultivées.* Paris: Gallimard.

Lafforgue, G. 1905. *L'Elevage à la Nouvelle Calédonie.* Paris: Chalamel.

Le Borgne, J. 1964. *Géographie de la Nouvelle Calédonie et des Iles Loyauté.* Noumea: Ministère de l'Education, de la Jeunesse et des Sports.

Leenhardt, M. 1937. *Gens de la Grande Terre.* Paris: Gallimard.

Lombard, D. 1974. La vision de la forêt à Java (Indonésie). *Etudes Rurales, No. Spécial, Agriculture et Sociétés en Asie du Sud-Est* 53–56, 473–85.

Marx, K. and Engels, F. 1974. *Lettres sur les sciences de la nature, et les mathematiques.* Translated, and with an introduction, by Lefebvre, J.-P. Paris: Editions Sociales.

Rousseau, J. 1972. *Des Colons qui Amènent avec eux leur Ecologie.* Paris: Klincksieck.

Sarasin, F. 1917. *La Nouvelle Calédonie et les Iles Loyalty: Souvenirs de voyage d'un Naturaliste.* Bâle: Georg.

Sauer, C.O. 1952. *Agricultural Origins and Dispersals.* New York: American Geographical Society.

Vavilov, N.I. 1951. The origin, variation, immunity and breeding of cultivated plants: selected writings of N.I. Vavilov. *Chronica Botanica* 13, 1–366.

Virot, R. 1956. La végétation canaque. Thèse présentée à la Faculté des Sciences de l'Université de Paris pour obtenir le titre de Docteur de l'Université, Paris.

Yen, D.E. 1977. Hoabinhian horticulture? The evidence and the questions from northwest Thailand. 567–99 in *Sunda and Sahul. Prehistoric Studies in Southeast Asia, Melanesia and Australia*, Allen, J., Golson, J. and Jones, R., eds. London: Academic Press.

7

Landscape of Conquest: Frontier Water Alienation and Khoikhoi Strategies of Survival, 1652–1780

Leonard Guelke and Robert Shell

In the seventeenth and eighteenth centuries European settlers ousted the Khoikhoi and San from much of the land they inhabited in south-western Africa using a strategic combination of technology and bureaucracy. The settlers possessed a powerful new fighting technology in the form of firearms and horses that enabled them to hold and defend lands taken from the Khoikhoi. The Dutch East India Company legitimised settler occupation of Khoikhoi land by granting them exclusive use of lands they acquired in freehold or on loan. The settlers took advantage of this permissive policy and their connection to the Cape Town bureaucracy to acquire choice watered land in the interior. These lands and the water resources and pasture they contained were denied to the Khoikhoi pastoralists who found it increasingly difficult to sustain themselves in a land in which access to limited water resources was necessary for survival. In a slow, non-catastrophic process the Khoikhoi were gradually squeezed out of the lands they had once occupied as European settlers alienated the springs and permanent water courses. The survivors of this process often became clients of European settlers and applied their skills in animal husbandry to the invaders' livestock instead of their own.

The Khoikhoi and San (Khoisan) were the first pre-colonial peoples of southern Africa to experience the impact of European-directed settlement and territorial expansion. In the mid seventeenth century, when the Dutch founded a fort and garden at the Cape, Khoisan pastoralists and hunter-gatherers were in sole possession of vast stretches of south-western Africa. Some one hundred and thirty years later European colonists and their slaves had displaced the Khoisan peoples from most of their lands and incorporated the survivors into a colonial settler economy. This paper seeks to provide a geo-political perspective on how and why this happened by analysing the relative strengths of settlers and indigenous peoples on the ground in frontier conditions. Our interpretation stresses the relationship of people to water and land and seeks to show that European settlers by acquiring and

---

* We would like to thank Richard Elphick, Shamil Jeppie, Robert Ross and Anthony Whyte for reading earlier drafts.

legitimising their control and access to scarce surface water resources disrupted Khoisan patterns of life and effectively destroyed their traditional societies.

The decline of the Khoikhoi and the San has attracted a good deal of scholarly attention.[1] The question of Khoikhoi decline has been most thoroughly examined by Richard Elphick in a detailed study of the Khoikhoi that argues for a secular cyclic ecological decline.[2] Both Shula Marks and Robert Ross have written articles on the subject: the former emphasising Khoisan resistance to Dutch expansion, the latter questioning whether the 1713 smallpox epidemic was a major factor in the destruction of the Khoikhoi.[3] This study re-examines all these interpretations of Khoikhoi decline placing emphasis on the nature of European-Khoikhoi interaction on the level of the individual farm. We have made use of the rich archival materials dealing with the geographical and economic foundations of Dutch-Khoikhoi relations, including extensive records relating to settler acquisition and control of land.[4] Our conclusions differ substantially from earlier interpretations and in the penultimate section of the paper we contrast and then synthesise our own views with those of Elphick, Marks and Ross.

The area of south-western Africa that the Khoisan had inhabited for centuries before European colonisation is a rugged, immense and largely arid land of imposing mountains and vast plains. The higher mountains and coastlands of the south-west and south-east receive more than 20" of rainfall per year. In those areas of the south-west receiving ample amounts of rainfall there are typically extended periods of drought in the summer. Only along the south-east coast can rainfall be relied upon throughout most of the year. As a result of the generally low and unreliable seasonal rainfall permanent sources of fresh water are scarce in almost all areas, and especially in the interior where rain-bearing clouds seldom penetrate. Most of the various groups of Khoikhoi pastoralists who inhabited this land secured a livelihood for themselves by practising an extensive form of transhumant pastoralism. They followed the rains with their cattle and sheep setting up their kraals in different places as they sought to utilise effectively the resources of their diverse territories. The mainly San hunter-gatherers were even more mobile in search of game and *veldkos* on which they depended for a livelihood.

The Cape Khoikhoi, numbering perhaps 50,000 people in the mid-seventeenth century, were divided into several autonomous groups each with its own political

---

[1] G.M. Theal, *History of South Africa* (London, 1922), III, pp. 475-77. Theal's emphasis on the disastrous impact of the smallpox epidemic of 1713 has found support among later historians such as W.M. MacMillan, J.S. Marais, P.J. Van der Merwe and Monica Wilson.

[2] R. Elphick, *Kraal and Castle: Khoikhoi and the Founding of White South Africa* (New Haven and London, 1977).

[3] S. Marks, 'Khoisan Resistance to the Dutch in the Seventeenth and Eighteenth Centuries', *Journal of African History*, 13, 1 (January 1972), pp. 55-80; R. Ross, 'Smallpox at the Cape of Good Hope in the Eighteenth Century', Mimeo, p. 13.

[4] These materials are found in the Cape Archives (hereafter CAD): Old Cape Freeholds (hereafter OCF), vols 1-3; Old Stellenbosch Freeholds (hereafter OSF), vols 1-2; and Receiver of Land Revenue (hereafter RLR), vols 1-32.

leader and roughly demarcated territory.[5] The members of these groups migrated in bands within their territories in complex patterns that involved the periodic dispersal and reunification of people and their livestock. These patterns were of a seasonal character, but as Elphick points out 'were never regular enough to enable the Dutch to forecast the location of a particular tribe without hesitation'.[6] Individual Khoikhoi owned their own stock, but the land and water that sustained them were the common property of the group. The Khoikhoi peoples made good use of the resources of the areas they inhabited by grazing their cattle and sheep in a given area for a short period before moving on. This practice gave the veld ample time to regenerate itself before it was used again. The Khoikhoi practices, however, notwithstanding their ecological merits turned out to be vulnerable to encroachment by European colonists, because choice springs and sites of permanent water were often left unoccupied as Khoikhoi bands moved about their territories. The Khoikhoi might have left water sites unattended, but their actions were not prompted by any lack of appreciation of their value. That the Khoikhoi valued water is made abundantly clear from the multitude of words they used to describe it. Nienaber and Raper noted that the presence of water (on the basis of the number of words in Khoikhoi languages to describe its various occurrences) was the most important geographical element in Khoikhoi descriptions of landscape.[7] The idea that individuals could gain exclusive control of land or water was alien to their thinking, and their experiences with other Khoikhoi and San would have done little to prepare them for the remorseless privatisation of these resources by settlers.

When in 1652 Jan Van Riebeeck, the military commander of a small Dutch expeditionary force, took possession of Table Valley in the name of the Dutch East India Company (hereafter VOC) he did so as an experienced colonial official on behalf of a trading company with half a century of successful and quite ruthless involvement in colonial undertakings in the Indian Ocean basin. This involvement included some prior political alliances with the Khoikhoi.[8] In planning to place a small settlement at the Cape, the VOC directors assumed that the land was there for the taking not because it was 'empty' -- they had traded with the Khoikhoi long before 1652 -- but because they never intended the settlement to go beyond the Cape Peninsula. The famous hawthorne hedge was the proscribed border for the small number of market gardens needed to serve the outward and homeward bound fleets. At the outset, there was no specific colonising vision of settling people on the land, only a limited way station, as Cornelis de Kiewiet aptly pointed out, a micro-managed 'cabbage patch' on the way to the Indies.

The task the VOC entrusted to Jan Van Riebeeck looked simple enough. He was to establish a fortified base at the Cape, see to its sea and land defences, begin trade

5 The 50,000 Khoikhoi inhabiting the area south of the Orange River occupied an area of about 130,000 square miles (see Elphick, *Kraal and Castle*, p. 23). The overall population density was well under one person per square mile.

6 Elphick, *Kraal and Castle*, p. 58.

7 G.S. Nienaber and P.E. Raper, *Toponymica Hottentotica* (Pretoria, 1977), pp. 58-61.

8 R. Raven-Hart, *Before Van Riebeeck* (Cape Town, 1967), II; Eric Axelson, *South-East Africa, 1488-1530* (London, 1940), pp. 12-22.

with the Khoikhoi for cattle and sheep and establish intensive agriculture, especially market gardens with orchards and vegetable patches.[9] It was a threefold mission that was easier to plan than to accomplish. Van Riebeeck soon found that the Khoikhoi were often less than willing traders and that intensive agriculture on unimproved ground and the building of fortifications demanded much more labour than was available.[10] Within the first month, Van Riebeeck had asked the Directors for slave labour. He sought to remedy the trading problem by sending out trading expeditions into the interior and to solve the agricultural shortfall by creating in 1657 a class of vrijburghers (free independent farmers, hereafter, burghers) on their own land.

The VOC directors envisaged that the primary role of the Khoikhoi would be that of trading partners and considered that such a relationship could be best achieved by recognising the cultural, legal and political autonomy of the various Khoikhoi clans. The VOC ostensibly treated the Khoikhoi as 'free' peoples, although that freedom was nowhere defined in the Dutch period (1652-1795).[11] At first they were not coerced to labour in the colony. The official VOC view of Khoikhoi autonomy, though, did *not* include a respect for the territorial integrity of their land and did not prevent Company officials from alienating the Cape peninsula in the name of the VOC. More importantly the local officials, many of whom farmed in a private capacity, did little to prevent the more aggressive burghers disrupting the lives of Khoikhoi by hunting and trading on their land.

The Van Riebeeck initiatives were premised on Asian colonial precedents. If the Khoikhoi would not trade freely, ways would be found to force them; if agriculture languished, new lands and coerced labour would revive it. Yet in spite of a developing colonial vision, Van Riebeeck's thinking remained fixed on the essentially conservative ideal of European intensive farming, combining crops and livestock in a clearly proscribed area. He was greatly impressed by the potential of the Cape Peninsula for realising this vision, estimating that the area could support thousands if not tens of thousands of farmers.[12] Initial land grants were therefore small, averaging twenty-nine acres and located in the very choicest traditional Khoikhoi pasturage, along the Liesbeeck Rivier.

Several Khoikhoi watched apprehensively as the VOC's surveyor, Peter Potter, measured the land grants. Although notions of private landed property were alien to their way of thinking, they were in no doubt that the European intruders were taking *their* land. When it became clear that the Europeans were there to stay, they took up arms and went to war against the Dutch in a futile effort to regain the land they had lost. When peace was concluded in 1660 they continued to object, as Jan Van Riebeeck candidly recorded in his journal:

---

[9] L. Guelke, 'Freehold Farmers and Frontier Settlers', in R. Elphick and H. Giliomee (eds), *The Shaping of South Africa Society, 1652-1840* (Middletown, 1989), p. 69.

[10] Guelke, 'Freehold Farmers', p. 70.

[11] CAD, M142 (6) Anon., 'Laws Relating to Free Blacks'. Historical summary of all laws relating to blacks, compiled between 1825 and 1828, 11 pages.

[12] H.B. Thom (ed), *Journal of Jan Van Riebeeck, 1651-1662* (Cape Town and Amsterdam, 1954), I, pp. 35-36.

They strongly insisted that we had been appropriating more and more of their land, which had been theirs all these centuries, and on which they had been accustomed to let their cattle graze, etc..... They therefore strongly urged that they should again be given free access to this land for that purpose. At first we argued against this, saying that there was not enough grass for their cattle as well as ours, to which they replied: 'Have we then no reason to prevent you from getting cattle, since if you have a large number, you will take up all our grazing grounds with them? As for the claim that the land is not big enough for us both, who should rather in justice give way, the rightful owner or the foreign intruder? ... [E]ventually they had to be told that they had now lost the land as the result of the war and had no alternative but to admit that it was no longer theirs, the more so because they could not be induced to restore the stolen cattle which they had unlawfully taken from us without any reason. Their land had thus fallen to us in a defensive war won by the sword as it were, and we intended to keep it.[13]

The end of the first Dutch-Khoikhoi war in favour of the VOC was a political reality that some of the Khoikhoi who had fled the Cape peninsula were now ready to accept and they sought permission to return to their homes. The local VOC officials questioned them about their decision:

On being asked why they wished to come back to the Cape now and make peace, they had replied that the Cape was their birthplace and their own country with an abundance of fresh water, that their hearts continually hankered after it, and that at the Saldanha Bay [an arid northern Cape bay] the ground was everywhere barren and brackish, moreover Oedasoa refused to have them near him at the best places and rivers, and had asked them to arrange to live peacefully and quietly with us in their country.[14]

This incident provides one of the earliest indications that access to water was a critical factor in the Khoikhoi pastoral economy and placed severe limits on the number of kraals that could be sustained in any given area. In this case it was instrumental in forcing a group of Khoikhoi to make an accommodation with the VOC, and do something they had not been prepared to do earlier.

When Jan van Riebeeck left the Cape in 1662 the VOC had established a self-contained, if not self-sufficient 'fortress' colony in the northern Cape peninsula. The VOC protected it on its landward side with a series of forts which were partly connected by the hawthorne hedge. These steps were designed to keep autonomous Khoikhoi out of the colony, which was conceived as a clearly defined territory with a demarcated frontier. The northern peninsula provided the invaders with a secure territorial base from which neighbouring Khoikhoi lands could be threatened. Neither the VOC nor the burghers acknowledged Khoikhoi jurisdiction in the hinterland. When it suited them the VOC sent exploration parties into the interior and put pressure on Khoikhoi clans to trade with them. These expeditions often

[13] Thom, *Journal*, II, pp. 95-96.

[14] Thom, *Journal*, III, p. 77.

created political tensions among the Khoikhoi peoples who adopted different strategies for dealing with these unwanted visitors.[15]

The hard-working and self-sustaining small holders Van Riebeeck envisaged as the basis for a flourishing colonial agriculture did not materialise. A slave-owning elite soon formed in Table valley, which quickly dominated the market for fresh vegetables and fruit. The remaining burghers sought to make a living as best they could in an economic context that made the labour intensive cultivation of the land unprofitable.[16] Several burghers resorted to a wide variety of non agricultural activities to maintain themselves including trading livestock, wine, selling wood to foreign ships, lime-burning, salt-collecting, running boarding houses, hunting and fishing.[17] Other burghers, especially the younger generation, could not be contained within the old boundaries and ventured onto Khoikhoi lands in search of first game, trade, then pasture and finally, watered freehold land.

In 1672 and again in 1673 Khoikhoi of the Cochoqua clan, seeking to re-establish control of their pasture attacked European hunting parties that entered their territory.[18] These actions precipitated a strong European response in the form of punitive military expeditions and ended some years later in 1677 in the capitulation of the Cochoqua.[19] The Khoikhoi who remained in the South-western Cape continued their traditional pastoral way of life, but were unable to resist further European occupation of their territory. The Khoikhoi retained internal political control of their affairs, but permanently lost that most fundamental aspect of political authority - the ability to secure the integrity of their own territory and its resources.

The defeat of Cochoqua leader, Gonnema, gave Europeans continued access to Khoikhoi territory on favourable terms. Although this access left most of the Khoikhoi in possession of their land, it established effective European military hegemony over the south-western Cape. The political inequality between European and Khoikhoi was made clear in terms of the way they both 'controlled' territory. Europeans had the power to move onto and use Khoikhoi land as they wished, but the Khoikhoi remained in European controlled areas on terms set by the VOC. In a sense, the frontier beyond the Cape flats was 'closed' before a single permanent white settler had established residence in the area, because the Khoikhoi were no longer able to control access to the lands on which they lived.

A critical component of European military strength was the armed soldier or settler mounted on a horse. This combination of fire power and mobility made a devastating impression on people used to fighting or defending themselves with spears and bows and arrows. A well-educated Cape official, J.G. Grevenbroek,

---

15 Elphick, *Kraal and Castle*, pp. 117-174.

16 Guelke, 'Freehold Farmers', pp. 69-73.

17 K. M. Jeffreys et al., *Kaapse Plakkaatboek* (Cape Town, 1948), I-IV, passim. The plakkaats restricting activity are the most illuminating. For a full listing of all legal occupations in the cape of 1731, see Algemeen Rijksarchief, Collectie Rademacher Inventory, Number 507, 'The de la Fontaine Report'.

18 De Wet et al., *Resolutions of the Council of Policy* (Cape Town, 1957 onward), (28 November 1672), 1, pp. 102-3 (hereafter *Resolutions*).

19 *Resolutions*, I, pp. 190-191.

described in vivid chauvinist detail the reaction of the Cape Khoikhoi to armed horsemen:

> In the early days of the Colony the natives thought that our men who lived next to them could vomit fire. They stood by in amazement, while on the discharge of our guns they saw the largest wild animal, a lion, or an elephant, laid low by one leaden bullet as if struck by lightning. Our cavalry likewise they regarded as so many Centaurs, half horses half men; at the sight of them they began to shake; and they held their strength in such awe that any native wife or maid would prostitute herself to them, hoping, like a second Thalestra, to conceive an Alexander; the woman's husband perhaps would urge her to it, and she would be all the more ready because firmly convinced that the virtue of the sire would be implanted in any child conceived from one of our men. Poor woman! All because she had never heard of powder and shot.[20]

The above passage leaves little doubt about the reality of European power.. However, it seems more likely that fear of the destructive capabilities of European weapons had more to do with Khoikhoi women having sexual relations with European men than their expectations of producing superior children. In later years the awe with which Khoikhoi regarded firearms would have lessened, but the possession by European settlers of a deadly fighting machine combining mobility and fire power must have continued to pose a military threat to peoples who had to rely on their traditional weapons in courageous attempts to stave off European occupation of their lands.

The defeat of the peninsular Khoikhoi under Gonnema provided additional land for European settlement. Local officials, desirous of increasing agricultural production, made new lands available to free burghers on a first-come-first-served basis. The settlers responded quickly and occupied choice bottom lands along the Eerste River and its tributaries in the newly formed District of Stellenbosch. They claimed much larger areas than the early, modest peninsula farms, but large areas of common grazing lands in areas away from the rivers were not alienated. A similar process occurred in the Tijgerberg hills near Cape Town, where a dozen or so burghers established large farms with access to spring water leaving the areas between the farms as common grazing lands for their stock.

The new settlements created increasing pressures for the Khoikhoi who had formerly used the pastures of these areas, notwithstanding the relatively small land area actually occupied in freehold farms and notwithstanding the Company assurances that pasture lands should be shared by all. There was certainly plenty of pasture land available in the Stellenbosch area and in the early years of settlement sufficient for all, particularly as Khoikhoi patterns of transhumance created only a seasonal demand on the land. It was not the quantity of land taken by the settlers

[20] J. G. Grevenbroek, 'An Elegant and Accurate Account of the African Race Living Round the Cape of Good Hope, Commonly Called Hottentots', in I. Schapera (ed), *The Early Cape Hottentots* (Cape Town, 1933), p. 295. The military advantages of cavalry in other areas of Africa are described in J. Goody, *Technology, Tradition and the State in Africa* (Cambridge, 1971), pp. 39 ff.

that was the problem, but rather the pattern and style of settlement and also the type of title.

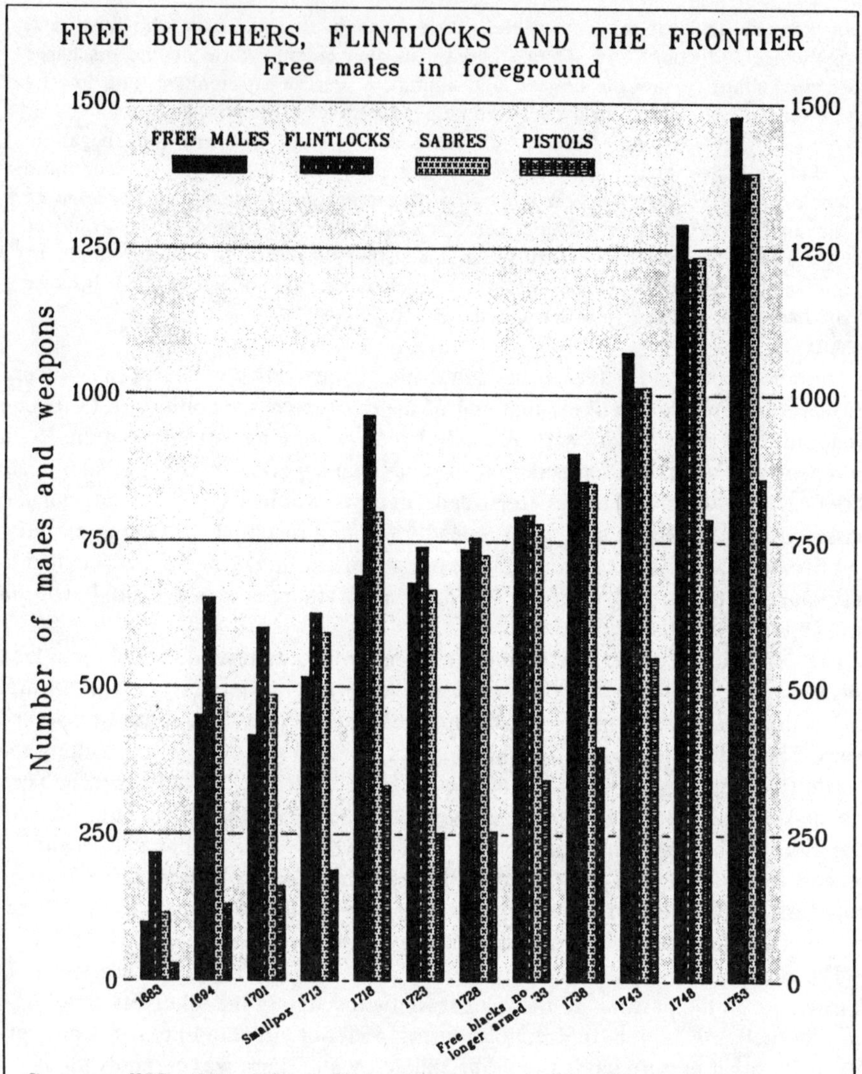

Sources: VOC 4018: 218; C 504, p.868; Beyers (B–H), pp. 240–245.

Figure 1

First, the land was granted to the settlers in 'freehold', that is in private ownership in the expectation that they would make permanent homes in the area and cultivate the land. This system of land tenure and the associated settlement excluded the Khoikhoi from lands that had once been theirs and placed the control of the system of property allocation in European hands. The Khoikhoi were technically a 'free' people, but this freedom did not give them access to land that had been alienated from them. In the Receiver of Land Revenue books, there is only one example in the Dutch period of a Khoikhoi laying claim to property, namely Adam Kok, described in the source as a 'Hottentot', who obtained his farm De Stinke Fonteijn (The Stinking Fountain) in 1751 and held onto it until 1771, when it went to Hugo Lambregts, a settler.[21] Although under no known restriction regarding property ownership, the Khoikhoi at the Cape never owned or even claimed a single piece of freehold land, an astonishing piece of evidence, illustrating the near total Dutch 'legal' hegemony over the landscape.

Second, the settlers of Stellenbosch, taking advantage of the open-ended land policy of the VOC, took up all freehold land along the rivers. The map of freehold grants is also a map of the permanent water ways of this area. When VOC commissioner Van Reede visited the Stellenbosch area in 1685 he was concerned that future freeholders would not have access to water.[22] Van Reede did not comment on the implications that the pattern of land holding must have had for the Khoikhoi of the area. If potential future settlers were to be denied access to water, the Khoikhoi must have been, to their great disadvantage, in that predicament already.

Since the early days of the colony, skirmishes and frontier violence had characterised burgher-Khoikhoi relations. To avoid expensive involvement in the interior, the VOC encouraged the burghers to arm themselves and to shoot well: the VOC instituted annual militia exercises and shooting competitions. By 1722, even the free blacks were forced into their own militia company. Every household, with the exception of those headed by widows, was heavily armed with arquebuses, sabres, flintlocks and pistols[23] (Figure 1). Burghers used their military capability to hunt down runaway slaves, retrieve stolen cattle, avenge Khoikhoi raids, or harass Khoikhoi who came to close to their land or water. It is hardly surprising that the Khoikhoi complained about their situation to a sympathetic Van Reede, who recorded:

> They say ... that they were being more and more oppressed both by the freeburghers and the Honourable Company not only because the best grasslands of our nation were being turned into grainland but also because the mountains and veld were being occupied by livestock by [sic] Dutch people, who

[21] CAD, RLR, 12, entry 201 (30 March, 1751).

[22] H. A. Van Reede, 'Joernaal van zijn Verblijf aan de Kaap', *Bijdragen en Medelingen van het Historisch Genootschap* (gevestigt te Utrecht), 62 Jaargang, pp. 122-123.

[23] The annual *opgaaf* was part muster: no less than three types of weapons had to be declared. We know from the probate documents that burghers often had more weapons than they declared on the muster.

> prohibited them [the Khoikhoi] from coming thereabouts with their animals. Should their oxen and cows happen to graze too close to the free peoples' grain fields they were shot dead by them, or [by] their people [i.e. slaves].[24]

The burghers were using the VOC legal system, their own slaves and guns to restrict the Khoikhoi from using the resources near 'their' lands. The Khoikhoi were obliged to get close to the settlers' grain fields because they needed to water their livestock. The burghers clearly felt that they were justified in shooting any animals that strayed onto their property. Whether the Khoikhoi understood the meaning of private land ownership is a moot point; they quickly learned that they could not get too close to cultivated land. Since the location of these lands denied them access to water and nearby pasture, they were obliged to wander into the drier interior, setting off a slow domino effect, where Khoikhoi displaced Khoikhoi; a process moreover, which accelerated wherever the carrying capacity of the land was low. The burghers, having secured a decisive legal and geographical advantage over the Khoikhoi, gradually squeezed them out of much of the common lands the two groups were supposed to share.

As their access to water resources dwindled, the Khoikhoi found it increasingly difficult to support themselves in their transhumant way. Many Khoikhoi, who resolved to stay in the land of their birthplace, were compelled by necessity to work for burgher farmers at harvest time, or to tend their cattle. By 1688 the Khoikhoi were so prominent in the ploughing or harvest work force that Van der Stel wrote that it was just like the descent of migrant workers from Westphalia upon the Netherlands.[25] Grevenbroek, who lived in Stellenbosch, was impressed by the new Khoikhoi farm worker. He wrote:

> The natives on this side of the mountain enthusiastically hire out their labour for a modest wage, and toil more submissively than Spartan helots. They are apt in applying their hands to unfamiliar tasks. Thus they readily acquire the veterinary skills to cure scab in sheep, and they make faithful and efficient herds. They train oxen for use in ploughing; and if put in charge of a wagon, coach or cart, they are found exceedingly quick at inspanning or outspanning or guiding a team. Some of them are very accomplished riders, and have learned to break horses and master them ... They chop wood, mind the fire, work in the kitchen, prune vines, gather grapes, or work the wine press industriously ... Their wives and daughters make reliable washerwomen and busy chars. They wash plates and dishes, clean up dirt, gather sticks from the fields round-about, light the fires, cook well, and provide cheap labour for the Dutch.[26]

These activities which began as part time activities, ended for many Khoikhoi in their becoming fully dependent client workers, we are tempted to say, even serfs.

---

24 Van Reede, *Joernaal*, p. 128.

25 Algemeen Rijksarchief, The Hague, C502: Uitgaande Brieven, p. 39 (26 April, 1688).

26 Grevenbroek, 'Account of the Hottentots', pp. 271-273.

The settlers of Stellenbosch, secure on their freehold land, developed a viable form of extensive agriculture dependent on slave and client labour, large land holdings and extensive tracts of open grazing lands.[27] The principal crops included wheat and other cereals, while some settlers cultivated vineyards, an agricultural import perfectly suited to the Cape's Mediterranean climate. The livestock of the burghers were the issue of cattle and sheep breeds of the Khoikhoi, which foraged for themselves in the rough vegetation of the country. In 1687 the Governor, responding to settler concerns about the ability of Stellenbosch to support additional farmers and their livestock, stopped giving out freehold land and looked elsewhere for the land on which expansion would occur. By this time the settlers had gained effective control of the Stellenbosch District, but they continued to share its pasture lands with the Khoikhoi, who were still present in considerable numbers and lived independently in their own settlements, even if some of them had to supplement their incomes by working on settler farms.

When, in 1687, Governor Simon van der Stel opened new lands for settlement in the upper Berg River Valley he abandoned the first-come-first-served policy of earlier years in favour of one in which settlers received rectangular grants of re-surveyed land at right angles to the Berg River.[28] Notwithstanding this planned approach, the basis of the agricultural and settlement strategy was not altered. Van der Stel drew on his Stellenbosch experience in establishing a standard 60 morgen farm size and made sure that the freehold grants comprised but a small proportion of the total land area, thereby ensuring the continued success of extensive agriculture. The Khoikhoi of the Upper Berg River Valley continued, as in Stellenbosch, to live side by side with the settlers, but were somewhat better off to begin with than those of Stellenbosch in terms of the easier access they had to water thanks to the new farm geometry.

If the Drakenstein Khoikhoi were somewhat better off to begin with than those of Stellenbosch, they did not long remain so. The growth of settler activity and especially their flocks and herds put increasing pressure on the available pasture. Kolbe described the Drakenstein landscape some twenty years after its first settlement by Europeans:

> The Farms on this River [the Mountain (Berg) River] are at about Half an hour's distance from one another; yet the Inhabitants complain, that they are too near one another; urging, that between their Plantations there is not Pasturage sufficient for their Cattle, tho' Grass grows in plenty every where on both sides of the river. The reader will gather from hence the vast Numbers of cattle belonging to those Planters.[29]

By the early eighteenth century the European colonists had consolidated their hold on the lands near Cape Town, lands suitable for mixed farming and livestock husbandry. Yet the VOC still could not contain the burghers within fixed

[27] Guelke, 'Freehold Farmers', pp. 73-74.

[28] CAD, Old Stellenbosch Freeholds, I and II.

[29] P. Kolbe, *The Present State of the Cape of Good Hope* (London, 1731), II, p. 49.

boundaries of the districts. Individuals and groups of colonists invaded Khoikhoi land to hunt, trade and secure additional pasture for their livestock. There were some VOC attempts to regulate frontier activity, but in a thinly settled land there were too many opportunities for individuals to take things (mainly guns) into their own hands. The successful burgher revolt of 1706 which resulted in the recall of the highest echelon of local VOC officials improved economic opportunities for burghers, who no longer had to compete with wealthy VOC officials with privileged access to a small market. Notwithstanding strict *plakkaats* against violence and illicit trade in the interior, the frontier area became an unsettled zone where Khoikhoi were increasingly vulnerable to settler violence and coercion.[30]

In an attempt to regulate this frontier activity, but also to raise revenue for local administration, the VOC began issuing salt collection permits and hunting and grazing permits to colonists for the seasonal use of frontier lands (Fig.2). These documents, extensively used by P.J. van der Merwe in his detailed study of the wandering interior farmers, are a key source for the understanding the decline of the Khoisan people. The mundane wording of these permits is in strong contrast to their significance in displacing the indigenous people, viz. in 1699 one such licence read: 'to Hans Henske ... over the Klein Berg River, around the Red Sands, 1 load of Eland, Hartebeest or Rhinoceros' or in 1700, 'to Abraham Diemer, over the Red Sands, in the Land of Waveren, next to the Breede river, 1 load of salt'.[31] Such salt collection privileges and hunting trips (*togts*) served as informal surveys of the land and often matured into land grants. In 1714 the VOC began charging a fee for the use of grazing permits, and at the same time gave the permit holders permission to grow small amounts of cereal. The grazing permits fostered frontier expansion on a model that confirmed individual access to and control of land, but did so on a vastly different scale: the standard loan farm size was 6,000 acres, fifty times the size of the old freeholds.

Moreover, the farmers used the ancient Dutch medieval burgher right of *opstal* to gain de facto ownership of their grazing permits or loan farms as they came to be called. In the Netherlands this right entitled a landless burgher to erect a building on or cultivate another's land provided that former lived on the property that was being used. Moreover, any improvements could legally be sold to the next tenant, although the ground itself remained the property of the original owner. In South Africa the VOC retained ownership of the frontier lands it leased to colonial settlers, but the *opstal* could be sold and its tenure became so secure that price of the *opstal* came to reflect the value of the entire property. It is ironic that a system originally intended to favour impoverished burghers in medieval Netherlands should later become one of the main legal engines for European colonial expansion in South Africa at the expense of its original inhabitants.[32]

---

30 Jeffreys, *Kaapse Plakkaatboek*, I, p. 262.

31 Leibbrandt, *Requesten*, I, p. 109 and p. 117.

32 *Encyclopedia van Nederlandsch-Indië*, s. v. 'opstal' III, pp. 174-175. The system was also used throughout the VOC possessions in Indonesia.

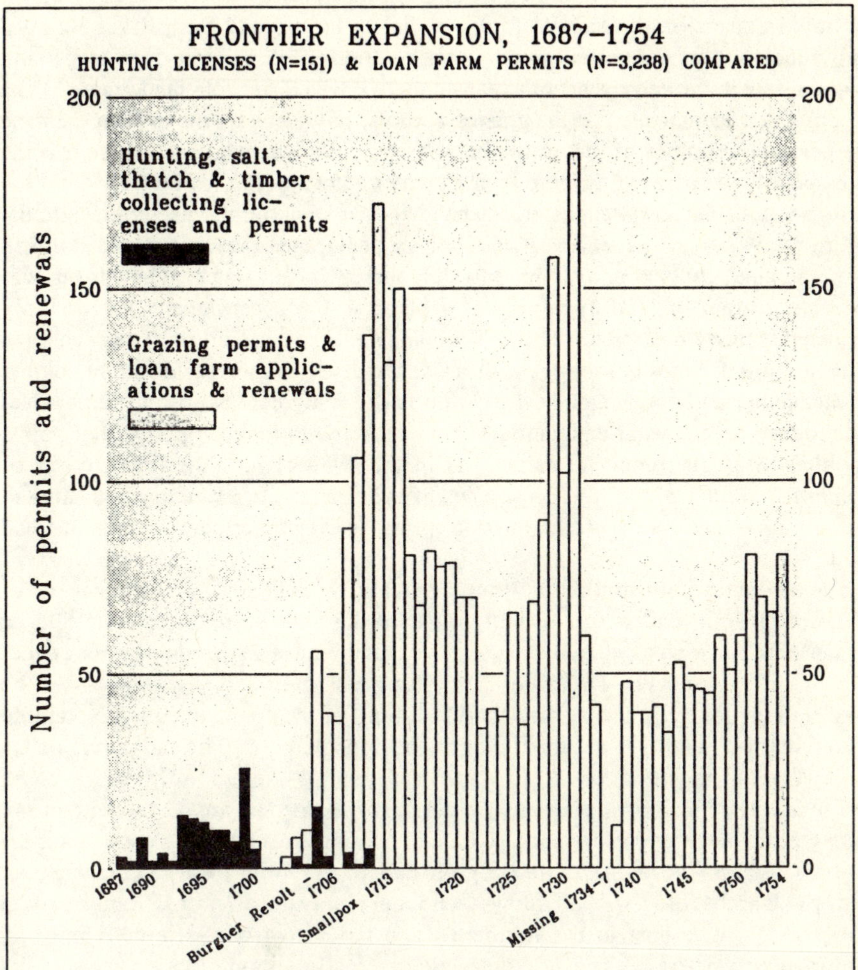

Notice that frontier expansion PRECEDES 1713 smallpox epidemic, but follows burg. revolt 1706

SOURCES: CAD, Receiver of Land Revenue Books, RLR, Volumes 1–13.

Figure 2

The movement of the pastoralist *trekboers* into frontier areas paved the way for the rapid expansion of freehold farms in the Swartland and Berg River Valleys. Individual settlers converted grazing and hunting privileges into loan farms and occasionally in the early years of expansion into coveted freehold land grants.[33] The specific areas demarcated on the grazing permits, salt licenses or loan farms became the permanent homes of increasing numbers of colonists, who used these permits to expand the borders of the colony at a greater velocity. The loan farm met all the basic needs of the *trekboers* as they expanded their stock farming activities into the interior, always at the expense of Khoisan pasturage: they secured a hold on strategic land, containing a spring or watercourse. C. Hall states: 'The custom was to issue a grant of land in respect of a place where a permanent water-supply rendered habitation possible'. The *landrost* and *heemraden of* a district were charged with settling disputes involving water as was the case in the very different context of the Netherlands. Although in the period under study here almost all settlers had full control of the water emanating or flowing through their property, Hall points out that the Government remained, as in the Netherlands, the legal owner or *dominus fluminis* of the water resources of the country. In a country with limited rainfall, only land on which there was a spring or even a periodic watercourse had value.[34]

The colonial administration of water is evident in the letter of the younger son of Christoffel Smit who wrote in the twilight of the loan farm system. In 1806 he requested a 'loan on ordonnance [sic] at the usual rent of 22 Rixdollars per annum [for] the ... place called De Holle Vlei ... as the said ground is without water, he asks further, for the grant, at half the usual rent, of the fountain called Noodhulp [Need help] ... and that the fountain may be considered as part and parcel of De Holle Vlei (signs with a cross)'.[35]

The holder of land on a permanent water supply gained *de facto* command of not only the surrounding pasture but also gained hunting advantages by commanding a point at which the game of the region had to converge. Carmel Schrire, an historical archaeologist, has shown with her analysis of faunal remains of a company frontier post in the northern Cape the extent of the inroads into the hunting ecosystems of the Khoikhoi which a small platoon of company soldiers made in a short period of sixty years. Although these men were supplied by ship with food, they *preferred* hunting as a way to obtain food.[36] It is not difficult to imagine what kind of impact groups of similarly armed settlers in search of game would have had on the natural ecosystems on which the Khoikhoi and San depended for their livelihoods.

---

[33] This practice was common in the first two decades of the eighteenth century, but did not again become a possibility until 1743 when new policies were adopted allowing settlers to convert a portion of a loan farm into freehold tenure. Relatively few settlers actually availed themselves of this opportunity.

[34] C. G. Hall, *The Origin and Development of Water Rights in South Africa* (Oxford, 1939), pp. 7-26.

[35] Leibbrandt, *Requesten*, 3 (no. 50 of 1806-7), p. 1173.

[36] C. Schrire, 'Excavating Archives at Oudepos 1, Cape', *Social Dynamics*, 19 (1990), pp. 11-22.

The movement of white settlers to the north and east continued in the 1720's and 1730's and was largely unchecked until the easternmost *trekboers* encountered the numerous Xhosa agriculturists in the vicinity of the Great Fish river in 1779 (Fig. 3). The huge area of South Africa that approximately 1,000 independent settlers and their families occupied in the period from 1717 to 1780 encompassed large areas of arid land.[37] The *trekboer* strategy involved seeking out and settling the better watered areas. In the south settlers were able to advance eastwards without having to contend with extreme arid conditions all the way to the Great Fish River. In the arid north settlers sought out the islands of better watered land to the north and east of the Great Karroo along the Great Escarpment. They settled the Roggeveld, the Nieuwveld and Sneeuwbergen and then turned south east as they entered the area of summer rains. There was much land that the *trekboers* did not occupy, but most of it was inferior land lacking permanent sources of water. The number of European settlers was small and the land they occupied was vast. In this setting the strategic nature of that occupation becomes critical for an understanding of how a relatively small number of European settlers was able to occupy and gain control of a land occupied by many thousands of Khoisan pastoralists and hunter gatherers.

In establishing a hold on a strategic spring, the *trekboers* deliberately limited their mobility, but gained certain access to whatever resources were available. Moreover, many *trekboers* took out more than one loan farm simultaneously, vastly increasing their domination of the land with very few men. The *trekboer* pastoral economy rooted individuals to one or more specific areas and was certainly ecologically less efficient than the full blown nomadism of the Khoikhoi. The *trekboer* livestock often used the same grazing land for longer periods than the Khoikhoi had and did much harm to the natural grazing resources of the country. Yet the advantages of having enduring control over a permanent water source proved to be the decisive secular advantage in securing the unpromising landscape entirely for themselves.[38]

As the *trekboer* population increased and more and more strategic land was privatised, the lot of people without land, the Khoikhoi, but especially the San (Bushman) hunter-gatherers who were dependent on an even more land extensive way of life, became ever more difficult. Such people had to work around the privately controlled loan farms, or migrate in the hope that water sources were available elsewhere on unoccupied land. There were a few Khoikhoi who eked out a bare subsistence in this way, but the process of privatisation effectively destroyed the viability of the grand transhumance systems and without them, the ways of life of the pre-colonial Cape disappeared.

---

[37] This figure is based on the RLR volumes. Confer Guelke, 'Freehold Farmers', p. 85.

[38] P. J. van der Merwe, *Die Trekboer in die Geskiedenis van die Kaapkolonie* (Cape Town, 1945), p. 73.

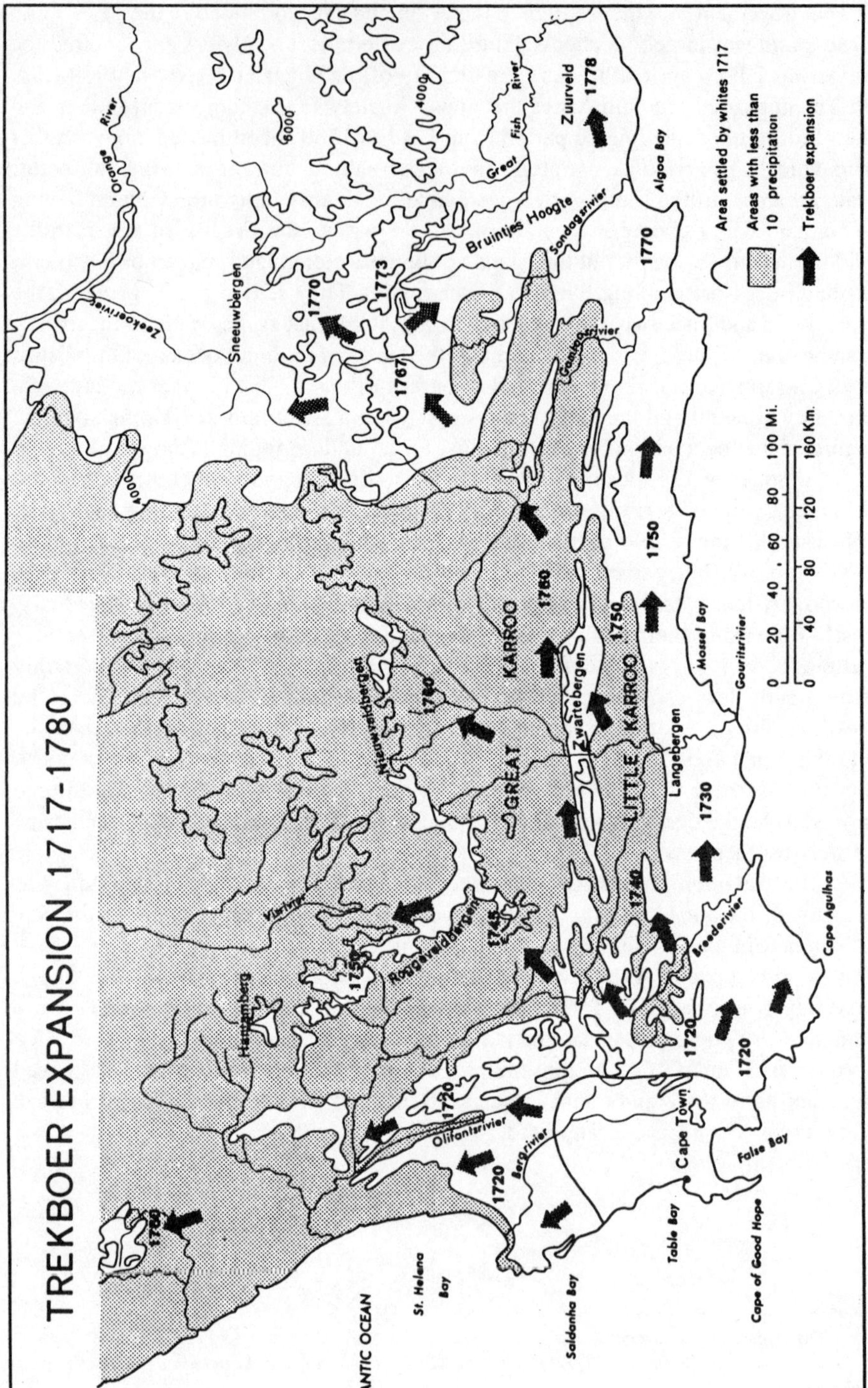
TREKBOER EXPANSION 1717-1780
ATLANTIC OCEAN
St. Helena Bay
Saldanha Bay
Table Bay
Cape of Good Hope
False Bay
Cape Town
Cape Agulhas
Mossel Bay
Algoa Bay
Gouritsrivier
Breederivier
Bergrivier
Olifantsrivier
Langebergen
Zwartebergen
Roggeveldbergen
Nieuwveldbergen
Hantamberg
Sneeuwbergen
Bruintjes Hoogte
Sondagsrivier
Gamtoosrivier
Great Fish River
Orange River
Zeekoerivier
Zuurveld
GREAT KARROO
LITTLE KARROO
1720
1730
1740
1745
1750
1760
1767
1770
1773
1778
1780
4000
6000
Area settled by whites 1717
Areas with less than 10 precipitation
Trekboer expansion
0 20 40 60 80 100 Mi.
0 40 80 120 160 Km.

Figure 3

The well-known observer of eighteenth century Cape life, Otto Mentzel has provided a detailed description of how European settlers displaced Khoikhoi people. He wrote:

> The Hottentots are, as it were, the bloodhounds who smell out the most fertile lands. When their kraals are discovered in such places several Europeans or Afrikanders soon appear and, by gifts, flattery and other forms of cajolery, wheedle the Hottentots into granting permission for them to settle alongside. But as soon as the pasture land becomes too scanty for the cattle of these newcomers and the Hottentots, the latter are induced by trifling gifts to withdraw and travel further inland. They prefer to remain in close proximity to the white man.[39]

In the above somewhat contradictory passage Mentzel has undoubtedly captured the essence of how the displacement process worked. It would have been entirely natural for a settler or settlers to make use of Khoikhoi knowledge by occupying areas they had already recognised as valuable in terms of water and pasture, register the land with the VOC and end up denying Khoikhoi use of the lands they once possessed. At this point a Khoikhoi family or group would face the difficult choice of either taking their chances on their own by moving further into the interior or accept the reality of European power and become the dependent clients of some *trekboer*. This choice and its different outcomes probably explains why Mentzel wrote both about Khoikhoi withdrawal inland and simultaneously noted their 'preference' to live in close proximity to the white man.

As European colonists acquired more and more spring land they were better able to dictate the terms of existence to the Khoikhoi. The tipping of the balance of geo-political power must have appeared obvious in the micro regions around a spring. After perhaps several heart-rending displacements, accommodation even to virtual serfdom would have made a good deal of sense to a people with hierarchical traditions in which clientage and even slavery was practised. Lichtenstein provides the best example of an indigenous person making such a choice:

> At this garden we noticed a Hottentot, whom Rossouw singled out for us. He was one of the tribe [*Stamme*] of the Great Namaquas and distinguished from the Hottentots in being both bigger and stronger. In earlier years of his youth he was in a war between his own people [Nation] and the Dammaras, a neighbouring group somewhat to the north[40] who captured him and among whom he lived for a number of years. According to this people's custom he was circumcised, and as a sign of his captivity, the middle upper *incisor* teeth [*Schneidezähne*] were knocked out [*ausgeschlagen*]. He showed us this and indicated that on account of this, if he were captured again, his life would be forfeit [emphasis added]. He therefore preferred living among the Christians to

[39] O. F. Mentzel, *A Complete and Authentic Geographical and Topographical Description of the ... Cape of Good Hope* (Cape Town, 1925), p. 36.

[40] According to A. Morris this was invariably a custom of Northern rather than Southern groups (private communication).

> running this chance, and making his escape, engaged in his present service, with which he was extremely satisfied.[41]

The burger's farmstead offered a relatively attractive and reasonably secure subsistence to people on the brink of starvation and most Khoikhoi eventually opted for that dependence over a barren, insecure and waterless independence away from their homes.

The importance of water as a basis on which individuals controlled land is well illustrated by the names of loan farms. Ignatius Maree held the first loan farm in 1709, De Langefonteijn (The Long Fountain).[42] In the period from 1704 to 1754 over 3,238 loan farms applications were issued or renewed. An increasing percentage of these had references to water in their names (Fig.4). Farms were, and still are, named after their most important asset, their water. Thus, Brakkefonteijn (Brack Fountain); Droogevallei (Dry Valley) Elandsfontein, (Buck-Fountain), Geelblommetjefontein (Yellow Flower Fountain), Klaarstroom (Clear Stream), Riviersonderend (River-without-end), Soebatsfontein (Beg-for-Fountain) and even simply Watergang (water course). In a thirsty land water was the one commodity that everyone needed and people expressed its importance in the most permanent way: the names of their farms. Nearly all have lasted to the present day, evidence of the continuing importance of water in that changing setting.

The Khoikhoi found it impossible to maintain their traditional transhumant way of life as their access to water and pasture was curtailed. The gradual economic and social disintegration followed the Khoikhoi's loss of the political control of their land. The Khoikhoi increasingly became clients of the burgher farmers and frontier *trekboers*, living permanently under their domination. The Khoikhoi retained some social independence despite their client status, but this vestige of their independence did not last long. Surviving Khoikhoi began mingling with the slave population and eventually lost their language.[43]

The San hunter-gatherers reacted quite differently to the Khoikhoi as European settlers encroached upon their lands. Unlike the Khoikhoi pastoralists the hunter-gatherer communities had no traditions of clientage and little knowledge of animal husbandry. Their first response to European settlement was to retreat into the interior, but as early as 1715 some San reacted violently to European expansion and attacked settler farms in the Land of Waveren (present day Tulbagh). The Europeans responded to this attack by forming a body of armed volunteers or

---

[41] H. Lichtenstein, *Reisen In Südlichen Africa* ... (Berlin, 1811), I, 91; cf. We are grateful to Mr Deeke for pointing out that during the nineteenth century travellers often referred to the 'Hereros' as 'Damara', thus adopting normal Nama usage. Deeke to Shell, 17 March 1989 (private communication). In a note to Shell A. Morris also made the crucial point that the Dama already practised tooth mutilation. This practice would have made tooth mutilation a sign of incorporation rather than as a physical mark signifying discrete slave status. A. Morris, 'Dental Mutilation in historic and prehistoric South Africa', *Quarterly Bulletin of the South African Library*, 43, 3 (March 1989); pp. 132-4.

[42] Leibbrandt, *Requesten*, I, 215.

[43] R. Elphick and R. Shell, 'Intergroup Relations: Khoikhoi, Settlers, Slaves and Free Blacks, 1652-1795', in Elphick and Giliomee, *Shaping*, pp. 229-30.

commando to deal with the attackers.[44] This pattern of attack and counter attack was to become a familiar element of frontier life for San and settler as the San

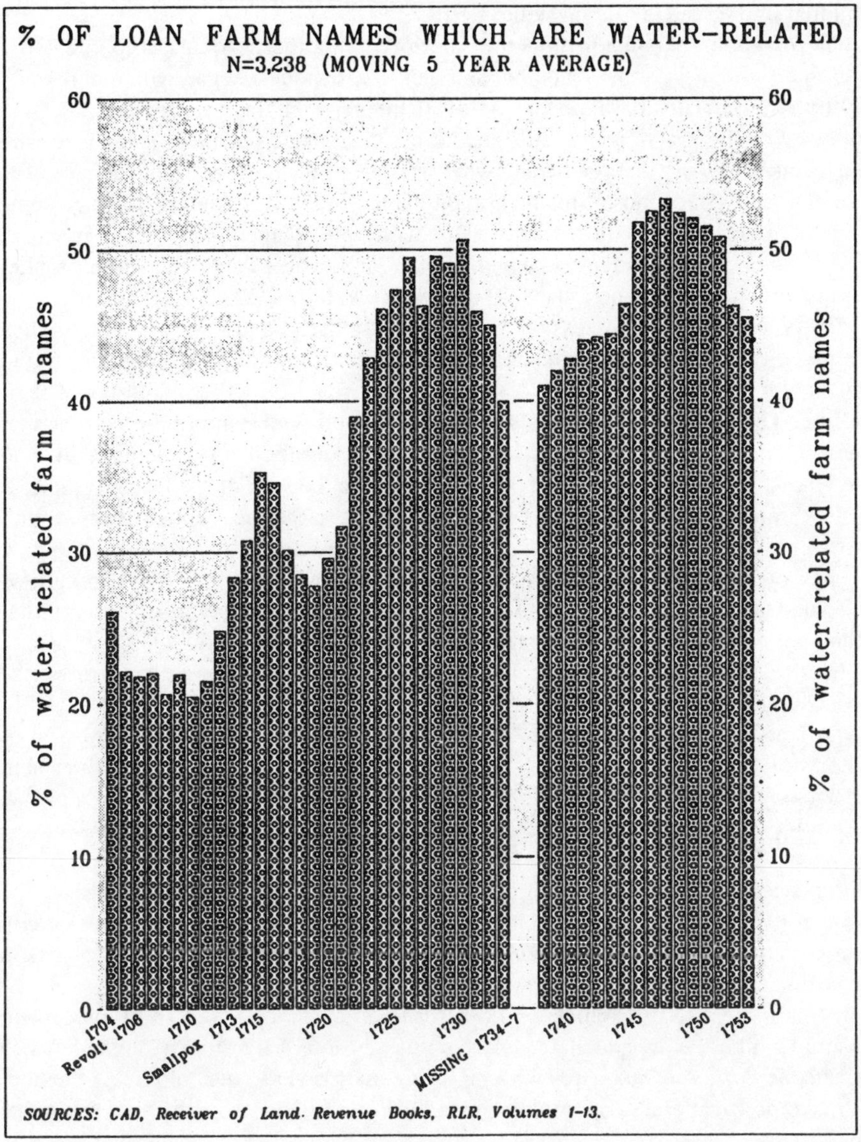

Figure 4

44 R. Elphick and V. C. Malherbe, 'The Khoisan to 1828', in Elphick and Giliomee, *Shaping*, p. 25.

struggled to survive the loss of their hunting grounds. It was an unequal struggle and the San suffered terrible losses at the hands of these armed commandos, which routinely executed the male San they managed to entrap and pressed women and children into forced labour on settler farms.[45]

The *trekboer* expansion into the interior of Southern Africa amounted to a piecemeal conquest of the Khoikhoi and San. The conquest began with the inability of the Khoikhoi to maintain the political integrity of their territories; this was followed by the loss of some cattle and access to strategic resources and ended with the Khoikhoi losing their cattle and what was left of their social independence. The rate of social and economic disintegration varied. Many Khoikhoi maintained some kind of independence on the lands still available to them, or moved onto the frontier away from the areas occupied by *trekboers*, where they met another frontier of hostile inhabitants, other clans. There was no exit.

The colonisers took possession of the land in a literal and symbolic sense. They occupied choice land and remade the landscape with buildings, fields, crops and objects familiar to them - the ox-wagon, ploughs and the other materials of their culture. Their physical presence went hand in hand with naming the land and its hills, mountains and rivers with familiar words from their own language. A land the Khoikhoi and San had called their own became the land of the Dutch settlers. It is now impossible to recreate the landscape as it was before Van Riebeeck. South African historical maps are the maps of conquest, with the names of conquerors. Names such as Tafelberg, De Caabse Vlakte, De Paarl, De Berg Rivier redefined this land and gave it a Dutch-African identity, although some of the original Khoikhoi place names such as Grabouw and Karoo were retained.

A critical process of European expansion involved European control of the procedures for legitimising the occupancy of land. The ability of Europeans to acquire and to defend their title to their lands made their strategic occupation of Khoikhoi territory possible. The Khoikhoi were powerless to stop European settlers encroaching on their land because they had no significant mechanisms, military or legal, to defend or entrench their rights, once the they had allowed the invaders to land. They were also in no position to use the 'deeds' system of the Europeans to register their own claims to land. Perhaps they were excluded at the door of the Company offices, but this is conjecture. As noted above, there is only one example, the exception that proves the rule, of a Khoi-descended person acquiring title to a loan farm in the period of Dutch rule.[46]

In his major study Elphick sets out five factors that needed to be present to ensure Khoikhoi independence: (1) secure possession of livestock (2) maintenance of satisfactory living standards without losing manpower to the colony (3) freedom to make political and economic decisions (4) secure and exclusive occupation of

---

45 Nigel Penn, 'The Frontier in the Western Cape' in John Parkington and Martin Hall (eds), *Papers in the Prehistory of the Western Cape* (Oxford: BAR International Series 332, 1987), pp. 475-476. It is not possible to do justice to the history of the San here. We have included only the essential outlines of their relationships with frontier settlers and as a contrast with the Khoikhoi.

46 See note 18 above.

traditional pasture and (5) retention of traditional culture.[47] Elphick argued that Europeans assaulted these conditions of Khoikhoi independence together, but concluded that 'the critical point was not land but the possession of cattle and sheep'.[48] We disagree. There is no question that the Khoikhoi were weakened by European harassment and by the unequal terms of frontier trade, but many retained considerable herds and flocks, which they could regenerate if they were allowed to husband them in their traditional way. Their growing inability to water and pasture their animals became the ever worsening problem. European colonists, in gaining control of springs and pasture lands, made it impossible for the Khoikhoi to maintain their livestock, and just as importantly prevented the Khoikhoi, who had lost some or all of their stock, from ever re-establishing themselves as independent graziers. This process started slowly, first in the high rainfall areas of the peninsula in the seventeenth century, but increased most rapidly after 1706 when the prime peninsular land was occupied and burgher settlement started to spread into the low rainfall areas where the carrying capacity of the land was substantially less.

Although Elphick identified a gradual internal cycle of decline, he, in common with many historians (with the exception of Robert Ross), places some emphasis on the smallpox epidemic of 1713 in dealing a final and fateful blow to a weakened Khoikhoi society:

> Thus, for all the nearby Khoikhoi except the Hessequa, the effects of the 1713 plaque were catastrophic. A traditional society, already in precipitous decline, had been virtually destroyed and the people who had lived in it had barely escaped annihilation.[49]

Robert Ross has questioned the importance of the 1713 smallpox epidemic as an explanation for the decline of the Khoikhoi.[50] He pointed out that the Khoikhoi were, as nomadic peoples, in a good position to avoid infection, provided that no one in the *kraal* became infected, and in the vastness of southern Africa, avoiding infection would not have been unduly difficult. Ross also questioned whether the Khoikhoi were particularly susceptible to the smallpox virus and concluded they were not. The Khoikhoi lived in a part of the Old World and were in contact with other African people, a situation very different from that of New World and Pacific peoples. In addition, it seems probable that as a pastoral people, close to cows, many of them, and especially the women who did the milking would have been exposed to cow pox and acquired some immunity to smallpox.[51]

Ross, using census data for other groups (the Khoi were not enumerated until 1798) estimated the probable Khoikhoi losses from smallpox:

---

47 Elphick, *Kraal and Castle*, p. 237.

48 Ibid., p. 238.

49 Ibid., p. 234.

50 Ross, 'Smallpox', pp. 3-5.

51 Khoikhoi women seem to have survived the smallpox epidemic in greater numbers than Khoikhoi men.

> It would thus seem exceedingly unlikely that smallpox killed more than, say, thirty percent of the Khoikhoi population of the Cape in 1713. This of course, represents a terrifying loss of life, by any standards. It is, however, the sort of situation from which a healthy population, with reasonable means of subsistence can recover fairly quickly ... provided that it does not suffer such calamities recurrently. Now, the Khoikhoi seem to have remained free from such diseases for the rest of the century, but, on the other hand, were not able to regain their old strength because their land was increasingly taken out of their control by the advance of the European farmers and stock ... far into the interior.[52]

Although we are in essential agreement with Ross' assessment of the historical significance of the 1713 smallpox outbreak and his judgement that the Khoikhoi were unable to recover because they were losing their land, we have added to his analysis a geo-strategic explanation of why and exactly how a relatively small number of heavily armed burghers were able to occupy the lands of the far more numerous Khoikhoi, regardless of epidemiological events. The settlers also suffered losses during the smallpox epidemic, but their population recovered within a few years of the outbreak.

Whatever the impact of the smallpox epidemic on the Khoikhoi of the south-western Cape, it is clear that most Khoikhoi to the north and east were not touched by the disease. And yet the decline and subjugation of these people by marauding colonists continued without pause. Europeans were able to gain a stranglehold on the lands they occupied, undercutting the very basis of Khoikhoi livelihood. The exact mechanism of the process is set out above, and we believe it provides a plausible non-catastrophic explanation, beginning around 1706, of the gradual decline of the Khoikhoi as independent peoples. Why did the Khoikhoi put up such a weak resistance to European encroachment of their lands after the defeat of Gonnema in 1678? Shula Marks has sought to chronicle that resistance and has assembled considerable evidence to show that Khoikhoi, as opposed to the San, did actively resist the advance of the European frontier.[53] This resistance, however, proved feeble and Khoikhoi accommodation to the settlers' society was the order of the day. The pattern of accommodation and sporadic resistance, it seems to us, is better explained by the gradual way in which Europeans infiltrated Khoikhoi land one spring or *spruijt* at a time.

---

52 Ross, 'Smallpox', p. 3.

53 Marks, 'Khoisan Resistance', pp. 70-73.

8

Fuelwood in Colonial Brazil: The Economic Consequences of Fuel Depletion for the Bahian Recôncavo, 1549–1820

Shawn W. Miller

Among the various tasks that constituted the daily routine of Brazil's sugar economy, collecting fuel from the colony's plentiful forests was among the most extensive. In addition to preparing land for planting, cutting cane at harvest, and refining it at the mill, African slaves through four centuries had the additional burden of gathering the crucial energy source that fuelled the sugar mill. This article is in part a study of that resource's depletion and of sugar production's detrimental effect on Brazil's Atlantic forest. But it is more immediately an examination of the impact that the forest's retreat had on the fortunes of the colonial capital, Bahia, located in northeastern Brazil.[1] Fuel's increasing scarcity increased labor and capital costs related to fuel supply, exacerbated elite social conflict, multiplied petitions to the Crown, and eventually dictated the adoption of more efficient firing technology. Moreover, activities that vied with the sugar mill's furnaces for fuel contributed to the contest, in which all colonists participated, for one crucial natural resource: wood.

Discovery of the Forest

At the height of the Bahian sugar harvest, a view from the bluffs of the Bay of All Saints contained a great deal of smoke, for all along the northern shoreline, to a few kilometers inland, a considerable array of fires pushed arching columns into the prevailing southeasterly winds. The simple process of boiling the water and impurities from sugarcane's viscid juice, an essential stage in sugar refinement and one requiring large quantities of wood fuel, was in full swing. Ashes to ashes, the cooling clouds of smoke came finally to settle on what remained of Brazil's once primeval forests. With rarely a pause, the fires of the sugar mills burned as many as nine months of the year, more than twenty hours of the day and six days of the week, as long as there was fuel to feed them. And to the smoke-filled

2 Shawn W. Miller

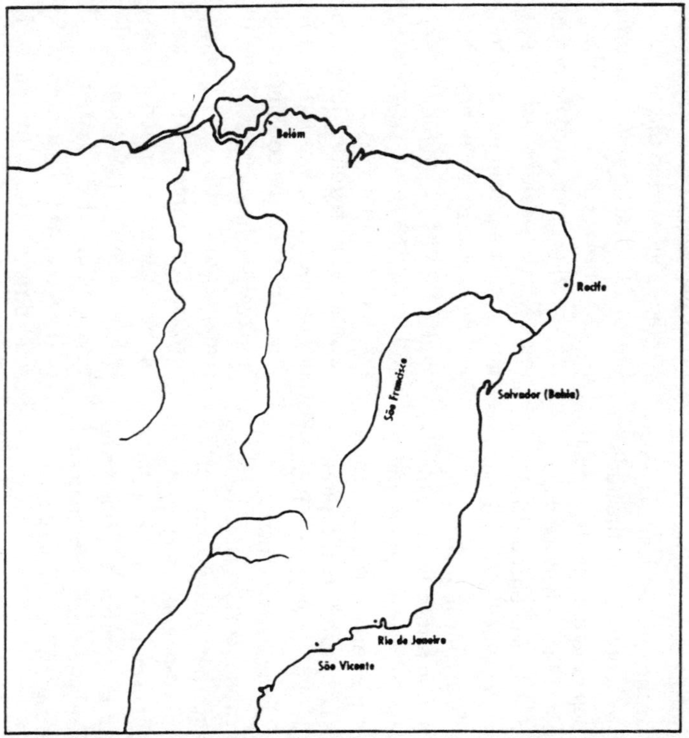

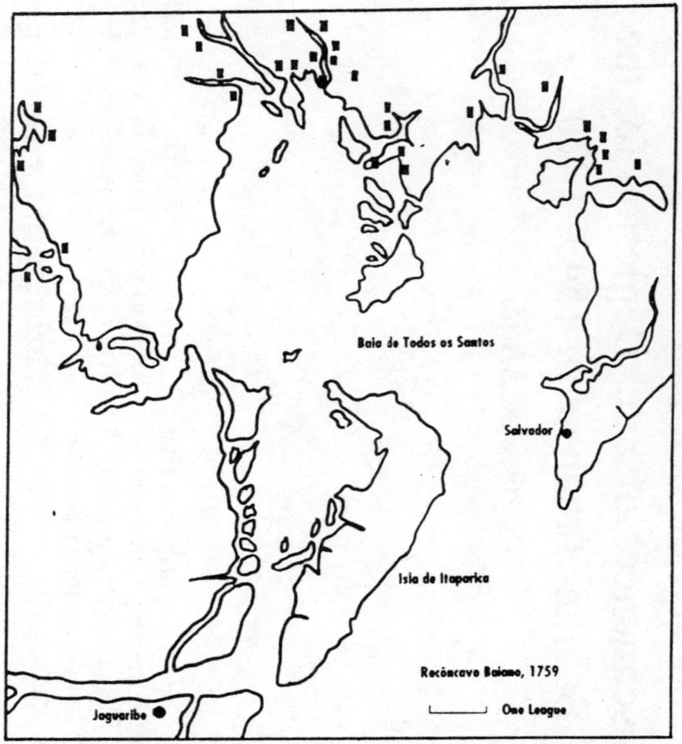

Figure 1 (left) *Colonial Brazil, drawn by author.* **Figure 2 (right)** *The Bay of All Saints, 1759. Salvador, the colonial capital, was the point of embarkation for sugar exported to Europe. Sugar mills clustered along the bay's northern arms and estuaries. The map in figure 2 was drawn by the author, after a 1759 map by José António Caldas, whose original utilized a mushroom of rising smoke to represent each mill.*

eyes of the early Portuguese, the fuel capacity of the Brazilian forest was boundless.

Portuguese, Italian, and Spanish explorers, the people who initiated the expansion of Europe, came from an environment that had long been stripped of its trees. In fact the entire Mediterranean seaboard, except a few isolated and coveted pockets, had been bereft of forests for centuries. Growing populations and the growing material demands for ships of discovery, war, and trade continued to make European wood resources increasingly scarce and costly, thus explaining the general amazement European sailors expressed at the sight of the New World's plentiful forests. When the Portuguese, during the earliest stage of their quest for the African route to the East, came across one particularly heavily forested island, the first they would colonize, they named it "Madeira," meaning "wood." Poet-adventurer Luís de Camões wrote a century and a half later that the island was more renowned for its namesake than for any other particular fame.[2] In 1500, en route to the spices of India, Pedro Alvares Cabral bumped into the South American continent and christened it "Land of the Holy Cross." The name that finally stuck originated in the red dyewood commonly called *pau-brasil* (brazilwood), South America's first trade contribution to the Portuguese colonial empire. To the doughty sailors of the fifteenth-century Mediterranean, wood was neither unremarkable nor unappreciated.[3]

The first important Brazilian export was wood, but the more important second commodity, cane sugar, could not have made its unrivaled contribution to colonial development had Cabral encountered a land of fewer trees. As if to claim parity of worth with precious metals, the nature of sugarcane dictated that it be processed by fire; and unless the purifying flames were applied soon after harvest, the touch of the refined product would be significantly lessened. To make this rapid procedure possible, large quantities of firewood had to be within easy reach of the sugar mill's furnaces. Before the mid-eighteenth century, wood and its by-product, charcoal, were the only practical sources of fuel. Any attempt at sugar production without a ready stockpile of forested land would not succeed no matter how favorable other environmental factors such as climate and soil.

Despite the delight of early arrivals to Brazil's tree-blanketed northeast, the novelty of the bounty quickly receded. Residents of wood-rich regions often perceive timber, lumber, and firewood to be as common as oxygen in the air. But, in the way that an oxygen shortage causes desperate burning in the lungs, rapid destruction of Bahia's virgin forests forced millers and planters, shipbuilders and carpenters, brick makers and tanners, even common folk living in the city, to grasp for their share of a steadily diminishing resource.

Fuel Requirements at the *Engenho*

The Bahian sugar mill, known as the *engenho*, was simply the combination of apparatus by which planters turned cane into varying qualities of crystallized sugar. The engenho pressed juice from the cane, boiled it, purged it, graded it, and then boxed it for transport to the city of Salvador where it awaited the near-annual convoy arrival for shipment to European markets. The mill also represented a concentration both of capital (in the form of machinery, buildings, and slaves) and political power in the hands of the *senhor de engenho*, a proto-capitalist who not only held extensive cane plantings, but also held the sugar region's essential mechanical means of production. In the Recôncavo, a fertile crescent of land surrounding the Bay of All Saints, life, work, religion, trade, planter aspirations, and natural resources centered on the engenho, a quasi-urban, quasi-industrial entity on the rural, colonial landscape. The sugar mill fueled Bahia's economic and demographic growth, and Brazil's forests fueled the mill.[4]

Estimates of the expense of fuelwood as a commodity in sugar production vary according to the sources and periods analyzed. One study indicates that in 1752 firewood amounted to 10 percent of finished sugar's selling price; as a percentage of cost, the figure would probably be greater.[5] Another source reveals that fuel costs in Bahian engenhos were often much higher (see table 1).

Although fuel expenses never dominated engenho account books, the cost of fuel generally stood at about one-fifth of total operating costs. Fuel costs were probably even greater than some numbers in the table suggest. The mill could either employ its own slaves, oxen, and salaried labor in gathering firewood or it could purchase fuelwood, sometimes more cheaply, from those who cut the forests independently. Engenho Sergipe in 1611 and Engenho São Caetano recorded little or no fuel expenses yet both must have used fuel, so fuel-associated costs were not always properly categorized. These two mills ostensibly paid little to outside fuel suppliers, but they would still have engaged their own capital in the collection of local firewood. The higher transportation and livestock costs of the mills at São Caetano and Buranhaem, whose reported fuel expenditures are about half the average, support this contention. For any given volume of cane refined the mill required at least half as much fuel. Hence, the average oxcart or boat servicing mill that did not purchase fuel from outside sources could be expected to haul large quantities of firewood and to do so over greater distances than it did either fresh-cut cane or finished, boxed sugar. For many mills, transportation, livestock, and equipment costs unavoidably included fuelwood-related expenses, especially in cases where the mill used its slaves and other capital to gather wood, making total cost of fuelwood acquisition greater than the table reflects.

Labor in the Forest

Getting fuel to the fire was a labor-intensive operation. Initially, mill owners found abundant forests at their front doorstep, but the double

Table 1 Distribution of Bahian engenhos' annual expenses, 1611–1822, as percentage of total expenses

Mill	Years	Fuel	Salaries	Slaves	Food	Medicine	Equip.	Trans.	Livestock	Misc.
Sergipe	1611–12	1.4	27.0	4.3	16.0	1.3	30.0	3.0	–	4.7
Sergipe	1634–35	26.0	33.0	4.1	6.0	–	24.0	3.0	–	3.1
Sergipe	1643–52	20.0	26.0	8.2	7.9	–	35.0	1.1	0.2	–
Sergipe	1669–70	18.0	14.0	13.5	3.8	–	17.2	–	–	–
Sergipe	1707–16	20.1	20.5	18.7	13.1	0.8	8.3	8.0	0.8	10.5
Lages	1711–1800	19.0	14.4	14.6	30.0	2.5	13.5	1.0	3.2	1.4
S. Caetano	1726–1800	–	13.4	12.3	30.5	0.6	14.7	10.0	15.4	3.1
Avg. Mill	1751	11.8	20.6	18.9	11.3	3.1	28.0	–	9.6	1.7
Buranhaem	1796	8.5	12.1	10.7	21.4	3.9	11.6	5.9	18.8	7.2

A dash signifies that account books for particular years do not utilize the category in question and may, or may not, include them under another category. The author carried over to the table any errors in the original documents.

Source: Stuart B. Schwartz, *Sugar Plantations in the Formation of Brazilian Society: Bahia, 1550–1835* (New York: Cambridge University Press, 1985), p. 222. Data that Schwartz consulted for Engenho Sergipe are from wider sources than data published as *Documentos para a história do açúcar* (Rio de Janeiro, Brazil: Instituto do Açúcar e do Alcool, 1956) and to which many researchers have had access. Schwartz's data include information drawn from the Arquivo Nacional da Torre do Tombo, Lisbon, Portugal, and the Archivum Romanum Societas Iesu, Rome, Italy. Statistics for Engenho Sao Caetano are from the Arquivo Distrital de Braga, Coleçao Sao Bento, Portugal. Statistics for Engenho Buranhaem are from caixa 46, pacote (1802), Arquivo Nacional do Rio de Janeiro, Rio de Janeiro, Brazil.

onslaught of land clearing for agriculture and timber felling for fuel and construction soon pushed this important resource far from the mills' easy reach. The competitive advantage switched from mills having local fuel sources to mills located at the water's edge where firewood from other locations could be easily delivered. In many instances fuel traveled long distances over water to the mill. The forests of Jaguaribe, located more than seventy kilometers south of mills on the bay's north shoreline, were a primary fuel source. The Jesuit André João Antonil suggested in 1711 that Jaguaribe's forests alone could supply sufficient fuel for all the mills of the Recôncavo at water's edge.[6] Just how crucial Jaguaribe's forests were to sugar production was made evident in 1671 when a general Indian rebellion expelled the colonists from many southern areas including Jaguaribe, Cachoeira, Camamu, and Maragogipe. Author and inventor Juan Lopes Sierra noted the impact of expulsion: "Everything dealing with provisions was missing throughout the south, and the northern part lacked the necessary items for the sugar mills such as wood, forms, bricks, tiles and crates. Because of this the mills had to close. With production halted, commerce ceased."[7] Even before the 1670s, many mills at the bay's shoreline depended on fuel sources far from their base of operation. José da Silva Lisboa, the colony's inspector of the royal forests in the late-eighteenth century, added as sources of fuelwood Itaparica and its neighboring isles located northeast of Jaguaribe in the southern part of the bay—"a voluminous article of commerce" that the islanders sold to sugar mills as well as to the citizens of Salvador.[8]

Fuelwood, from its felling to its fiery destruction, might require as many as four modes of transportation, consuming significant outlays of labor, time, and capital. The first and most dangerous movement was from vertical to horizontal: the actual felling of the tree. Male slaves, largely those newly arrived from Africa, throughout much of the year felled by ax centuries-old hardwoods, under the watchful eyes of foremen. In Jaguaribe the work commenced in July, for August was the beginning of the annual cane harvest, which often stretched into May of the following year.[9] It is possible that fuelwood harvesters in the North practiced the same mass felling technique that future coffee planters in the South employed. A large number of trees on the side of a hill, all standing in close proximity and often tethered to one another by limbs and lianas, would each be cut to within inches of felling and then, finishing off the largest, all felled together with a domino effect. "It was the foreman's task to decide which was the master tree, the giant that would be cut all the way through, bringing down all the others with it. If he succeeded, the entire hillside collapsed with a tremendous explosion, raising a cloud of debris, swarms of parrots, toucans, songbirds, and, from the woodsmen, a shout of joy and relief."[10] If the foreman failed, the woodcutters entered a precarious and unsettled environment to topple what remained of the wavering widow-makers.

If the trees fell any distance from a waiting oxcart they were moved in their entirety by any of several methods, depending upon the terrain. On relatively level ground the limbed logs were rolled on poles or skidded behind oxen; however, in most other cases, whether it was uphill or down, slaves dragged the heavy timber with ropes and chains to points where oxen could be hitched to the logs. During the rainy season the slippery mud made for easier pulling, but once the trees were at roadside, bucked into manageable lengths, and loaded into oxcarts, the mired roads threatened the life expectancy both of the oxen and the carts. One of the colony's English residents, Henry Koster, maligned the oxcarts, which he asserted too frequently turned over due to their narrow wheel gauge; some planters worked their tongue-dragging oxen literally to death in the harsh task of hauling firewood.[11]

At land's end workers removed the firewood from oxcarts and carefully stacked small boats with the proper proportion of large, medium, and small pieces for efficient firing. From there the fuel sailed straight to the mill, provided the mill was at bayside. If not, there was at least one more transportation break.

Antonil described a number of independent persons who divided this immense labor among themselves.[12] Professional woodcutters worked their own slaves, but may not have owned oxen and carts for transporting their goods and instead hired independent teamsters for that purpose. Men called *arrais* often privately owned and managed the sailing vessels, but millers also employed sailors, slaves, and their own boats in order to buy fuel direct from the woodcutters and avoid paying freight charges, which escalated as the forests receded. Antonil recommended that the senhor de engenho have at least two vessels so one could return while the other set sail, testimony to the distances traveled. Mills that still had locally accessible firewood probably worked their own human capital, when available, to feed the furnaces.

Both cane and fuel were measured by the *tarefa*, a unit of volume equal to that necessary for a day's production at the mill. The tarefa of firewood equalled eight cartloads, and the tarefa of harvested cane was twenty-four cartloads.[13] While initially each mill probably consumed only one tarefa (eight carts) of fuel per day, by 1583 Fernão Cardim reported that Bahia's increasingly powerful mills required as many as twelve carts for a sixteen-hour day of milling; in 1710, Antonil explained that between twenty-five and thirty carts of cane were milled per day, requiring twelve to fifteen carts of fuel; the academic Vilhena, writing in 1802, calculated that a mill converted to an efficient furnace would consume between twelve and sixteen carts of fuel each twenty-four hours. Vilhena complained that ignorant and obstinate senhores de engenho nevertheless still used antiquated furnaces that consumed no less than one cartload per hour.[14]

In the harvest of 1650–51, the Jesuit-owned mill Sergipe de Conde milled sugar 224 days out of the total 310-day harvest (stoppages and the observance of religious holidays account for the 86-day difference).[15] Sixty years later Antonil reported that *engenhos reais* like Sergipe de Conde consumed in their six fiery furnaces 2,500 cartloads of firewood per harvest.[16] This means the mill consumed eleven carts per day.[17] Lay planters did not observe Catholic holidays as rigorously as did the Jesuits, meaning that their average number of milling days and total consumption may have been higher. Even if the mills observed by these contemporaries were somewhat larger than the average, a reasonable figure for the colonial period would be eight to ten carts of firewood per day per mill. Since employers expected every ax-wielding laborer to stock one full cartload of fuel per day, this would suggest that every operating mill employed between eight and ten laborers in cutting, dragging, and stacking firewood. It required forty oxcart trips from the forest edge to fill the hold of a large sailing bark, which had a capacity of five tarefas. Whatever the number of carts making those trips, two skilled teamsters manned each. As many as four sailors and a master manned the boats themselves, which "follow one right after the other, without halt."[18] The number of hands required for 36 Bahian mills in 1583, 63 mills in 1610, 146 mills in 1710, and 180 mills in 1758 illustrates Vilhena's simple statement that "[i]n the woods significant numbers are required to cut [and haul] firewood."[19] Based on these estimates, in 1758 Bahia's mills consumed more than 3,300 cubic meters of fuel per day (18.3 cubic meters per mill per day), involving close to 4,000 men in its harvest and transport during much of the year. This labor force drew from a total population of 72,833 in the sugar-producing Recôncavo. During one year's harvest, the Recôncavo's mills consumed approximately 750,000 cubic meters of fuel. None of these calculations include fuel consumption for firing clay, distilling rum, cooking meals, tanning leather and furs, or producing charcoal. Sugar production alone pushed the forest back at a breakneck pace.

Labor at the Engenho

The life of the sugar mill stemmed from its consuming center. Fire was so integral to the workings of the engenho that Brazilians employed the epitaph *"fogo morto"* ("the fire is out") to identify inoperative mills, no matter the reason for their expiration. One colonial cartographer indicated Bahian mills on his map with the mushroom-like symbol of rising smoke. Antonil described the sugar mills' multiple furnaces as "truly devouring mouths of woodland, a prison of perpetual fire and smoke, a living image of the volcanoes Vesuvius and Etna, and almost of Purgatory and Hell themselves."[20]

Hell it was for the poor African souls allotted the task of keeping the fires stoked; tending fire frequently came as an assignment for medical and penal reasons. The sick, who might benefit from a good sweat, and the chained, intractable slave were both sent to the mouths of the furnaces in hope that the heat of the flames and the humid Bahian summer would remove from their constitutions whatever it was that adversely affected their productivity. One contemporary believed that stoking the fires along with hydrating clay constituted the worst jobs at a mill because these were the only individuals who could not secretly dip into the sweet product of their labors as a way to fend off hunger. A fire tender likely had much more with which to concern himself. Besides the skin burns that radiant heat and ballistic embers caused there was the possibility of falling into the furnace, although the construction of some mills minimized this danger. An insubordinate slave at the fire door was chained as much to keep him from running off as to deny him even the freedom of taking his life in the flames, an event that occurred on at least one occasion.

Building the fires was a job that required some architectural skills. The average six furnaces, one per kettle, were arrayed along the boiling house's outer, basement walls. Since the floor of the furnace could be a significant distance from the bottom of the kettle on the floor above, a structure of firewood spanned the gap (see figure 3). Using a long pole called a *trasfogueiro,* workers placed large logs crossways on top of each other, supporting gradually smaller pieces up to the top. In the early stage of boiling, the more fire the better, but the process became more delicate as the semi-purified juice moved to new receptacles, and the fire tender had to pay strict attention to his superiors. Too little heat and the juice would not boil; too much and the kettleman had to throw cool water in the vessel to keep the precious fluid from boiling over. If the foreman above and the slave below maintained proper temperatures, the resulting sugar was of higher quality and could be finished more quickly.[21]

Some smaller fires were apparently designated to burn softer hardwoods, including white mangrove and cashew among others, for although these woods had lower heat potential they rendered ashes that were essential in making lye *(decoada)* to further purify the cane juice. Important enough to be saved from one harvest to another, workers raked ash from the furnace floor and placed it in a heated brick recess from which it could be taken as necessary. Once the ashes had been shoveled into smaller, perforated tubs, boiling water was filtered through the ashes and then mixed with the juice boiling in the cauldrons.[22] The ashes of particular tree species were regarded as essential to the processing of sugar for they were key to its purity and whiteness, the qualities that fetched the highest prices. Senhores de engenho included this fact in arguments for preservation of the forest.[23]

Competition for a Dwindling Resource

As an essential, expensive, and increasingly scarce commodity, competing mill owners frequently argued over fuelwood. Although blood ties often connected these lords of agriculture, such links did not include the bonds of charity. The removal of one

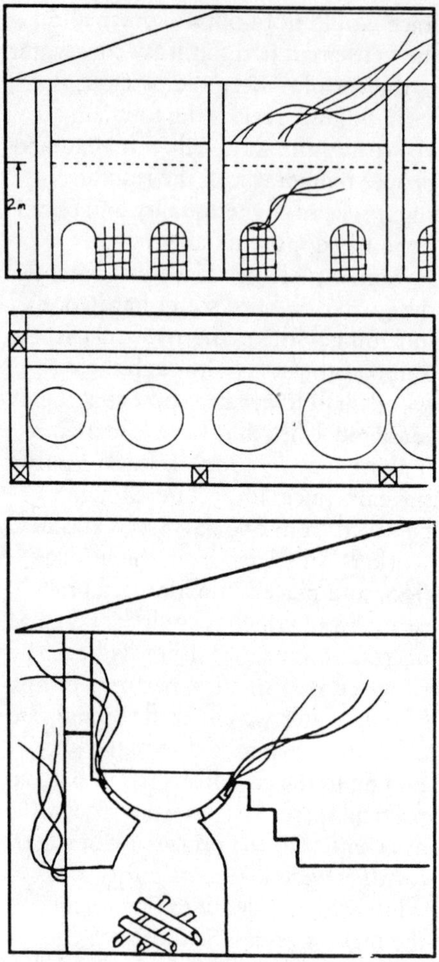

Figure 3 *Plan of a flueless eighteenth-century furnace house drawn in three elevations: front, side, and top. All drawings by the author.*

Top *Front view of the furnace house, showing the grated fire door of each furnace and the open terrace above. Note the great distance between furnace floor and kettles.*

Middle *Top view demonstrating the arrangement of the large copper kettles over each furnace. Larger mills utilized six kettles, the juice passing from the largest to the smallest during processing.*

Bottom *Cut-away side view of the furnace house, showing the internal dimensions of the furnace, the absence of a flue, and the working floor.*

length of timber or the trespassing of one wayward, cane-trampling ox could throw cousins into sometimes bloody dispute.[24] Foremen who oversaw plantings and cuttings on the fringes of the mills' property had the added responsibility of defending both crops and fuel from sneaking neighbors. A 1690 altercation in which familial ties had no mollifying influence took place when two Jesuit priests of the Engenho Pitanga discovered a group of men from António da Rocha's mill in Jesuit woodlands. When the priests attempted to obstruct the depredations the mob verbally insulted them and beat them with the stolen goods.[25]

As pockets of forested land diminished in size and increased in distance from the mills, the scale of the fuelwood conflict escalated. By the mid-seventeenth century, however, mill owners turned to more formal means than theft to acquire what remained of an increasingly costly natural resource. In 1679 Pedro Garcia Pimental received special attention from the Crown concerning his exclusive right to exploit the woodlands that encircled his mill at Cahicabo. The prince-regent, soon to become Dom Pedro II, learned that certain heirs to those lands were planning to construct a mill not far from Pimental's, "which would be greatly prejudicial to both as it is not possible that the site has enough firewood for them, and from which act damage to the conservation of that state would ensue." Reminding his royal officials about related legislation his predecessors had passed, the regent forbade construction of any mill within a one-league circle of Pimental's mill and ordered that no tree could be removed from the woods of that prescribed area for any reason, unless those doing the clearing sold the wood as fuel to the Pimental mill at the current market price. This order insured that the surrounding forests would be used only for fueling the mill.[26]

In July 1680 the regent received a formal list of grievances from the municipal council at Salvador concerning the deleterious effects the large number of mills had on the sugar industry as a whole. The council confirmed that there was not enough fire wood. Once cut, forests could not be replenished in less than twenty years, and trees required considerably more growing time than that if mills were to continue burning fuelwood in the sizes and quantities they were accustomed to using. Although the council and residents were aware of royal decrees setting minimum distances for new mills from neighboring mills, arguments over what that distance was and whether the space had to be wooded caused serious quarrels and lawsuits. The council's recommendations to the Crown fell firmly behind the senhores de engenho. Existing sugar mills, they said, should have precedence over upstart neighbors in the exploitation of firewood within one full league of their furnaces, for "it is more useful...to conserve one [engenho] for many years than to lose two in a brief time."[27]

On 3 November 1681 the regent, paraphrasing much of the council's letter as justification, prohibited construction of new mills within one-half league of existing mills.[28] That he cut in half the suggested one-league law is evidence that the king received conflicting pressure. Less than three years later the Salvador city council, for reasons unexplained, reversed its position. Their letter dated 5 August 1684 claimed that the woodlands of the Recôncavo were such that mills could often be established closer even than half a league without the slightest damage to production. The opposite was the case, they argued, for if one mill owner could hoard many acres of unused forests, less sugar would be produced and at greater cost.[29]

The desire of the planter class to escape their dependence on the senhores de engenho, whose monopoly over milling services promoted occasional high-handedness, was a primary stimulus to sugar mill proliferation. Despite the huge initial outlay in capital (slaves, waged employees, milling machinery, and transport vehicles) and the difficulties inherent in mill ownership, many planters had both the wealth or credit and the ambition to pursue milling operations on their own land. Hence the ongoing battle over how many mills could be built in a limited area was in part a contest between rival classes; senhores de engenho attempted to bar planters from acquiring any means of production beyond the land's inherent fertility. Declaring a dearth of firewood when new mills were to be built too close to their own was a means to that end.

The Impact of Fuel Scarcity

Production figures seem to support the mill owners' appeals. Brazil's three hundred working mills squeezed two million *arrobas* of finished sugar from the 1650 harvest. Nevertheless, in 1670, even with an increase of two hundred mills, the colony produced about the same volume.[30] While mill proliferation might have promoted a more equal distribution of income among the landed elite and ostensibly expanded productive capacity, actual production stagnated. Fuel scarcity was not necessarily a limiting factor in Brazilian sugar production at this point, but fuel shortages were significant enough that senhores de engenho somewhat successfully invoked this concern as a weapon in defending their interests. Planters during 1650–70 invested in processing capacity rather than in expanding cane plantings.

Statistics from the 1650–51 harvest at the Jesuit-run Engenho Sergipe de Conde, however, shed light on fuelwood scarcity's impact on sugar refining. Refining was interrupted on eighty-three days of the 310-day harvest period from August to May. Sixty of those interruptions were due to Sabbaths and religious holidays; the remaining twenty-three days of inactivity were divided among various causes: bad weather accounted for three days, five days were devoted to mechanical repairs, a lack of cane stopped the mill once. Fuel shortage was the reported cause for shutting down on fourteen separate days.[31] Insufficient firewood, in this particular mill and during this particular year, accounted for 61 percent of unwanted, reported production stoppages. There is no way to discern what caused the fuelwood shortages. Potential causes included bad weather, insufficient labor, lack of foresight on the part of the sugar master, and inadequate carts and boats for transportation. But as the distance of the forest from the mill increased, as a growing population demanded more wood for construction material, cooking, and shipbuilding, the complications in keeping sufficient amounts of fuelwood at the mill increased. Fuel supply became far less reliable than cane supply.

The firewood shortage at the mill had less to do with overall forest reserves than with inability to move sufficient amounts of fuel to areas of sugarcane production. As late as 1807 the author Rodrigo de Brito commented that talk of wood being in short supply was fiction. Cutting wood was just part of plantation expansion. Yet he conceded the bigger reality: the price of wood would continue to increase as the distance between mill and forest grew more quickly than the forests could be replenished.[32]

Despite senhores de engenho's statements suggesting the contrary, the debate among planters, millers, and other interested parties over fuel reserves was not about the availability of fuel but its accessibility. Bahia's forests stretched for miles into the interior; there was theoretically enough wood for everyone many times over. The problem was getting access to that far-flung wealth. Trees any distance inland were as good as nonexistent. Unlike some areas, such as the United States where relatively light conifers could float down even unnavigable rivers, Bahia's predominantly non-floating hardwood trees had to be shipped from the river bank, thus limiting woodsmen to the distances they could navigate. Even if there had been more conifers in the forests of northeast Brazil to float downriver, these would not have served as well for fuel.

Nature produced a wealth of forest resources, and generally the only cost to those who needed those resources was labor. The double benefit from minimizing labor and using the cheapest modes of transportation suggests that forests receded from the edges of navigable water more quickly than they receded from the mills themselves. Hence, most mills at water's edge, even before they exhausted the woods on their own lands, bought wood more cheaply from those who cut and hauled it from the banks of the bay and rivers. Attempts to limit the construction of new mills to a distance of one league may have had the additional intent of barring aspiring planters' access to water. Mills in the interior, however, while relying heavily on expensive oxen and manpower to move their cane, wood, and finished sugar, also had advantages. Interior, untouched lands were reportedly more productive, and distances to fuel sources were invariably shorter.[33]

Even when true scarcity threatened the viability of the sugar trade from time to time, senhores de engenho were loath to invent, adapt, or adopt new milling methods. It took centuries before a critical shortage of fuelwood forced millers to take note. Bahian millers then benefited from the experience of their competitors in the West Indies, who suffered from a far earlier and more severe depletion of fuel resources and therefore devised more efficient and alternative means of fueling their mills.

Part of human existence and survival involves predation on the surrounding environment by the economic system hunter, gatherer, or capitalist. And despite the importance of certain cultural and religious factors in conditioning the intensity and extent of human impact, the major determinant is population density. Even before the expansion of its culture, Europe's economic philosophy—whether in farming, mining, irrigation, or logging—had been to exploit natural resources to the ends that technology permitted. Native American cultures also exploited nature; the difference resided first in numbers, but also in the exploiter's intentions. Where Brazil's indigenous peoples were unanimously satisfied with daily subsistence, Europeans imported the idea that economic growth was a worthy end in its own right. Colonists could not, with satisfaction, simply produce for themselves what was meet for the day at hand. Bahia's sugar mills, like those of the West Indies or other Brazilian provinces such as Pernambuco and Rio de Janeiro, would single-handedly attempt to satisfy the world. But the growth of sugar production could not be sustained without technological progress. As has happened many times since in Bahia, technology ran up against scarcity, and either new technology had to be developed to leap scarcity's recurring obstacles or sugar interests would have to face the agony of economic stagnation. Nature's limitations had the final word: her many gifts were finite. Colonial Brazil's sugar producers faced only one option, to use what remained more efficiently.

In the bounteous land of colonial Bahia efficiency was a concept almost unheard of when it came to using natural resources. Planters and millers alike were interested in improved technology, but improvement meant expanding sugar production, and this was accomplished by adding more slaves, more acreage, and by squeezing more cane juice with more powerful mills. Although firewood amounted to a considerable portion of mills' total costs, little was done to use it prudently until the early-nineteenth century.

The Roots of Fuel Scarcity

The most extensive firewood destruction occurred even before the first cane cuttings were planted, when planters cleared the land. José da Silva Lisboa explained that the *massapé,* the black, unctuous soil that turned to a quagmire in rain, also produced the best cane yields, attracted the highest land prices, and had the best quality of trees. This exceptionally fertile soil was both the origin and result of the trees it anchored. The accumulation and decomposition over many centuries of leaves, downed timber, and other organic material on top of a layer of clay formed a coveted soil type for the perennial sugarcane. Silva Lisboa claimed that in some places this ground had been planted profitably for over sixty years without lying fallow or adding fertilizer; but he also observed that once the fertility diminished, a planter had to wait years for the land to recover.[34] Most planters simply moved to virgin, fertile forest instead.

Land clearing and fuelwood gathering were not complementary jobs. As Lisboa noted, "The planting of cane commences with the cutting and felling of the trees, if the plantation is made on new ground. When the forest is virgin, composed of woods that are enormously thick and large, if there is the possibility for them to be sawn, they are made into planks for sugar crates; otherwise all is reduced to ash....The smaller, remaining firewood is piled into mounds called *covairas* and fire is set to them until all the wood is consumed."[35] Forested land was burned over so that it would produce to planter expectations, although the practice wasted expensive firewood. Nevertheless, because of this burning the mills' fuelwood resources disappeared as cane plantings expanded into new areas.

Figure 4 *Large team of oxen hauling a cartload of logs, not necessarily for fuel. Note the smoke rising on the horizon. Drawing by Percy Lau reproduced with permission from the Instituto Brasileiro de Geografia e Estatistica, Rio de Janeiro, Brazil.*

The condition and form in which firewood was prepared for the furnace also contributed to the inefficient use of wood resources. Contemporaries observed that it was common to burn green or unseasoned wood. It is generally recommended that timber cut for firewood should be stripped completely of its bark and left to dry in the open air at least two to three months. At the peak of the harvest, when milling often halted for occasional interruptions of firewood supply, the mill did not wait even an extra day beyond the delivery date for fuel to dry. The caloric potential of green timber, containing from 60 to 90 percent moisture, might be compared to that of a wet, tightly rolled newspaper. A fresh-cut log with the minimum 60 percent moisture content provides less than 80 percent of the useful heat created by an air-dried log of 20 to 30 percent moisture content.[36] Every degree of heat that the furnace applied to boil moisture from the steaming, hissing wood represented energy that could not be applied to the cane juice it was intended to boil. Mills that stockpiled quantities of firewood during the off-season may have had fairly dry wood to fuel the furnaces during the early part of the harvest, but such stockpiles would have been exceptional.

In addition to burning green wood, mills burned wood in large pieces. Vilhena estimated that many logs put on the fire wholly intact were as large as .66 meters in diameter; this corroborates Sampaio e Melo's claim that some fire logs were so enormous that a man could not put his arms around them.[37] A piece of wood burns more efficiently as the ratio of oxygen-exposed surface area to volume increases. The circumference of a large round log exposes the smallest surface area possible, thereby impeding the critical processes of drying and pyrolysis.[38] Split firewood would obtain better heating and fuel-use efficiencies. However, mill operators likely considered splitting an operation too labor intensive due to the daily volumes involved.

The inefficient construction of the furnaces themselves added considerably to resource waste. Until the late-eighteenth century, mill operators largely ignored the fuel-saving claims of furnace inventors. In 1656, Juan Lopes Sierra claimed his new furnace would save one-third the fuel that other furnaces consumed and offered his patented plans for a price of 100 *milréis*. He reminded millers of the costs involved in gathering firewood, including the great distances traveled and the number of slaves required.[39] Most mill owners, either out of ignorance or a lack of faith in the claims of such entrepreneurs, did not invest in either the plans or the new construction (see figure 3, p. 185).

Throughout the colonial period, chimneys and flues that would direct the application of heat were almost unknown. Smoke, heat, and flame escaped through any opening in the furnace. As late as 1802, Vilhena described operating mills in which the fires were 2.2 meters below the kettle bottoms. Newer mills, at the time of his writing, had reportedly cut that distance in half, also cutting wood consumption. Fourteen years later, Koster described the sugar refining process as sloppy: "Ovens [are poorly made,]...enormous quantities of fuel are consumed, and the Negroes who attend to ovens are soon worn out." He thought the operations slovenly, with owners paying little attention to the "minutiae of business."[40]

During the first half of the nineteenth century millers showed greater interest in furnace innovations, suggesting that the costs of fuel finally had a recognizable impact on prodigal practices at mills. The Bahian Recôncavo had few rivals with such profitable conditions for growing sugar, and for this reason Brazil was late to adopt specific improvements. The extremely fertile *massapé* soils; proximity to the source of African slaves; long harvest season; vast forest reserves; a large, sheltered bay; and an extensive system of navigable rivers on which to transport goods and people cheaply gave Bahian sugar planters sharp advantages over their West Indian counterparts. Yet these advantages may have been deterrents to progress. Sampaio e Melo, in 1812, explained:

> The quantity of firewood that abounds in this country has been cause for the inhabitants not to study the merits of economizing it; but now that the forests are lengthening themselves from the *engenhos*...it is necessary to study this branch of physics which in the northern countries has made great improvements even in the kitchen's fires....If abundance has been the cause of our tardiness, necessity begins to cause our industry. Firewood has increased greatly in price in the past twenty years in this region; and if we do not imitate the industry

of the inhabitants of Barbados and Martinique, our sugar will not be able to compete in price with theirs in Europe.[41]

Through much of the colonial period, nature's abundance was the strongest argument against attempts at conservation.

Buying into the Engenho da Ponta in 1806, Manoel Ferreira da Camara, well-educated, liberal, and a colleague of Brazil's leading politician José Bonifácio de Andrada e Silva, revamped his furnaces in such a way that he saved two-thirds of the average fuel consumed.[42] Although there is no record of the actual changes made to the boiling apparatus, Ferreira may have adopted the Jamaica train system that had been earlier developed and accepted in the West Indies.

The Jamaica train, or any of its many variously named versions, had several advantages over older furnaces. As opposed to the old Spanish train configuration in which each kettle had its own fire, the Jamaica train had only one fire and its heat was directed by means of a flue over an array of kettles. Although the process was slower than the old process, significant savings in firewood convinced senhores de engenho to adopt it. More importantly, the Jamaica train burned bagasse—the dried cane husk wrung of its juice—as well as wood, or a mixture of both. Prior to this, mills either threw bagasse into the water, clogging shipping lanes, or spread it over empty fields to be dried and burned as a fertilizer. Although its use created new problems for the mill and even influenced the varieties of cane grown (some were more volatile than others), millers had an alternative fuel to the increasingly costly wood.[43]

Other Fuel Consumers

If sugar mills had been the only firewood consumers in the Recôncavo the intensity of local conflicts and deforestation might have been less notable. But there were other demands on the forest that competed directly with people whose interests were tied to sugar. Salvador's city dwellers faced the same fuel scarcity and price increases as their rural neighbors. In 1672 city residents could not gather even a small bundle of firewood near the city because private parties owned all the nearby wooded land. The city council suggested that wooded properties near the city should be bought, even if a specific tax had to be levied, to remedy this inconvenience.[44] Estimated late–nineteenth-century per capita annual fuel consumption was approximately two cubic meters, and it is likely that this consumption rate had changed little since the late-colonial period.[45]

The mills' ancillary operations themselves were often the most serious competitors for declining fuel resources. Most mills took advantage of sugar processing by-products to make rum for consumption both locally and abroad. In 1759, José Antônio Caldas counted seventy-one distilleries in the Recôncavo providing three different grades of spirits.[46] Some, especially clerics, wrote of the moral consequences that this drink had on the population; others warned against being too liberal with alcohol for slaves. However, relative to the furnace house, these operations and their demands for firewood were small.[47]

Another fuel contender at the mill was the kiln for ceramic production. The Bay of All Saints had an ample natural supply of clay, and besides the well-known ceramic sugar forms whose conical shape gave the name "Sugar Loaf" to more than one mountain, the engenho needed a great deal of brick and tile. These items were used in the walls and extensive roofs of the furnace and purging houses, the chapel, and the master's *casa grande*. Mill owners constructed the furnaces themselves using brick and mortar for the lower walls, but the bricks in the furnace's upper reaches were set in clay because intense heat made mortar friable.

Gabriel Soares de Souza, writing in 1587, gave the impression that all mills had their own wood-burning brick kiln, but there was some debate about the cost effectiveness of these kilns. Some people concluded that the kiln required too much large firewood, which included that of the mangrove in whose roots bedded oysters, slaves' favorite food. Antonil suggested that the pottery works required the labor of six slaves and were so expensive that buying from other sources would be advisable. Antonil concluded that "if the senhor de engenho has many people, much firewood and mangroves in excess for oystering, he can also have a pottery house, and this workshop will serve to increase the greatness, utility and convenience of the mill."[48]

Other sectors of Bahia's economy also relied upon the forests for fuel. Even before the arrival of the first royal governor in 1549 there was a lime kiln at Bahia. In 1550 Afonso Jorge received $800 *réis* for his employment there, and Estevão Fernandes was paid the sum of 5$160 réis for supplying it with firewood.[49] Again in 1587, Soares de Souza wrote of three lime producers on the island of Itaparica where quarried chunks of limestone were placed in vaulted kilns and pulverized by the heat from large wood fires burning "day and night." When oyster shells could be easily accumulated, these were the preferred source of lime because they required less fuel per pound of product.[50] As the colony grew, demand for lime increased.

Although there are scant references to charcoal and its use in Bahia, there is evidence of a sizable trade from as early as 1549. Charcoal, used in smelting and blacksmithing, was made by burying closely stacked wood underground or in earth-covered piles and setting to it a slow, controlled burn. In this process of distillation wood lost half its volume and half its total heat potential but gained the advantage of burning both hotter and cleaner. Between April 1550 and August 1552 five named individuals delivered a total of 2,051 bags of charcoal to the city of Salvador's smithies.[51] According to an established contract, the city paid each *carvoeiro* (one who produces charcoal) thirty-five réis per bag. Of the five named, some were also listed as lumber dealers and canoe makers, suggesting that the requirements of gathering and processing wood for charcoal allowed sufficient time for other related economic activities.[52]

In the early-eighteenth century

leather goods became a major factor in Bahia's yearly exports, requiring large amounts of the forest's organic fuel. Salvador's hinterland, especially that bordering the cattle trough of the San Francisco River, produced numerous herds used for traction (turning mills, hauling carts), meat, tallow, and shoe and crude leather. In 1757 Bahia exported over ninety-five thousand pieces of leather and 14,585 bales of tobacco tightly sewn into sizable hides. So important were trees to the leather industry that in 1760 cutting mangroves for fuel was prohibited by law in many provinces because such activity caused harm to hide processors who relied on the tree's bark for tannic acid. Anyone who cut those trees that had not been peeled of their bark faced a large fine and three months of imprisonment. Cattle hides, which like cut sugarcane decayed if left untreated, were turned into durable leather by boiling them in a tank of water with the mangrove bark. In 1746, Pernambuco had twenty-seven separate tanning operations employing more than three hundred slaves, many of them undoubtedly peeling bark and cutting firewood. Although there are no similar records for Bahia, the numbers could be even greater since Salvador annually exported more leather than any other port.[53]

In the early-seventeenth century, Bahia established the colony's first whaling industry, and although overfishing of the bay and local coasts quickly removed the captaincy from its preeminent position, the processing of whale oil remained a significant employment until about 1825. Even in 1802, Vilhena recorded at least two large whale oil refiners in Bahia, one at the Ponta da Armação das Baleias on the northern tip of the island of Itaparica, and another on the beach at Itapoan on the Atlantic coast. The German scientist Karl Friedrich Philipp von Martius gave the impression there were smaller operations in the vicinity as well.[54] The fuel necessary for reducing wale blubber, at least in the case of the Itaparica armação, relied on the same sources as did sugar mills.[55]

Shipbuilding, although never requiring large amounts of firewood, elicited complaints from colonists who saw Salvador's royally administered shipyard as a detrimental competitor for highly valued hardwood species *sucupira* (*Bowdichia virgilioides* H.B.K). By 1587 Soares de Souza counted more than fourteen hundred watercraft on the bay, one hundred of them more than ten meters long, a number that still insufficiently served the needs of commerce.[56] Most of these boats and the best oxcarts, milling machinery, and bushings were made of *sucupira* and had to be frequently repaired and replaced. Hence shipbuilding competed for firewood indirectly by competing for the materials required to build small boats and carts that transported fuel to the mills. Planters and millers also often hauled timber to the shipyard, which lacked its own means of transporting primary materials. In 1704 Salvador's municipal council informed the king that the practice of building large ships in Bahia (six galleons in the past seventeen years) had caused millers, sugar planters, and tobacco planters "general ruin." As they had done previously, they petitioned the king to move the royal shipyards south to Ilheus, Camamu, or Rio das Contas where wood was still abundant and there was less danger of damaging the local sugar economy.[57]

Bahia also exported a significant amount of lumber and a variety of wood products. Although most of these were species specific, wood in the holds of Portuguese ships bound for Europe, the West Indies, or other parts of Brazil could not be used to fuel the fires of sugar mills. Little is known about the extent and value of the Brazilian lumber trade outside of brazilwood, but the inventories of goods shipped out on fleets bound for Lisbon provide some indication. Of the twenty-nine items that José António Caldas listed as cargo on the fleet of 19 November 1757, twelve were wood products, ten were packaged in wood (crates and barrels of varying sizes), and four were processed by the application of heat from wood fires. Whatever the total commercial value of the lumber trade, the volume was substantial. Trees left Brazil in significant numbers as sawn lumber, protective packaging, and spent heat (see table 2).[58]

Even Portugal used wood from Bahia's forests to fuel her fires, at least by the late-eighteenth century. The same fleet in 1757 carried 146,200 billets of firewood to Lisbon, and in 1796 Bahia exported at least 14,875 pieces to Lisbon at the price of two réis each. Much of the latter shipment seems to have been destined for the royal kitchens and probably served as ballast on the voyage.[59]

Table 2 Exports of timber, containers, and fuelwood-dependent goods from Bahia to Portugal, 19 November 1757 merchant convoy

Timber products	
Unmilled flitches (dozens)	680
Planking (dozens)	130
Beams	235
Ship timbers	1,000
Carriage switches	37
Vine trellis poles	1,897
Billets of firewood	146,200
Planks of *vinhatico*	180
Planks of *piguia*	28
Roof timbers	615
Logs of cabinet wood	
Jacarandá	356
Sebastiao de Arruda	3,068
Wooden containers (varying sizes)	
Barrels	1,271
Crates	12,756
Fuelwood-dependent products	
Units of sugar	12,751
Bales of tobacco	143
Units of leather	89,177
Furs	5,058

Note: Quantities not otherwise designated are counted in units of "each."

Conclusion

In the Bahian Recôncavo, as in all of Brazil's sugar producing regions, the extraction of firewood was an operation extensive in scale and of widespread concern. Even in a tropical climate where domestic heating had almost no impact, fuelwood was among the colony's most crucial economic products. Through most of the colonial period, fuel acquisition for Brazil's fuel-dependent enterprises was generally never more than a nuisance. But by the end of the eighteenth century senhores de engenho, tanners, whalers, and distillers alike began to experience the crisis West Indian colonies had suffered a century before. Although Brazil's sugar industry commenced a full century before its West Indian competitors, the sheer extent of Brazil's Atlantic forests averted the exigency for two and a half centuries. Firewood's increasing scarcity impinged upon the scale and direction of sugar mill proliferation, multiplied the costs and complexities of production for all fuel-dependent entrepreneurs, and even dictated the articulation of the terms in which social conflict played out.

The impressive scale of fuel demand also affected the colonial division of labor. The planting, caring for, and harvesting of sugar cane on the colony's coastal lowlands abetted the importation of millions of slaves to Brazil, the majority of all Africans who arrived in the Americas.[60] But not all slaves received a machete or hoe. The ax played as large a role as any tool in the ongoing attempt to make the colonial venture a success.

In his 1944 study of the state of the sugar industry in Bahia, Adrião Caminha Filho marveled at the colonial mansions that had been financed by the old sugar mills and concluded that "without doubt, the status of the past mills was far more prosperous than that of the present, modern sugar factories." In addition to exhausted soils and cane blight the "lack of fuel is one of the most serious difficulties of the local industry," and many mills were forced to stop constantly as a result. Echoing his predecessors' oft-repeated concerns, Caminha Filho declared that "there are no longer any forest reserves in the Recôncavo, and the situation is so grave and offers such serious prognostications for the industry that it requires a very careful study."[61] Brazil's eminent sociologist and geographer, Gilberto Freyre, writing about the same time, reminded those concerned that "the warning call of the first governor of Pernambuco against the devastation of the forests remained a voice crying in an ever increasing desert."[62]

The Recôncavo's wood resources continued to decline through the nineteenth century as they had done the previous three centuries. The succeeding generation of planters inherited the legacy of their parents and made little or no effort to preserve or renew accessible sources of fuel. Despite the crucial relationship between the colony's prosperity and nature's well being, smoke-filled skies continued to obscure the vision of a less promising future.

Notes

1. The official name of the city is Salvador (Saviour), but in their correspondence kings and viceroys referred to the city as Bahia (Bay), also a common practice today. I refer to Bahia in this article, except where there is ambiguity as to whether I mean the city or the province Bahia. Bahia remained the capital until 1763, when Rio de Janeiro became the capital.

22 **Shawn W. Miller**

2. Luís de Camões, *Os Lusíadas* (Porto, Portugal: Porto Editora, 1992), chap. 5, v. 5. However, sugar production had nearly completed Madeira's deforestation by the time of Camões's passage.
3. For the causes and state of deforestation in Portugal, see José Bonifácio Andrada e Silva, *Memoria sobre a necessidade e utilidade do plantio de novos bosques em Portugal. . .* (Lisboa, Portugal: Academia Real de Sciencias, 1815). The author is unaware of contemporary related research for Portugal or Iberia. For southern Europe in general see J. V. Thirgood, *Man and the Mediterranean Forest: A History of Resource Depletion* (New York: Academic Press, 1981).
4. Students of the colonial sugar economy have often mentioned the vast quantities of fuelwood that sugar production required, but rarely have they elaborated. See Stuart B. Schwartz, *Sugar Plantations in the Formation of Brazilian Society: Bahia, 1550-1835* (New York: Cambridge University Press, 1985), for the most comprehensive economic study in English to date. Schwartz's research substantiates fuel's importance but does not address its social and environmental implications. Vera Lucia Amaral Ferlini, *Terra, trabalho e poder: o mundo dos engenhos no Nordeste colonial* (São Paulo, Brazil: Editora Brasiliense, 1988), touches on the fuel issue but emphasizes what occurs above the flames. José Wanderley Araújo Pinho, *História de um engenho do Recôncavo, 1522-1944* (São Paulo, Brazil: Companhia Editora Nacional, 1946), is an early source with an excellent description of furnace innovations, but suffers from understatement of fuel's total cost.
5. Wanderley Pinho, *História de um engenho* (1946; reprint, São Paulo, Brazil: Companhia Editora Nacional, 1982), p. 245, n. 8. Pinho arrives at a figure of 10 percent by dividing the selling price of seven loaves of sugar (23$800) by the cost of the one tarefa of firewood (2$500) consumed in its refining. Colonial Brazil's basic money of account was the *réal* (plural *réis*). One thousand réis was called a *milréis*. The dollar sign's location three places from the decimal point was simply a means of punctuation, much as American English uses the comma in numbers beginning in the thousands. Thus, ten réis would have been written $010; one thousand réis would have been written 1$000. Brazilian coinage often referred to in primary sources (e.g., tostão, vintem, cruzado) was made up of réis, but the actual value changed from one century to another.
6. André João Antonil [pseudonym of Giovanni António Andreoni, S.J.], *Cultura e opulência do Brasil, por suas drogas e minas . . .* (São Paulo, Brazil: Companhia Melhoramentos, 1976), bk. 2, chap. 8.
7. Juan Lopes Sierra, *A Governor and his Image in Baroque Brazil: The Funeral Eulogy of Afonso Furtado de Castro do Rio de Mendonça*, trans. Ruth Jones (Minneapolis: University of Minnesota Press, 1979), p. 43.
8. José da Silva Lisboa, "Carta muito interessante para o Dr. Domingos Vandelli," (1781), *Annaes da Biblioteca Nacional do Rio de Janeiro* 32 (1910): 496.
9. Antonil, *Cultura e opulência*, bk. 2, chap. 7. Indian labor may have also been used for fuel cutting (as it was in timber extraction for lumber), but the author found no mention of it in Bahian sources.
10. Warren Dean, "Coffee Dispossesses the Forest," *With Broadax and Firebrand: The Destruction of the Brazilian Atlantic Coastal Forest* (Berkeley: University of California Press, forthcoming), chap. 8.
11. Henry Koster, *Travels in Brazil*, ed. C. H. Gardiner (Carbondale: Southern Illinois University Press, 1966), p. 169; and Luiz dos Santos Vilhena, *Recopilação de notícias Soteropolitanas e Brasílicas Contidas em XX cartas*, 3 vols. (Bahia, Brazil: Imprensa Official do Estado, 1921), 1:184.
12. Antonil, *Cultura e opulência*, bk. 2, chap. 8.
13. The oxcart had a capacity of 2.04 m^3 (1.7 x 1.5 x 0.8 meters); hence, the *tarefa* of fuel equaled approximately 16.3 m^3. See Antonil, *Cultura e opulência*, bk. 2, chap. 8; and Koster, *Travels in Brazil*, p. 169, for dimensions. A tarefa of fuel in 1711 cost 2$500 réis, or about sixteen times the value of a woodcutter's daily wage ($160 réis).
14. Fernão Cardim, *Tratados da terra e gente do Brasil* (Rio de Janeiro, Brazil: J. Leite & Cia., 1925), p. 320; Antonil, *Cultura e opulência*, bk. 2, chap. 6 and 8; Vilhena,

Recopilação de notícias, 1:184.

15. Stuart B. Schwartz, "Colonial Brazil, c. 1580-c. 1750: Plantations and Peripheries," *The Cambridge History of Latin America*, ed. Leslie Bethell, 7 vols. (Cambridge, England: Cambridge University Press, 1987), 2:434-35.

16. Antonil is probably describing Sergipe de Conde, with which he was most familiar. See Antonil, preface to *Cultura e opulência*, p. 213. An *engenho real* was a mill run by waterpower rather than animal traction.

17. The daily usage rate is calculated by dividing 2,500 carts of firewood by 224 days of milling.

18. Antonil, *Cultura e opulência*, bk. 2, chap. 5.

19. Vilhena, *Recopilação de notícias*, 1:185. For the number of mills at Bahia see Cardim, *Tratados da terra e gente;* Antonil, *Cultura e opulência;* Vilhena, *Recopilação de notícias;* and Schwartz, "Colonial Brazil," 2:431.

20. José António Caldas, *Notícia geral de toda esta capitánia da Bahia*, Edicão Facsimilar (Salvador, Brazil: n.p.). Antonil, *Cultura e opulência*, bk. 2, chap. 8.

21. For descriptions of fire handling at the mill see Antonil, *Cultura e opulência,* bk. 2, chap. 8; and Vilhena, *Recopilação de notícias,* 1:184, 186-87.

22. See Antonil, *Cultura e opulência,* bk. 2, chap. 8; and Koster, *Travels in Brazil,* p. 166.

23. João Rodrigues de Brito, *Cartas econômico-políticas sobre a agricultura e commércio da Bahia* (Bahia, Brazil: Imprensa Official do Estado, 1924), p. 97. Brito quotes Manoel Ferreira da Camara's statement that when the proper woods could not be secured for making lye the fabrication of sugar suffered immeasurably.

24. Antonil, *Cultura e opulência,* bk. 1, chap. 3.

25. Royal Order to António Luís Gonçalves da Câmara Coutinho, Lisbon, 7 December 1690, vol. 1, Royal order #92, Ordens Régias, Arquivo Público do Estado da Bahia, Salvador, Brazil.

26. *Documentos Históricos,* 110 vols. (Rio de Janeiro, Brazil: Biblioteca Nacional do Rio de Janeiro, 1928-55), 27:260-61. The Portuguese colonies, like those of Spain, were ruled and administered in every branch of colonial life by the crown and its multitudinous appointed officials. Affairs from the momentous to the trivial merited their attention. During this period Pedro ruled as regent, after deposing by signed agreement his brother Affonso VI, becoming Pedro II after Affonso's death in 1683. The best evidence suggests that the one-league law commenced in 1609 (see Brito, *Cartas econômico-políticas*, p. 96). It met with little success.

27. *Cartas do Senado: Documentos históricos do Arquivo Municipal do Salvador Bahia,* 5 vols. (Salvador, Brazil: Prefeitura do Municipio, 1951-62), 2:88-89.

28. Pinho, *História de um engenho*, p. 220.

29. *Cartas do Senado*, 2:128-29.

30. Stuart B. Schwartz, "Free Labor in a Slave Economy: The Lavradores de Cana of Colonial Bahia," *The Colonial Roots of Modern Brazil*, ed. Daniel Alden (Berkeley: University of California Press, 1973), p. 189. An arroba equals 14.7 kilograms.

31. Schwartz, "Colonial Brazil," 2:434-35. Although not critical to my argument, there is a small discrepancy between the data provided in the text of "Colonial Brazil" and that suggested by the graphic representation on the facing page. I have relied on the latter.

32. Brito, *Cartas econômico-políticas*, pp. 31-32.

33. Both Antonil, *Cultura e opulência*, pp. 101-102, and Vilhena, *Recopilação de notícias*, 1:174, discuss the pros and cons of waterside mills and those inland. The waterside mills' soils were exhausted, forests were depleted, and subsistence agriculture was inadequate.

34. Silva Lisboa, "Carta muito interessante," p. 499. Schwartz, *Sugar Plantations*, p. 107, discusses soil and climate.

35. Silva Lisboa, "Carta muito interessante," p. 499.

36. See *Fuelwood and Charcoal Preparation: An Illustrated Training Manual on Simple Tools and Techniques for Small-scale Enterprises* (Geneva, Switzerland: International Labor Office, 1985), p. 47, for the suggested time to allow wood to dry before burning. Percentage of moisture content is determined by comparing the timber's oven dry weight (0 percent moisture) with its initial green weight. The equation is [1 — (oven dry weight ÷ initial green weight)]. For this equation and average percent moisture in fresh cut timber, see

Fuelwood and Charcoal Preparation, p. 46.

37. Vilhena, *Recopilação de notícias*, 1:196; Manuel Jacinto de Sampaio e Melo, *Idade de Ouro do Brasil*, no. 76 (22 September 1812), cited in Pinho, *História de um engenho*, p. 241; and Antonil, *Cultura e opulência*, p. 116, all describe the excessive size of firewood.

38. See David A. Tillman, *Wood Combustion: Principles, Processes, and Economics* (New York: Academic Press, 1981), pp. 17-33. Tillman also discusses wood's moisture content.

39. *Atas da Camara: Documentos Históricos do Arquivo Municipal do Salvador Bahia, 1625-1700* (Bahia, Brazil: Prefeitura Municipal do Salvador, 1944), 3:311.

40. Koster, *Travels in Brazil*, p. 165.

41. Sampaio e Melo, *Idade de Ouro*, in Pinho, *História de um engenho*, pp. 236-37.

42. Brito, *Cartas econômico-políticas*, pp. 81, 96.

43. For a comprehensive account of the numerous furnace innovations at Bahian mills in the nineteenth century see Pinho, *História de um engenho*, pp. 227-41.

44. *Atas da Camara*, 5:81. The king granted lands *(sesmarias)* to the citizens of Salvador on which pasturing and wood gathering were permitted, but they were inconveniently located in relation to the city.

45. An incident in Rio de Janeiro in 1800 provides some evidence. The Brazilian viceroy charged the English 25$800 réis for the fuel needs of four hundred fifty French prisoners that the English captured on the high seas. This charge was for 3,775 billets and 256 bundles of firewood for thirty-two days. Based on this example, the prisoners' consumption was approximately 120 wood pieces per capita per year, keeping in mind the greater efficiency likely exhibited due to the institutional nature of keeping prisoners. See Viceroy Conde de Reyerde to Captain Bulteel of the English Navy, código 68, vol. 16, f. 280, pp. 307-308, Arquivo Nacional do Rio de Janeiro, Rio de Janeiro, Brazil.

46. Caldas, *Notícia geral*, p. 445. Antonil, *Cultura e opulência*, appendix, stated the trade value of Recôncavo rum as 30,000 *cruzados* in 1710, a significant portion of Bahia's total exports. A *cruzado* was a gold coin worth 480 réis at the time.

47. The Jesuit Antonil, among others, makes frequent mention to the consequences of permitting slaves to drink. See Antonil, *Cultura e opulência*.

48. Gabriel Soares de Souza, *Tratado descriptivo do Brasil* (Rio de Janeiro, Brazil: Typographia Universal de Laemmert, 1851) in *Revista do Instituto Histórico e Geográfico do Brazil* (Rio de Janeiro, Brazil: Imprensa Nacional, 1853), 14:356; Antonil, *Cultura e opulência*, bk. 3, chap. 4.

49. Edson Carneiro, *A cidade do Salvador* (Rio de Janeiro, Brazil: Organização Simões, 1954), p. 108.

50. Soares de Souza, *Tratado descriptivo do Brasil*, 14:296, 355-56.

51. There is no record of the size of these bags of charcoal.

52. *Documentos Históricos*, vols. 37 and 38.

53. See Francisco Augusto Pereira da Costa, *Anais Pernambucanos*, 2d ed. (Recife, Brazil: Governo de Pernambuco, 1983-87), pp. 203-205; Caldas, *Notícia geral*, following p. 442.

54. Vilhena, *Recopilação de noticias*, 1:238. Johann Baptist von Spix and Karl Friedrich Philipp von Martius, *Viagem Pelo Brazil* (Rio de Janeiro, Brazil: Imprensa Nacional, 1938), pp. 273-76.

55. *Documentos Históricos*, 27:209. See also Myriam Ellis *Aspectos da pesca da baleia no Brasil colonial* (São Paulo, Brazil: n.p., 1958), pp. 55-70.

56. Soares de Souza, *Tratado descriptivo do Brasil*, 14:151. Diego de Campos Moreno claimed in 1612 there were more boats in Bahia than in the rest of Brazil combined; see Diego de Campos Moreno, *Livro que da razão do estado do Brasil*, published in *Hispanic American Historical Review* 29 (August 1947): 533.

57. For a sample of the municipal council's correspondence concerning the royal shipyard and its effects on the sugar economy, see *Cartas do Senado*, 3:4-5, 38-39, 42-43; and 5:98-101.

58. Caldas, *Notícia geral*, pp. 442-43. Millers reported sugar crates, made of softwoods that would not alter sugar's taste or color, as increasingly expensive. Ambrósio Fernandes Brandão, *Diálogos dos grandezas do Brasil*, ed. José A. Gonçalves de Mello, 2d ed. (Recife, Brazil: n.p.,

1966), p. 159, was acquainted with individuals who with slave labor built and sold up to two thousand chests per year (*Diálogos*, p. 159). Sacks did not replace crates until the mid-nineteenth century.

59. *Balança Geral do Comércio de Portugal, 1796*, 11, 4, 8, folha 43, Biblioteca Nacional do Rio de Janeiro, Rio de Janeiro, Brazil; and Caldas, *Notícia geral*, following p. 442.

60. Philip Curtin, *The Atlantic Slave Trade: A Census* (Madison: University of Wisconsin Press, 1969), table 34, p. 119.

61. Adrião Caminha Filho, *A cana de açucar na Bahia* (Bahia, Brazil: Tipografia Naval, 1944), pp. 5, 36.

62. Gilberto Freyre, *Nordeste: Aspectos da influência da cana sobre a vida e paisagem do Nordeste do Brasil*, 3rd ed. (Rio de Janeiro, Brazil: José Olympio, 1961), p. 54.

9

Ax or Plow? Significant Colonial Landscape Alteration Rates in the Maryland and Virginia Tidewater

David O. Percy

Imagine standing on the deck of a small vessel gently sailing the relatively calm waters of the great Chesapeake Bay some three hundred and a half centuries ago. Having survived the real or imagined terrors of an ocean crossing, the vista before you is a shore covered with a verdant mantle of great trees stretching to the horizon and, perhaps, beyond. This is America—a land so different from home. Anticipation mixed perhaps with relief and a little dread about the future is high. Is America the land of unfettered opportunity or one which had limits like home?

The first impression of this new land is the abundance of seemingly fertile land. Perhaps, the tales by men who had been to this new land were true. Even if the rhapsodic passages were discounted as what in a latter age would be considered advertising "hype," one could imagine the possibilities upon first sight. Listen to George Percy: "The same day [April 26, 1607] wee entred into the Bay of Chesupioc directly, without any let or hinderance. There wee landed and discovered a little way, but wee found nothing worth the speaking of, but faire meadowes and goodly tall Trees, with such Fresh-waters running through the woods, as I was almost ravished at the first sight thereof."[1] Fifty years later, George Alsop, an indentured servant, wrote his parents: "That this Country of Mary-Land abounds in a flourishing variety of delightful Woods, pleasant Groves, lovely Springs, together with spacious Navigable Rivers and Creeks, it being a

1. Observations of George Percy in Lyon Gardiner Tyler, ed., *Narratives of Early Virginia, 1606–1625* (New York: Barnes & Noble, Inc., 1907), 9–10.

67 Colonial Landscape

most healthful and pleasant situation."[2] At least for those new "Americans" in Maryland and Virginia, there seems to have been a general belief in a seemingly inexhaustible supply of fertile land awaiting the "civilizing" efforts of man.

Two major obstacles stood in the way of these transplanted Europeans and their African-American slaves. The first of these was a chronic shortage of labor. Although Maryland and Virginia planters attempted various solutions such as luring additional immigrants with liberal land and social policies: indentured servants, convict labor, and finally chattel slavery, the problem remained until late in the colonial period.

The second obstacle was that those soils deemed the most fertile were those covered with dense hardwood forests. This assumption arose from two logical conclusions. The first being that the more fertile the soil, the larger the plant that could be found growing in it naturally. Obviously, soils supporting hardwood trees of fifty, sixty or even a hundred feet in height must be very fertile indeed. As Captain John Smith noted in his *Description of Virginia:* ". . . the best ground is knowne by the vesture it beareth, as by the greatnesse of trees or the abundance of weedes &c."[3] The second conclusion was also based on empirical evidence both from Europe and America. When agricultural crops were initially planted on virgin forest soils, the yields were often spectacular. Contemporary observers believed that the soil "is so rich that if it be not first planted with Indian corne, Tobacco, Hempe, or some such thing that may take off the rankness thereof, it will not be fit for any English graine."[4]

These two obstacles combined to lead American farmers to fully utilize the soils' fertility. Rather than husbandmen of the soil most of them became miners of the land. Like miners they stripped the covering—the forests and extracted the resources—the fertility. But unlike miners, they believed, and to a degree their belief was well founded, that the soil would regain a measure of fertility sufficient to mine again if simply abandoned or to use a more modern term placed in long fallow. The surrounding woods did reseed the abandoned fields and in twenty, or more likely thirty, years the land would again produce adequate yields of tobacco, maize, and small grains. This field rotation of tobacco, maize, small grains and long fallow was adequate as long as each laborer had sufficient land. When population pressures increased the system fell apart.

Since the population was increasing geometrically rather than arith-

2. George Alsop to my Father at his House, Mary-Land, January 17, 1659/60, in Colman Clayton Hall, ed., *Narratives of Early Maryland: 1633–1684* (New York: Barnes & Noble, Inc., 1910), 378.
3. Smith's Description of Virginia, in Tyler, *Narratives of Early Virginia,* 82.
4. A Relation of Maryland, in Hall, *Narratives of Early Maryland,* 81.

metically, in a society where most people worked the land the changes in the landscape and eventually the attitudes were profound. The population growth in the Chesapeake colonies were little less than phenomenal during the colonial period. Taken in fifty-year increments, the population, excluding Native Americans who were declining in numbers anyhow, increased from 700 in 1650, to 19,500 in 1700, to 144,000 in 1750, and 1,150,000 in 1800.[5]

This rapid growth caused severe land shortages in the areas most desirable such as those having the best soils with adequate water supplies and easy access to transportation routes. As Father Joseph Mosley in Charles County, Maryland wrote to his brother in 1772: "It has been a fine poor man's county, but now it is well peopled, the lands are all secured, and the harvest for such is now all over."[6] While some such as Father Mosley saw the land as "all secured," others commented on nearby lands being thinly populated. Charles Varlo traveling from Annapolis, Maryland to Alexandria, Virginia in 1785 reported that "the country is only thinly inhabited, being woody, and the land light and sandy."[7] What had happened in a decade? The answer lies in perceptions. In addition to the different views arising from one observer being a resident and the other a tourist, there was a difference in looking at lands that were less desirable because they were more remote from transportation and had less access to water for personal, livestock and even agricultural uses.

While the state of agriculture was not very high among landowners, it may have been even less among tenants who could only expect to gain by getting as much as possible from the land. We know from the work of Gregory Stiverson, Allan Kulikoff, Lois Carr and others that the percentage of landholders in the population was declining as the eighteenth century progressed, particularly in the Tidewater counties. In southern Maryland and the northern neck of Virginia, there was a rise in tenancy until the American Revolution.[8]

5. Series Z, 1–19, Estimated Population of the American Colonies, 1610 to 1770, U. S. Bureau of the Census, *Historical Statistics of the United States: Colonial Times to 1957* (Washington, D.C.: U. S. Government Printing Office, 1960), 756; Series A, 123–80, Population, for States: 1790 to 1950, ibid., 12.

6. Joseph Mosley to Michael Mosley, S. J., Tuckahoe, June 5, 1772, "Letters of Father Joseph Mosley, S. J., and Some Extracts from his Diary. (1757–1786)," comp. Edward I. Devit, S. J., *American Catholic Historical Society of Philadelphia Records* 7 (1906): 208.

7. C[harles] Varlo, *The Essence of Agriculture, Being a Regular System of Husbandry, Through All Its Branches: Suited to the Climate and Lands of Ireland . . . with the Author's Twelve Months Tour thro' America* (2 vols.; London: The Author, 1786), II, 99.

8. Allan Kulikoff, *Tobacco and Slaves: The Development of Southern Cultures in the Chesapeake, 1680–1800* (Chapel Hill: The University of North Carolina Press, 1986), 131–35; Gregory A. Stiverson, *Poverty in a Land of Plenty: Tenancy in Eighteenth-Century Maryland* (Baltimore: The Johns Hopkins University Press, 1977), xii.

69 Colonial Landscape

By 1800 the population of Maryland and greater Virginia had grown to the point where there was no longer enough prime land for all potential laborers. If we assume that Arthur Karinen's proportion of taxables or those deemed capable of working in the fields remained constant at approximately one-third of the population throughout the eighteenth century and that each laborer required 50 acres of arable land, then by 1800 the population became too great for the amount of prime land available in Maryland and greater Virginia.[9] If we divide the 6,820,200 acres of prime land by a potential labor force of 383,333 we find that each laborer would have had just under 18 acres which is less than half the 50 acres considered necessary.[10]

Thus while the labor problem was solved after the end of the colonial period, much of the land had been already severely damaged. Even contemporary observers were aware of what 150 to 200 years of mining the soil had done. As James Garnett wrote in 1818: "[T]he circumstances alone of a new country, almost boundless, in various parts of which, fresh lands of great fertility, have been for sale during many years, at very low nominal prices, has greatly contributed to accelerate among our land killers, the exhaustion of our soil; and to prevent attempts at improvement, which appear, as if they would necessarily take up more time than any land-killer thinks he can spare, and be attended according to *his* arithmetic with more expense in most cases, than removing to, and settling new lands, which the God of nature has already fertilized to our hands."[11] Garnett was not alone in recognizing what had occurred and the causes for it after the fact.[12]

How had the condition of the land reached such a state? Did not the Chesapeake planters know better? The answer to the second question is easy. By in large, they knew that their system of agriculture was degrading the land both through erosion and draining the fertility. Their immediate practical needs however overroad considerations they may have had for

9. Arthur E. Karinen, "Maryland Population: 1631–1730: Numerical and Distributional Aspects," *Maryland Historical Magazine* 54 (Dec. 1959): 365–67.

10. Prime Cropland, Robert J. Mason and Mark T. Mason, *Atlas of United States Environmental Issues* (New York: Macmillan, 1990), 28, Table 4.1.

11. James M. Garnett, "Defects in Agriculture," *Memoirs of the Virginia Society for Promoting Agriculture: containing Communications on Various Subjects in Husbandry and Rural Affairs, 1818,* 54.

12. "Answer," *Agricultural Museum,* II (August 1811): 37; Julian Ursyn Niemcewicz, *Under Their Vine and Fig Tree: Travels Through America in 1797–1799, 1805 with Some Further Account of Life in New Jersey,* trans. and ed. J. E. Budka (Elizabeth, N.J.: Grassman, 1965), 88; Philip Vickers Fithian, *Journal & Letters of Philip Vickers Fithian 1773–1774: A Plantation Tutor of the Old Dominion,* ed. Hunter Dickinson Farish (New Edition; Williamsburg, Va.: Colonial Williamsburg, Inc., 1965), 88–89; Jacques Pierre Brissot de Warville, *New Travels in the United States of America, Performed in 1788* (New York: Augustus Kelley, 1970), 438.

preserving the land for future generations. Second, the large scale effects of their agricultural practices did not become so severe until the last half of the eighteenth century, when the population, particularly in the Tidewater counties, became dense. This denseness forced planters to till more highly erodable soils and to reduce, if not eliminate long fallow. But by then the agricultural practices which seemed relatively benign in the earlier years had become well entrenched.

The colonial agricultural practices grew out of adoption and adaptation of Native American swidden agriculture. The European settlers of the Chesapeake as elsewhere in colonial America found a thriving Native American agriculture which produced bountiful subsistence crops of maize, beans, and pumpkins and a crop which would form the basis for commercial agriculture in the area—tobacco. All these crops were grown in a modified row culture. As the early settlers quickly realized, that if they were to continue their existence in the New World, they would have to adopt Native American crops and practices.

The new settlers found that by adopting Native American practices they could insure their subsistence. They could avoid the laborious task of clearing the woods by girdling the hardwood trees in the fall after the trees had gone dormant. By peeling a one-foot strip of bark about three or four feet from the base, the sap would not rise in the spring. The planters then could cultivate the soil directly under the trees since no leaves would emerge. After a few years many of the dead tree trunks would blow down. Girdling, unlike cutting down a tree, generally killed the root systems as well, which prevented these girdled trees from sprouting.[13]

A small portion of the trees would be used for building, firewood, and fences. But of the average 500–700 hardwood trees per acre less than one-third of the trees on an acre or two would be needed during the first few years of settlement for all purposes. In subsequent years only about 10 percent of the trees on an acre or two would be required mainly for firewood.[14]

In addition to allowing planting the spring after girdling, this practice allowed colonial planters to make the most profitable use of their scarce labor. Instead of the approximately twenty days needed to clear an acre,

13. Thomas Anburey, *Travels through the Interior Parts of America* (2 vols., Boston: Houghton Mifflin, 1923 [1789]), II, 188–89; Hugh Jones, *The Present State of Virginia: From Whence is Inferred a Short View of Maryland and North Carolina,* ed. Richard L. Morton (Chapel Hill: The University of North Carolina Press, 1956), 77.

14. Estimates of lumbers and fencing materials cut from standing timber based on reconstruction projects at the National Colonial Farm, Accokeek, Maryland 1976–1986; Max George Schumacher, *The Northern Farmer and His Markets During the Late Colonial Period* (New York: Arno Press, 1975), 5; J. Thomas Sharf, *History of Maryland: From the Earliest Period to the Present Day* (3 vols., Hatboro, Pa.: Tradition Press, 1967[1879]), II, 75–76.

71 Colonial Landscape

the area to be planted could be girdled in a few days. The planter could remove deadfalls when he had time such as during the winter months. We assume that planters could be very selective in the trees for which they had a use. Only the straightest, knot and limb-free white oaks would be used for construction lumber. Chestnuts would be reserved for fences along with cypress and cedars. If they planned to use any of the white oak for firewood, they would down live trees rather than girdled trees, since it would be difficult to cut cured white oak into usable billets.[15]

By preventing the girdled trees from leafing out, the planters could use the virgin forest mold directly beneath the tree canopies for their cash crop—tobacco. With a hilling hoe or mattock the planters formed small hills about a foot high and three or four feet apart in each direction. One laborer could tend about 6000 tobacco plants or something on the order of one and a half to two acres in addition to providing for his subsistence. Generally, a planter could get at least two crops of tobacco from a particular plot of land and more if the soil was very fertile. Once the yields declined, the planter moved on to a new plot for his tobacco.[16]

For most of the colonial period the primary cultivation implement was the hoe. This was a practical measure since the Chesapeake planters were cultivating previously forested lands whose soils were interlaced with the tough roots of mature hardwood trees. Through the use of this hoe culture planters were merely scraping the forest mold into hills that served as their planting medium. Since there was a certain randomness to these fields of small hills interspersed with dying trees and some natural ground cover, the amount of erosion from early tobacco fields was probably small. In addition, the early settlers were planting on the flattest lands—the river terraces. The depth of the forest mold or soil of six to twelve inches was also a factor in retaining moisture to prevent erosion. Although these early planters were mining the soil's fertility with their tobacco, their cultivation practices and the amount of land being cultivated for tobacco made only slight alterations in the landscape.[17]

After the tobacco yields declined, the planter could and did use his old fields for maize. Because maize was a taller plant—particularly the Virginia

15. A Relation of Maryland, Hall, *Narratives of Early Maryland,* p. 79; Peter Kalm, *The America of 1750. Peter Kalm's Travels in North America, The English Version of 1770,* ed. and trans. Adolph B. Benson (2 vols., New York: Wilson-Erickson, 1937), II, 50.

16. The two best descriptions of colonial tobacco culture are William Tatham, *An Historical and Practical Essay on the Culture and Commerce of Tobacco* (London: T. Bensley, 1800) and Curtis Carroll Davis, " 'A National Property Richard Claiborne's Tobacco Treatise for Poland," *William and Mary Quarterly* 21 (January 1964): 93–117.

17. Horatio Sharpe to Cecilius Calvert, February 10, 1754, *Correspondence of Governor Horatio Sharpe,* I, 38 in *Archives of Maryland* (70 vols., Baltimore: Maryland Historical Society, 1883–), VI, 38; Avery Odelle Craven, *Soil Exhaustion as a Factor in the Agricultural History of Virginia and Maryland, 1606–1860* (Gloucester, Mass.: Peter Smith, 1965 [1926]), 28–29, 32;

Gourdseed variety commonly grown in the Chesapeake region whose stalks could reach 14 feet, the hills for maize or Indian corn were placed up to 6 feet apart in each direction. It was possible to raise maize after a crop of tobacco because of the former's deep tap root which took nutrients from a different soil level than the surface-feeding tobacco plants. After several years of repeated crops the yields of Indian corn too declined.[18]

During the early colonial period, the damage done to the soil from maize cultivation was probably more than that from tobacco. The former tobacco field now in Indian corn gradually became freer from the dead trees. Ambitious farmers pulled stumps as the root systems rotted. The distance between hills opened more bare earth to the effects of summer thundershowers. Both these factors allowed a more regular planting that in turn provided channels for erosion.[19]

Many Chesapeake planters would then use the former maize field for a small grain usually wheat. This was a European crop whose cultivation was very familiar. The land for wheat was well worked with a spade or even a plow. Some planters sowed wheat in a field of standing Indian corn in the transition year. Although much of the soil's fertility had been used by tobacco and corn crops, the difference in the time when wheat needed attention made the small yields acceptable to colonial planters. A planter could expect several crops of wheat before abandoning a field to long fallow.[20]

In the late colonial period on the Maryland's Eastern shore, the northern end of the Chesapeake, the northern neck of Virginia, the inland reaches of both colonies and sporadically elsewhere, planters chose to cultivate wheat as a cash crop rather than tobacco. Wheat as a cash crop was possible when sufficient seasonable labor was available for the har-

18. Entry for April 13, 1777, Nicholas Cresswell, *The Journal of Nicholas Cresswell: 1774–1777* (Port Washington, N.Y.: Kennikat Press, 1968 [1924]), 197; John Harrower to Capt. James Craigie, Belvidera, August 20, 1775, Edward Miles Riley, ed., *The Journal of John Harrower: An Indentured Servant in the Colony of Virginia, 1773–1776* (Williamsburg, Va.: Colonial Williamsburg, Inc., 1963), 111–12; Entry for April 24, 1775, ibid., 93.

19. Thomas Mann Randolph to Thomas Jefferson, Monticello, June 4, 1792, in Edwin Morris Betts, ed., *Thomas Jefferson's Farm Book: With Commentary and Relevant Extracts from Other Writings* (Princeton, N.J.: Princeton University Press, 1953), 229–30; [John Beale Bordley], *Summary View of the Courses of Crops, in the Husbandry of England and Maryland; with a Comparison of their Products; and a System of Improved Courses, proposed for Farms in America* (Philadelphia: Charles Cist, 1784), 22; [John Beale Bordley], *Sketches on Rotations of Crops and other Rural Matters. To which are Annexed Intimations On Manufactures; on the Fruits of Agriculture; and on New Sources of Trade, Interfering with Products of the United States of America in Foreign Markets* (Philadelphia: Charles Cist, 1797), 13–16; Craven, *Soil Exhaustion,* 35–36.

20. Bordley, *Sketches on the Rotation of Crops,* 11–12; Cresswell, *Journal,* 198; Lewis Cecil Gray, *History of Agriculture in the Southern United States to 1860* (2 vols., New York: Peter Smith, 1941), 198.

vest and adequate amounts of cleared land could be made or possessed. Since about six times the amount of cleared acreage was needed for wheat as compared to the partially cleared acreage for tobacco to obtain approximately the same monetary return, a planter had to have adequate labor to thoroughly work a piece of ground or had to be able to use animal powered implements such as plows.[21]

Because of the nature of cultivation, the most damage was probably done from small grains rather than the row crops during most of the colonial period. Both tobacco and maize were planted in the spring when the rainfall was usually gentler. Since wheat was planted on bare ground in late August or September, the torrential thundershowers of late summer caused both sheet and gully erosion. In addition the subsoils were disturbed through spading or plowing. Throughout most of the colonial period both tobacco and maize were planted in hills with only a small area around each hill disturbed. The irregular planting of these two crops caused by planting among stumps and girdled trees also acted to reduce the flow of water. Finally, the early settlers utilized the levelest lands along the rivers.[22]

After several years of wheat, the yields declined to point where further cultivation of the field was impractical. Most farmers simply abandoned the field to nature. Abandoned fields dotted the Chesapeake landscape. After twenty, or more likely thirty, years these "old" fields would have recovered enough fertility to be used for tobacco again.[23]

During the first hundred years after settlement, there was enough land for each laborer to allow this cycle of field rotation to be maintained. During the eighteenth century as the population grew, the period of long fallow was shortened and sometimes eliminated. Those who remained on the Tidewater lands would have to fertilize their soils and adopt practices that reduced the effects of erosion. The poorer lands once abandoned returned to forests.

With the exception of the forests surrounding the towns and the charcoal smelting furnaces, most of the forest was cleared for agricultural production. Although the hardwood forest had some resource value for

21. Paul G. E. Clemens, *The Atlantic Economy and Colonial Maryland's Eastern Shore: From Tobacco to Grain* (Ithaca, N.Y.: Cornell University Press, 1980); Carville V. Earle and Ronald Hoffman, "Staple Crops and Urban Development in the Eighteenth Century South," *Perspectives in American History* 10 (1976): 38.

22. The source of the commonly held belief that the planting of tobacco and corn were the primary causes of soil erosion in colonial Maryland and Virginia can be traced to the seminal work of Avery Odelle Craven, *Soil Exhaustion as a Factor in the Agricultural History of Virginia and Maryland, 1606–1860* (Urbana: The University of Illinois, 1926), 35.

23. John Taylor, "Memoir on Clearing Land," *Memoirs of the Philadelphia Society for Promoting Agriculture*, I, 328; Edward C. Papenfuse, Jr., "Planter Behavior and Economic Opportunity in a Staple Economy," *Agricultural History* 46 (April 1972): 304n.

74 agricultural history

American colonists in the Chesapeake colonies, for the most part it was a hinderance to the main business of life—farming.

While it is difficult, if not historically dangerous, to assign the roots of a particular practice or belief to a particular historical time, we suggest that the practices of soil conservation and of mining the soil were established in our colonial past. There were those who felt that the only way to maintain themselves and possibly to prosper was to mine the soil. When population pressures and the concerns about moving to the frontier caused some people to stay put, the need for husbanding the soil became important to those individuals. The promise contained in a "well-wooded land" was there for those who could extract the wealth from the soils below those great forests of oak and chestnut. The seemingly inexhaustible resource was the land. Few of them, like few of us, are wise enough to conserve what seems abundant now for some future generation. While the ax created an unkept appearance to the colonial landscape, it was the unwise use of the plow that eventually damaged the soils.

10

Human Influences on the Pine and Laurel Forests of the Canary Islands

James J. Parsons

THE effect of human activities on the vegetation mantle of the earth historically has tended to be especially severe on island ecosystems. The Canary Islands, which originally supported substantial stands of both coniferous and broadleaf evergreen forests, offer an instructive, but little-known example of this continuing interaction between man and environment. Located some 100 kilometers off the North African coast at 28°N, the Canary archipelago, structurally an extension of the Atlas Mountains, has been an isolated ecologic system since the Tertiary period. The extravagantly broken volcanic topography has created a remarkable diversity of niches, from sea level to the snow-capped summit of Tenerife (Pico de Teide, elevation 3,715 meters) and from the dripping cloud forests of exposed trade-wind slopes to the desertic lee sides where annual precipitation may be less than 100 millimeters.[1] Tenerife and Gran Canaria, the largest of the seven islands, have been termed with some reason "continents in miniature."

The Canary Islands lie on the subsiding eastern side of the semipermanent Azores anticyclone. This subsidence produces a warm, dry atmosphere aloft that is separated at an average height of 1,500 to 1,800 meters from a lower layer of moist, southward-streaming air.[2] From the high parts of Tenerife in the good visibility of the clean, dry air, it is usually possible to see the small cumulus-cloud layer that marks the upper part of this inversion extending to the horizon.

When these clouds rise against the northern slopes of the islands with high elevations, the result is the light showers that may account for more than half of the annual rainfall of approximately 800 millimeters. Elsewhere almost all precipitation occurs during short periods of disturbed weather in the winter

---

[1] Telesforo Bravo, Geografía general de las islas Canarias (Santa Cruz de Tenerife: Goya Ediciones, 1954), Vol. 1, pp. 206–225.

[2] Luis Ceballos and Francisco Ortuño, Vegetación y flora forestal de las Canarias occidentales (Madrid: Instituto Forestal de Investigaciones y Experiencias, 1951). A revised second edition of this work with superb color illustrations was published by the Cabildo Insular, Santa Cruz de Tenerife, 1976, reference on pp. 65–94. See also P. R. García-Prieto, F. H. Ludlam, and P. M. Saunders, The Possibility of Artificially Increasing Rainfall on Tenerife, *Weather*, Vol. 15, 1960, pp. 39–51.

half of the year, associated with the close approach of the polar front or with the development of small cyclones off the African coast. Thus the precipitation regime is suggestive of the Mediterranean region with nearly rainless summers, but milder temperatures due to latitude.

The high degree of endemism that characterizes the plant life of the Canary Islands is a product of both environmental diversity and isolation.[3] Of the almost 2,000 species recorded from the archipelago, approximately 30 percent are known only in these islands. Xerophytic shrubs dominate the low, dry eastern islands of Lanzarote and Fuerteventura where the stately Canary Island palm, *Phoenix canariensis*, is locally conspicuous as the only naturally growing tree. Similar semidesert plant associations with numerous *Euphorbiaceae* cover the low slopes of the higher islands to the west, while endemic succulents are significant where the Atlantic influence is strongest. The windward slopes on the western islands between approximately 500 and 1,200 meters elevation originally supported a diverse broadleaf evergreen forest (*monte verde*) dominated by the Canary laurel (*Laurus* sp.) and its relatives to which European plant ecologists applied the term *laurisilva*. This dense canopied assemblage has been largely destroyed on Gran Canaria and much depleted elsewhere to make way for terraced cropland and to provide firewood and lumber. The verdant and luxuriant laurel woodland was extravagantly described by early visitors. Today remnants of this broadleaf forest are of interest, especially to students of plant evolution. On Gran Canaria this forest survives only in two steep-walled barrancas. Less than 6,000 hectares, perhaps 10 percent of its original extent, remain in the other western islands.[4]

The life zones of the high-elevation islands are popularly defined in terms of altitude as "below," "in," and "above" the clouds, especially on the northern and eastern exposures. Upslope from the monte verde and reaching through the persistent clouds into the trade-wind inversion is the pine forest (*pinar*). These pine stands have been under heavy pressure from woodcutters since the earliest European settlement and, before then, from the fires and the domestic animals of the aboriginal Guanches. The pine forests have been reestablished in recent years over most of their original extent through a massive governmental reforestation program. Pine forests, in varying conditions and density, today cover some 70,000 hectares in the archipelago (Fig. 1). Nearly a third of these is on lands that have been reforested during the last thirty-five years.

On Tenerife pines form a continuous belt around the island between 1,200 and 1,800 meters, although scattered stands may occur down to sea level and upward to 2,200 meters (Fig. 2). Above the 2,000-meter contour, pines are generally replaced by xerophytic leguminous shrubs, especially *retama* (*Cytocytisis* sp.) and *codeso* (*Adenocarpus* sp.). The steep cinder flanks of the conical Pico de

---

[3] David Bramwell, The Endemic Flora of the Canary Islands, *in* Biogeography and Ecology of the Canary Islands (edited by Gunther Kunkel; The Hague: Dr. W. Junk, 1976), pp. 207–240; A. Machado, Introduction to the Study of the Canary Islands' Laurisilva with Special Reference to Ground Beetles, *in* Biogeography and Ecology of the Canary Islands, this footnote, pp. 347–412, especially 342–356; Per Sunding, The Vegetation of Gran Canaria, *University of Oslo, Natural History Series 29*, Oslo, 1972; and J. V. Malato-Beliz, Conservación de la naturaleza y recursos genéticos, *Botánica Macaronésica*, Vol. 1, 1976, pp. 67–82.

[4] Francisco Ortuño and Andres Ceballos, Spanish Woodlands (Madrid-Sevilla: INCAFO, 1967), p. 224.

Teide are devoid of plant life except for a pansy-like endemic, *Viola cheiranthifolia*.

On the island of Hierro (elevation 1,520 meters) and more especially on the better-watered La Palma (elevation 2,423 meters), there are still impressive stands of old-growth pines, but on Gran Canaria (elevation 1,980 meters), larger and drier than Hierro or La Palma, with its tangled interior topography, such stands are sparse and scattered except where they have been recently replanted. Only on Gomera (elevation 1,484 meters) and on the eastern desert islands were pine forests originally absent.

The Canary Island Pine Forest

The Canary Island pine, *Pinus canariensis*, is a familiar ornamental in Mediterranean and California gardens. Its closest relative, *Pinus roxburghii*, is found in the Himalayas, not in Europe or North Africa.[5] The Canary Island pine has one of the most restricted distributions of any of the more than 100 species of the genus *Pinus*. Mature, straight-boled trees normally reach a height of twenty to thirty meters. Much influenced by human activities, stand densities vary greatly, but a light, open forest with sparse undergrowth is most characteristic, especially on drier sites. Open or disturbed areas are usually invaded by shrubs such as heather, myrtle, broom, and rockrose, although a peasant tradition of using this growth for agricultural mulch has tended to hold them in check. The long, grayish-green needles, in fascicles of three and bunched in slightly drooping fashion near the ends of the branches, give the tree a distinctive appearance. On the ground the needles decompose slowly, emitting a pleasant fragrance in the afternoon warmth. The rough, reddish bark is fire-resistant and light enough to be used as buoys on fishermen's nets. The species is unique among the pines in its ability to sprout from the base and the trunk after fire or frost damage. The species is well adapted to shallow, rocky soils with small amounts of humus. In its growth habit and tolerance for drought conditions, it somewhat resembles the yellow pine, *Pinus ponderosa*, of the American West or some of the Mexican pines. Growth rates are rapid, especially when the trees are young. The largest specimens have a girth probably exceeding that of any European member of the genus, and in some cases take on almost sequoia-like proportions.

Individual pines of exceptional size have long been the object of special attention and veneration in the archipelago.[6] Shrines or hermitages often have been built alongside them, for example the *Pino de la Virgen* near El Paso (La Palma), the *Pino de Buen Paso* near Icod (Tenerife), and the *Virgen de Los Angeles* at Victoria de Acentejo (Tenerife). For festivals, pine boughs are traditionally brought from the mountains to festoon the paths of processions and to decorate streets and churches. *Nuestra Señora del Pino* at Teror on Gran Canaria is the *patrona* of that island and the subject of fervent devotion, although the tree in which the Madonna is asserted to have miraculously appeared in 1484 fell three centuries ago.

---

[5] Nicolas T. Mirov, The Genus Pinus (New York: Ronald Press, 1967), p. 74.

[6] Leoncio Rodriguez, Los arboles históricos y tradicionales de Canarias (2 vols.; Santa Cruz de Tenerife: La Prensa, 1946); and Ceballos and Ortuño, footnote 2 above, 2nd ed., p. 166n.

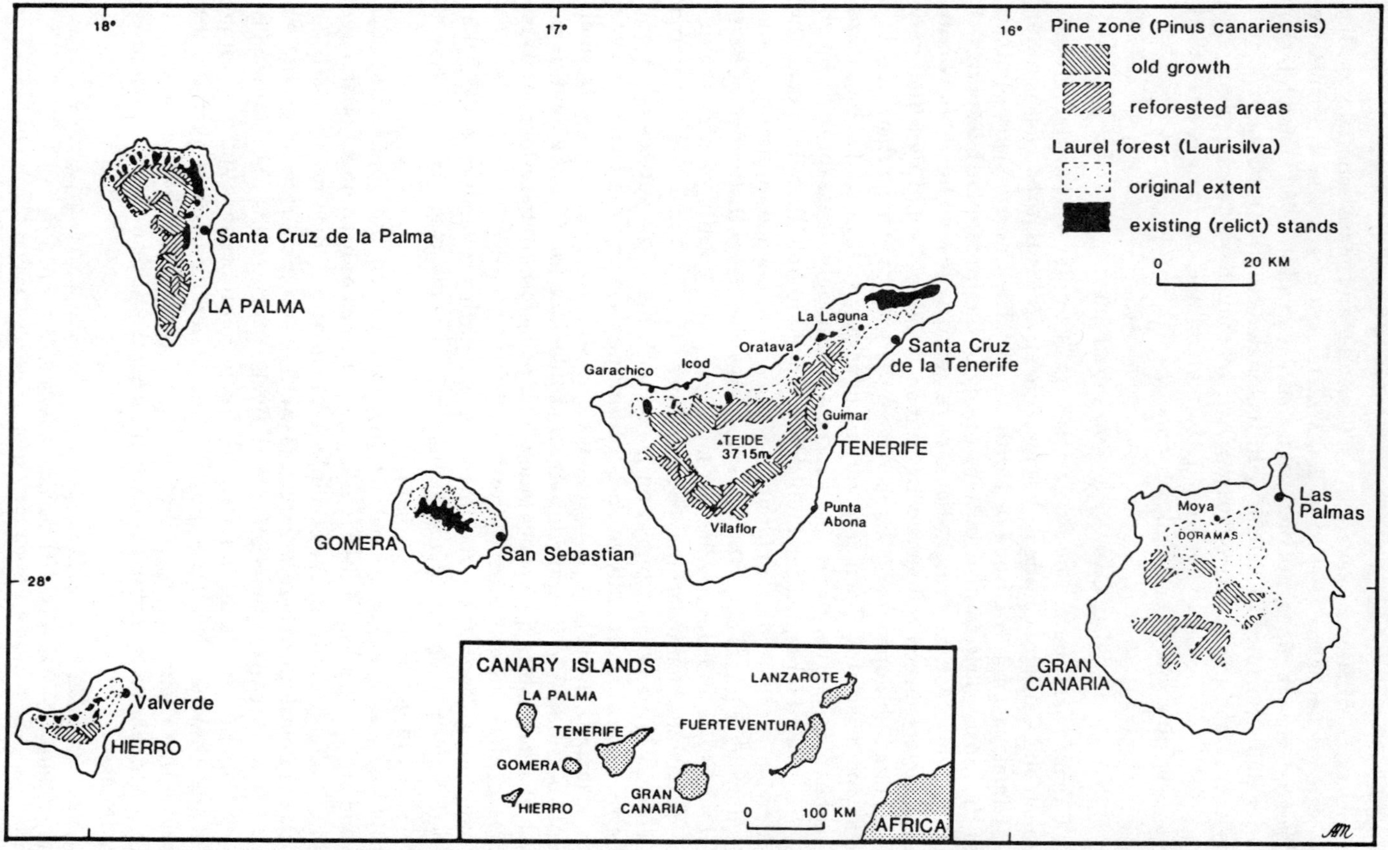

FIG. 1—The forest lands of the Canary Islands. *Sources:* Personal observations of the author, and adaptations from Ceballos and Ortuño, text footnote 2; Machado, text footnote 3; and Sunding, text footnote 3.

FIG. 2—Reforestation areas (*Pinus canariensis*) along Tenerife crest at elevations between 1,200 and 1,500 meters. *Carretera dorsal* is in the foreground, and snow-covered Pico de Teide is in the background.

The Cultural History of the Archipelago

There is uncertainty as to when the first Guanche peoples reached the Canary Islands from North Africa with goats, sheep, and pigs, but their arrival was definitely well before the time of Christ.[7] The Guanche were of Berber stock and full neolithic culture and apparently were isolated on the islands from other influences until the first arrival of Europeans in the fourteenth century. Livestock constituted the principal wealth of those cave-dwelling pastoralists, who also cultivated wheat and barley and gathered edible wild plants and shellfish. Remarkably, by the time of the European conquest they seem to have all but abandoned the art of boatbuilding and interisland travel so that the groups on each island were isolated and lacked communication with their neighbors.

The European presence in the archipelago dates from the activities of Catalan and Mallorcan missionaries on Gran Canaria between 1352 and 1391 associated

[7] Luis Diego Cuscoy, Los Guanches: vida y cultura del primitivo habitante de Tenerife, *Museo Arqueológico de Tenerife, Publicación No. 7,* Santa Cruz de Tenerife, 1968; Luis Diego Cuscoy, El Conjunto ceremonial de Guargacho, *Museo Arqueológico de Tenerife, Publicación No. 11,* Santa Cruz de Tenerife, 1979; and Ilse Schwidetsky, The Prehispanic Population of the Canary Islands, *in* Biogeography and Ecology of the Canary Islands, footnote 3 above, pp. 15–36.

with the sparsely documented bishopric of Telde.[8] The presence intensified in 1402–1403 with the conquest of Lanzarote, an early target of Andalusian slaving expeditions, by the Breton Jean de Bethencourt under the auspices of Henry III of Castile. By 1409 Fuerteventura and distant Hierro were occupied, as was Gomera, an important source of slaves, in 1447. These five islands were awarded as seignores by the Spanish Crown, but they offered little opportunity for economic gain other than from slaves and orchilla. They served as springboards for the conquest of Gran Canaria, Tenerife, and La Palma where the Guanche defenders were numerous and formidable. These three islands offered the promise of great wealth from the cultivation of sugar cane by irrigation and were proclaimed crown lands in 1477.

The Canary enterprise was in the hands of a small aristocracy until the conquest of Gran Canaria (1477–1484). The conquistador and first governor of that island, Pedro de Vera, was authorized by the Crown to distribute land and water rights to deserving noblemen, soldiers, and sailors according to their merits. So was the Adelantado Alonso de Lugo a few years later on Tenerife and La Palma. Persons who would build sugar mills and plant sugar cane, introduced from the Iberian Peninsula or from Madeira, were to be given the largest and best tracts. Genovese merchants, who had the necessary capital for such ventures, played a major role in establishing the activity. The technology for sugar making was brought by Portuguese from Madeira, already a "sugar island." To assure that the best lands and the water rights would not be monopolized, the sale of either land grants or sugar mills was forbidden.

With the completion of the conquest, councils were appointed to govern each of the crown-land islands. The councils were also authorized to make additional grants of unoccupied land. Much of it was forested or in the rough or drier parts of the islands. Some tracts close to settlements were set aside as commons for the use of local residents, especially for grazing and woodcutting. With the passage of time, most of these commons were distributed in allotments for cultivation.[9]

During much of the past 450 years, island society has been polarized between a subsistence agricultural-pastoral sector dependent on maize, potatoes, and domestic animals and a commercial export sector, long controlled by foreigners, that has gone through recurrent cycles of prosperity and depression. Orchilla dyestuff, sugar, wine, *barrilla* (*Mesembryanthemum*, an ice plant that was a source of soda ash), and now tomatoes and bananas have been the base of the export economy.[10] Tobacco and winter vegetables have made less contributions to it. Since 1950 a sun-and-sand-oriented mass tourism has engulfed

[8] Antonio Rumeu de Armas, El obispado de Telde, misioneros mallorquins y catalanes en el Atlántico (Madrid–Las Palmas: Biblioteca Atlántica, 1960); and Agustín Millares Torres, Historia general de las Islas Canarias (10 vols.; Las Palmas de Gran Canaria: Edirca, S. L., 2nd ed., 1977 [originally published 1873–1895]), Vol. 3, p. 31.

[9] A. M. Macías Hernández, La transformación de la propiedad agraria consejil en el paso del antiguo régimen al nuevo régimen, *Revista de Historia Canaria*, Anexo 1, 1978, pp. 44–47; and Germán Hernandez Rodriguez, Los montes de la Gomera y su conflictividad, *Aguayro* (Casa Insular de Ahorros de Gran Canaria), No. 84, 1977, pp. 31–34.

[10] Josef Matznetter, Die Kanarischen Inseln, Wirtschaftgeschichte und Agrargeographie (Gotha: Hermann Haack, 1958). On the "export" of people as a component of the commercial relations of the islands, see Coloquio de Historia Canario-Americano I, 1976 (edited by Francisco Morales Padrón; Las Palmas–Sevilla: Cabildo Insular de Gran Canaria, 1977).

the archipelago, which has brought both unprecedented prosperity and serious social and environmental ruptures.

The Canary Islands, with a population of 1.4 million, are an integral part of Spain and since 1929 have constituted the two provinces of Las Palmas (Gran Canaria, Lanzarote, Fuerteventura) and Santa Cruz de Tenerife (Tenerife, La Palma, Gomera, Hierro). The two provinces are almost equal in population, as are the provincial capitals of Las Palmas and Santa Cruz de Tenerife. The proposed entry of Spain into the European Economic Community and possibly into the North Atlantic Treaty Organization recently intensified concern for the always uneasy economic and political equilibrium of the archipelago.

Forests and Water

Water is the overriding problem in the Canary Islands today. "Mined" for more than a century from pervious lenses in the complex volcanic rocks that make up the islands and used mostly for irrigation, water is now in perilously short supply. These haphazardly distributed underground veins of water have been tapped since the nineteenth century by an extraordinary system of horizontal perforations (*galerías*), some as much as five kilometers long, on the middle and high slopes.[11] They are owned by public stock companies that sell water to banana and tomato farms below at increasingly higher prices. Shares in these speculative ventures are bought and sold through a "water exchange" (*Bolsa de Aguas*) in Santa Cruz de Tenerife. Extraction rates from underground reserves substantially exceed replenishment, and the galerías are drying up. There were 913 of them on Tenerife and 154 on La Palma in 1977. Many of these were abandoned, but new ones continue to be drilled in the hope of finding additional productive lenses of water. On Gran Canaria, where the more impermeable lavas make wells and small reservoirs feasible, galerías are less common than on the other islands. But on this dry island water shortages are especially acute. Seawater distillation, weather modification, and additional reservoir construction have all been resorted to, but the survival of commercial-irrigated agriculture depends primarily on a sharp reduction of water use. Mulching with either volcanic ash or organic matter and the new technique of drip irrigation provide partial answers.

The forests of the Canary Islands are dripping wet environments when shrouded in clouds. Moisture added to the soil beneath by "fog drip" often substantially exceeds that reaching the ground as rain. The legendary *garoé* or "rain tree" of the island of Hierro, probably a *til* (*Ocotea foetens*), is said to have condensed cloud droplets sufficient to support two permanent pools of water at its base until felled by a storm in 1612. This "horizontal precipitation" tends to be greatest where air movement is strongest and is especially pronounced in the pine forests.[12] In one extreme case, 5,090 millimeters were re-

---

[11] Hans Hausen, On the Ground Water Conditions in the Canary Islands and their Irrigation Cultures, *Acta Geográfica* (Helsinki), Vol. 12, No. 2, 1951.

[12] Luis Ceballos and Francisco Ortuño, El bosque y el agua en Canarias, *Montes* (Madrid), Vol. 8, 1952, pp. 418–423; Luis Ceballos, Consideraciones sobre la flora y la vegetación forestal de las islas atlánticas, *Anuario de Estudios Atlánticos*, Vol. 2, 1956, pp. 9–44; and Franco Kämmer, Klima und Vegetation auf Tenerife besonders im Hinblick auf den Nebelniederschlag, *Scripta Botánica* 7, Göttingen, 1974.

corded during twelve months in one Tenerife pine forest, while only 650 millimeters fell on a nearby clearing. Because the seasonal distribution of fog precipitation is similar to that of normal precipitation, the character of the winter-rain climate is not changed by this additional source of moisture. Tolerance of seasonal dryness is more decisive than fog interception to the distribution of pines, although the additional water added to the soil may range from 300 to 2,500 millimeters annually.

Horizontal precipitation, still little studied, is clearly a significant factor in the water economy of the archipelago. However, the same trees that induce this precipitation also transpire substantial amounts of moisture. Although they significantly reduce runoff rates, trees probably take more water out of the ground than they add to it under most circumstances. In mid-latitude forests transpiration rates have been shown to be substantially higher for pines than for mature broadleaf stands.[13] Data on this are still unavailable for the Canary Islands. In the popular mind there remains a conviction that there is a close relationship between declining availability of ground water and the destruction of the tree cover on the islands. This unsubstantiated belief has been the basis of the widespread support for forest conservation and replanting programs in recent years.

Early Exploitation of the Woodlands

From the Spanish establishment on the better-watered western islands, beginning with Gran Canaria in 1477, the woodlands were the object of intense exploitation. They supplied wood for fuel and construction, as well as pitch and livestock forage in an environment largely dominated by aridity. The woodlands were the object of insistent concern by the island councils that viewed them as both a source of income and a fragile resource demanding close supervision and control.

This preoccupation is evident from the minutes of early council meetings and from the numerous protective ordinances that were adopted.[14] Detailed restrictions on the use of the forests were enforced by a guard system with severe fines for violations. Every form of exploitation required a license, and the fee schedule was a topic of continued discussion. The export of lumber and firewood was closely controlled and at times prohibited, although there was always an active clandestine trade. Lanzarote and Fuerteventura, destitute of wood, imported it from the western islands. This trade was generally frowned on by local authorities, concerned about the threatened depletion of their own forest resources.

---

[13] W. T. Shank and J. E. Douglas, *Science*, Vol. 185, 1974, pp. 857–859.

[14] José Peraza de Ayala, Las ordenanzas de Tenerife (Madrid: Aula de Cultura de Tenerife, 2nd ed., 1976); Leopoldo de la Rosa, Catálogo del Archivo Municipal de La Laguna (Sucesor del antiguo cabildo de Tenerife), *Revista de Historia* (Universidad de La Laguna), various issues, 1944–1960; Acuerdos de Cabildo de Tenerife, *Fontes Rerum Canarium IV*: 1497–1507, 1949; *V*: 1508–1513, 1952; *XI*: 1514–1518, 1965; *XVI*: 1518–1525, 1970 (edited by Elias Serra Rafols and Leopoldo de la Rosa; La Laguna: Instituto de Estudios Canarias); Francisco Morales Padrón, Ordenanzas del Consejo de Gran Canaria, 1531 (Sevilla: Cabildo Insular de Gran Canaria, 1974); Libro Rojo de Gran Canaria (edited by Pedro Cullén del Castillo; Las Palmas: Tipografía Alzola, 1947); and Leopoldo de la Rosa, Evolución del régimen local en las Islas Canarias (Madrid: Instituto de Estudios de Administración Local, 1946).

Restrictions on the use of the forest were specific. The cutting of trees near springs was prohibited; none that were large enough for use in construction were to be taken for fuel wood; for each pine removed, ten new trees were to be planted; livestock were prohibited from entering the forest; fires in it were outlawed; night hunting in the dry season was illegal. There were explicit prohibitions on the taking of certain species such as *acevino* (*Ilex canariensis*). The monte verde of the Montaña Doramas of Gran Canaria was the object of particularly detailed restrictions that dated from the first years of the sixteenth century. The insular archives contain many volumes of litigation on the illegal activities in the forests. The excesses of the woodcutters and charcoal makers, the destructive fires, the incursions of livestock, and the perpetual shortage of reliable forest guards all contributed to what was early seen as an ecological disaster in the making.

Sugar manufacturing made the heaviest levies on the forests in the early years.[15] Tenerife and Gran Canaria each had twelve sugar mills in the 1560s, while La Palma had four and Gomera one.[16] Other evidence suggests as many as twenty-two for Gran Canaria somewhat earlier. The typical mill operated for six months of the year, with eight copper cauldrons continuously boiling cane juice. Substantial fortunes were made in the business in only a few years. Sir Francis Bacon wrote of how "being the first in an invention does sometimes cause a wonderful overgrowth of riches, as it was with the first sugarmen in the Canaries."[17] At 500 *maravedis* for one *arroba* (10 kilograms) of sugar, the capital investment in a sugar mill with a capacity of between 8,000 and 9,000 arrobas might be paid off in three years. The mills ground not only their own cane but also the production of smallholders on a shared basis. Each mill thus had fifteen to twenty independent suppliers. This practice was later to become the basis of the *colono* system of sugar production in Cuba. By the end of the sixteenth century, with new competition from the Americas and the North African coast, where *Canarios* were generally the sugar masters, sugar production was in sharp decline and wine was in the ascendancy. As sugar disappeared from the Canary Island economy, the drain on the deteriorating woodlands slowed but did not stop.

The gradual increase in the rural population, augmented by refugees from Lanzarote and Fuerteventura during drought periods, brought further pressures on the forest lands as subsistence agriculture pressed upward to the wooded mid-slopes of Gran Canaria, Tenerife, and La Palma. Additional titles granted on crown lands, together with clandestine occupations, led to widespread clearing, while livestock made increasingly destructive incursions into the remaining forests.[18]

---

[15] Miguel Angel Ladero Quesada, La economía de las Islas Canarias a comienzos del siglo XVI, *Anuario de Estudios Americanos*, Vol. 31, 1974, pp. 725–749; Guillermo Camacho y Pérez Galdos, El cultivo de caña de azúcar y la industria azucarero en Gran Canaria, 1510–1535, *Anuario de Estudios Atlánticos*, Vol. 7, 1960, pp. 11–60; and Maria Luisa Fabrellas, La producción de azúcar en Tenerife, *Revista de Historia* (La Laguna), Vol. 18, 1952, pp. 455-480.

[16] Thomas Nichols, A Pleasant Description of the Fortunate Islands Called Islands of Canaria (London: Thomas East, 1583).

[17] Quoted in Noel Deerr, The History of Sugar (London: Chapman and Hall, 1949), p. 155n.

[18] Antonio Manuel Macías Hernández, El motín de 1777 en Gran Canaria, *Anuario de Estudios Atlánticos*, Vol. 23, 1977, pp. 263–345.

George Glas, an English visitor, wrote in 1764 that on the high slopes above Oratava, Tenerife, where the clouds remained most of the day, "there was formerly a great abundance of stately pine trees; but being easy to come at they were almost all cut down by the inhabitants of the adjacent villages so that few now remain . . . but in other places of the island at the same altitude, and which are distant from any habitations, there are great numbers of them."[19] The effect of the fuel-hungry sugar mills would have been particularly heavy in those forests immediately above the best-watered and most productive lowlands where the cane plantings were concentrated. Accounts of visitors, including Alexander von Humboldt in 1799, who climbed the Pico de Teide, make no mention of a pine forest but instead describe a few named landmark conifers that stood as solitary guideposts along the route.

On the island of La Palma, Glas reported that rabbits had been the cause of the destruction of the high pine forests. Although the trees on the summits had been destroyed, they remained abundant in the cloud zone so that from a distance of two leagues "it appeared as a solid forest." Pines here were tall enough to serve as spars for large ships, but the timbers were heavy, and the broken terrain made the cost of transporting them to the coast almost prohibitive. The captain of an American sloop with a broken mast paid twenty-five pounds sterling to transport one tree to the harbor. "Nevertheless," Glas noted, "much wood is exported from here to the rest of the islands."[20]

The Destruction of Montaña Doramas

The history of the monte verde of Montaña Doramas on Gran Canaria exemplifies the continued pressures on the forests of the islands. Early travelers and poets wrote of its Arcadian verdure and its sparkling streams.[21] An observer in 1634 termed Montaña Doramas "one of the most grandiose possessions of Spain." For the historian José Viera y Clavijo, it appeared to be "a work of art, appreciated the more because it is not."[22] Glas thought that the forest made "a charming scene," the lofty boughs of its fragrant trees "so thickly interwoven as to exclude the rays of the sun." The rills that watered those shady groves, the whispering of the breeze among the trees, and the melody of the birds formed for him "a most delightful concert." A person in that "enchanting solitude," he reflected, was inevitably reminded of the descriptions of the Fortunate Islands by the ancients.[23]

The Montaña Doramas, named for a renowned Guanche chieftain, lay directly above the best sugar lands on the island. From the first years of Spanish control the forest was under heavy pressure. Many ordinances were passed by the council aimed at the conservation and the regeneration of the forest, but

---

[19] George Glas, Descripción de las islas Canarias, 1764, *Fontes Rerum Canarium XX* (La Laguna: Instituto de Estudios Canarias, 1976), pp. 79–80.
[20] Glas, footnote 19 above, p. 96.
[21] Alfredo Herrera Piqué, La destrucción de los bosques de Gran Canaria a comienzos del siglo XVI, *Aguayro* (Casa Insular de Ahorros de Gran Canaria), No. 92, 1977, pp. 7–10; and Mariano Nougues Secall, Cartas histórico-filisófico-administrativo sobre las Islas Canarias (Santa Cruz de Tenerife: Salvador Vidal, 1858), p. 380.
[22] José Viera y Clavijo, Noticias de la historia general de las Islas Canarias [1782–83] (edited by Alejandro Ciorenescu; Santa Cruz de Tenerife: Ediciones Goya, 1971).
[23] Glas, footnote 19 above, p. 65.

apparently to little avail.[24] There were prohibitions against the grazing of stock, the making of charcoal, and the setting of fires of any sort, the last under penalty of lashes, banishment, and even death. In 1533 when the forest was closed to any further exploitation for ten years, the sugar mills were directed to seek their firewood requirements elsewhere, chiefly in the less accessible pine lands. However, complaints caused the order to be rescinded. Soon thereafter the Crown ordered that during future council deliberations on matters related to forests, the members who owned sugar mills should be excluded from the sessions. The linkage of economic and political power was already evident.

The assault by settlers and woodcutters continued unabated, even as the sugar era neared its end. Already in the 1560s Thomas Nichols could write of Gran Canaria that "wood is the thing that is most wanted."[25] Increasingly wood was sought from the other islands as land clearing for agriculture impinged more and more on the remaining forests. A late eighteenth-century report referred to Montaña Doramas as "a picture of desolation." In 1802 fire further reduced by half what was left. Browsing livestock effectively prohibited regeneration. Already much of the area was in private hands and under cultivation. The remainder was a constant source of conflict among neighboring towns. With the proclamation of the constitutional system in 1820, influential citizens of the *ayuntamiento* of Moya requested distribution of the remaining Montaña Doramas as private property, on the argument that this was the only way to eliminate clandestine cutting and grazing. But the adjacent communities opposed such concessions and threatened armed intervention, if necessary, to prevent them.[26]

In 1831, as payment for salary arrears and for services rendered to the country, Fernando VII granted most of the remaining Doramas forest, some 1,000 hectares, to General Francisco Tomás Morales, celebrated field marshal of the defeated Spanish forces in the Americas.[27] Returning to his native islands in 1827, he was named governor of the archipelago. Deposed seven years later after a political dispute, he retired to develop his new property, which he named Hacienda San Fernando. An enigmatic figure, his name is still associated with the destruction of the legendary Montaña Doramas, although by his time it was only a caricature of what it had been in the past. He publicly proclaimed his innocence and good intentions with the assertion that he assigned guards to keep poachers and livestock from the remaining woodland and set severe penalties for sharecroppers who cut trees unnecessarily. Pascual Madoz wrote in his "Diccionario" that the forest of Doramas, "once the pride of the Canarios, has been reduced to some groups of trees which the proprietor condescendingly preserves but that are slowly disappearing."[28]

---

[24] Libro Rojo, footnote 14 above, p. lxiii.
[25] Nichols, footnote 16 above.
[26] Pascual Madoz, Diccionario geográfica-estadística-histórica de España y sus posesiones de ultramar (16 vols.; Madrid: P. Madoz, 1845–1850), Vol. 7, p. 308.
[27] Francisco Morales Padrón, Francisco Tomás Morales, último Capitán General de Venezuela, *Anuario de Estudios Americanos*, Vol. 33, 1976, pp. 641–712.
[28] Gunther Kunkel, Die Lordbeerwaldrelikte auf Gran Canaria, *Schriften des Geographischen Instituts der Universität Kiel, Vol. 39*, Kiel, 1973, pp. 121–129.

The Early Naval Stores Industry

There had been incursions into the pine forests of Tenerife for lumber and naval stores even before the Spanish conquest. By 1498, two years after the end of Guanche resistance, a tax of five maravedis per quintal (100 kilograms) of pitch (*pez*) was levied by the council. Later the sum was doubled. Licenses to take pitch were offered at public auction. The product was in demand not only for shipbuilding and repairs but also for caulking the wooden aqueducts that carried water across often considerable distances to the cane fields. An early prohibition on the export of pitch was soon relaxed. In an economy short of money, pitch was occasionally used as a medium of exchange, as were sugar and wheat. In the first years of the sixteenth century pitch was valued at 100 to 150 maravedis dockside. Twenty quintales might buy a cask of wine, 120 a healthy African slave.[29]

The authorities expressed continuing concern for the lifestyle of the pitch makers, many of whom were Portuguese. An ordinance in 1500 required that pitch makers must have a permanent residence and have "planted and fenced at least 800 vines" before they could be granted licenses. Later it was specified that they should be married. The council worried that the men who worked away from the towns for extended periods would not hear Mass, but no remedy for that situation was forthcoming.[30]

Pitch works were concentrated in the best stands of mature pines, above Icod on the northwestern coast and on the drier lee side of Tenerife. As early as 1500, it was explicitly forbidden to cut pines to make pitch in the Taora district (Oratava) "because they are for the sugar ingenios." There was less production of pitch on Gran Canaria, La Palma, and Hierro. The Tenerife archives at La Laguna contain an almost continuous record of council income from the pitch works of Icod and of Agache, between present-day Güimar and Abona, prior to 1651, when the record terminated.[31] Production peaked in 1593. In the nineteenth century, with the progressive destruction of the old-growth pines, references to pitch making were rare.

To obtain the resin-rich, red heartwood characteristic of old-growth Canary pines, it was the practice to fell the tree. The resinous wood was then cut into chunks and slowly cooked for twenty-four hours in three-chambered brick ovens. A charge of 1,600 kilograms was said to yield 200 to 240 kilograms of the black, lustrous pitch.[32] Because the works were in isolated locations, the remaining white sapwood was without value and was left to rot. The wastefulness of this primitive method of exploitation troubled Tenerife authorities. They made the cutting of trees to obtain pitch illegal in some areas. For the most part, such restrictions were unenforceable. Alternatively the heartwood

[29] Emma Gonzalez Yañez, Importación y exportación en Tenerife durante los primeros años de la conquista, 1497–1503, *Revista de Historia de Canarias*, Vol. 29, 1953, pp. 71–91, reference on p. 78; and Emma Gonzalez Yañez and Manuela Marrero Rodriguez, Protocolos del escribano Hernán García, La Laguna, 1508–1510, *Fontes Rerum Canarium VIII* (La Laguna: Instituto de Estudios Canarios, 1958).

[30] Gonzalez Yañez, footnote 29 above, pp. 79–80; and Acuerdos de Cabildo de Tenerife, 1508–1513, footnote 14 above, June 16, 1511.

[31] de la Rosa, Catálogo, footnote 14 above.

[32] Gaspar Frutuosa, Las Islas Canarias de "Saudades da Terra," 1590 (translated from the Portuguese; La Laguna: Instituto de Estudios Canarios, 1964).

might be hollowed from the lower trunk of a standing tree with an axe. This technique might yield one or two loads of the resin. Scattered old pines in the forest still bear evidence of the practices—their charred basal scars at least twice the height of a person. Ignited, these resin-saturated lesions, which also supplied torchwood for night fishing, might burn for three or four days.

Heartwood of Pine—A Prized Construction Material

The same resinous heartwood that produced the pitch of commerce was also prized as a building material. Valued for its rich, natural beauty and for its resistance to weathering and insect damage, the heartwood was considered "incorruptible." A memorial on the state of the carpentry trade on Tenerife in 1778 observed that the growing shortage of *tea*, as this wood was known, was having a damaging effect on the construction industry.[33] Builders, accustomed to using this material, found its cost prohibitive. They blamed the scarcity on the sorry state of the pine forests. The short supply of *viñatigo* (*Persea indica*), a wood favored by furniture makers, was also noted. The alternatives were either to bring in West Indian cedar or mahogany at much increased cost or to use inferior materials.

A distinctive "Canary style" of architecture, featuring half-timbering, exposed beams, and handsomely carved exterior balconies of natural wood evolved in the archipelago in response to the availability of this unique material.[34] So did a sophisticated furniture industry. Church interiors as well as the doors and the balconies of old townhouses or public buildings that have been exposed to the elements continuously for 300 or more years show no signs of deterioration. Today *tea* is rarely available on the market, and the demand for the wood has made the dismantling of old structures a profitable business. Buildings featuring the use of this dense, slightly translucent wood have become tourist attractions. A reviving nostalgia for the past, fueled in part by the new mass tourism, has further promoted public interest in the diversity of architectural uses to which *tea* has been put.

The Sociedades Económicas

The deteriorated state of the vegetative cover of the islands was a central concern of the *sociedades económicas de los amigos del pais* after their establishment in the late eighteenth century on Gran Canaria, Tenerife, and La Palma. In their publications and in the minutes of their meetings, these active, civic-improvement groups, so typical of the Spanish Enlightenment, reflected a preoccupation with the progressive destruction of the forests.[35] It was noted that the woodlands supplied construction materials, fuel wood, and naval stores, that they reduced runoff and erosion, and that "by attracting moisture

---

[33] Memoria sobre el estado actual del oficio de carpinteros de esta ciudad, 12 Diciembre, 1778, Archivo Municipal de La Laguna, *Industrias*, tomo 3.

[34] Fernando Martín Rodriguez, Arquitectura doméstica canaria (Santa Cruz de Tenerife: Aula de Cultura, 1978).

[35] Sociedad Económica de Amigos del Pais, Las Palmas, *Boletín*, Nos. 7–10, 1862, and Nos. 72–73, 1868; Sociedad Económica de Amigos del Pais, Las Palmas, *Annales*, 1872, pp. 44–45, and 1879, pp. 7–23; and Enrique Roméu Palazuelos, La Económica a través de sus actas, años 1776 a 1800 (La Laguna: Real Sociedad Económica de Amigos del Pais de Tenerife, 1970).

from the atmosphere [they] assured the flow of the springs on which agriculture depends."

The first session of the Gran Canaria branch of the society at Las Palmas in 1777 discussed the condition of the pine forests, unable to reproduce because of livestock depredations and fires set maliciously by stockmen to encourage new grass. It was lamented that the government had no control over privately owned forest lands where cutting, burning, and goat damage were also increasing. Five years later a report described "the sad spectacle of our once dense pinar" in which the actions of man and his livestock left no opportunity for the pines to regenerate.

Representations to the Supreme Council of Castile in 1788 brought the closing of a part of Montaña Doramas and the nearby Montaña Lentiscal, but bribery and intimidation of guards by "insolent woodcutters" only intensified. A committee returning from an inspection tour predicted that the useful forests of the island would be gone in a short time. The committee proposed, among other measures, that a reforestation program be initiated, "if necessary at the society's own expense."

Two inspectors, sent to the pine lands of the interior in 1836, returned to describe "a scene of mutilation" there. A few months later a "horrendous fire" had brought further extensive destruction. It was the goats, however, that were seen as the principal culprits. Gran Canaria, the inspectors reported, was being converted "from an island oasis crowned with perpetual green to an ugly skeleton floating in the sea." The rains were believed to have diminished and become less regular, and there was "more sickness at all seasons and an increase in rabid dogs." It was notorious that licensees were taking twice the authorized quantities of wood. The guards were irresponsible and completely ineffective in enforcing existent ordinances. There were not enough guards, and they easily fell under the influence of the exploiters. "We must tear the masks from these spurious sons of the island," the inspectors concluded, "to whom the care of our forest has been confided." On Tenerife and La Palma the situation was no different.

A report to the Gran Canaria branch of the society in 1878 stated that of the original 40,000 hectares of pines on the island only a few tracts remained, leaving the once forested land as "sad and ugly *páramos* supporting only rachitic stands of heather." At least eight different pine forests were described as having been destroyed during previous years. The society went to the highest authorities in Madrid but encountered only indifference. Royal orders that excluded livestock from state forests and authorized the replanting of the pine lands were ignored. The society was especially forceful in urging that state-owned forests should not be sold under the disentailment legislation (*desamortización*) of 1855 because of the special conditions in the archipelago. Instead such land continued to be sold, usually at a fraction of its true value.[36] On Gran Canaria the Crown had retained title to unalienated lands, in contrast with Tenerife and La Palma where most of the untitled woodlands had been ceded to the local governments in the 1830s.

---

[36] Juan José Ojeda Quintana, La desamortización en Canarias (1836 y 1855), *Cuadernos Canarias de Ciencias Sociales* (Las Palmas), No. 3, 1977.

The La Laguna branch of the Sociedad Económica heard its first account of the condition of forests on Tenerife and proposals for their restoration in 1778. Six years later the branch announced a prize of 200 *reales* or a medal for the best report "on the present state of the forests, the causes of their ruin, and possible means for their restoration." Neither these reports nor a study commissioned in 1794 on the same theme has been found in the society's archives.

There was repeated mention of the promise of reforestation of the pine belt. In 1813 and again in 1841 royal orders specified that such activities were the responsibility of local authorities. But, with few exceptions, there was little interest from the municipal officials who now controlled the woodlands of Tenerife. Those officials complained about the difficulties of obtaining supplies of planting stock and about the impossibility of keeping goats, sheep, and even camels from the forest during the dry season when the animals voraciously sought out the grayish-green pine seedlings. An observer wrote in 1858 that "if no remedies are taken these islands will soon be denuded and converted to unproductive crags . . . it is not enough to preserve the forest, it is necessary to replant it."[37] Because some of the original vegetation remained, he thought that "it would not take long to reclaim what has been lost if administrators would pay closer attention." The wealth of seedlings on the floor of the pine forest would reproduce if given protection. But this they did not receive.

On La Palma the records of the society are sketchier than those of the Tenerife and Gran Canaria branches, but the concerns were comparable to those of the other chapters. An 1859 classification showed 37,000 hectares of pines in the public forests of the island, an estimated 20 percent of the total for the archipelago.[38] They were in much better condition than those on the larger islands, although most were said to need "cleaning and thinning." On Hierro, 1,000 hectares of pine lands had been donated to the council by their owner, the count of Adeje.

Pine Needles as a Resource

The peculiar economy of the Canary Islands inverts the traditional scale of values of forest products. The minor items, not the logs or the lumber, provide the principal income.[39] And this pattern continues even today when charcoal and firewood are no longer of much significance. Pine needles (*pinocha*), sometimes mixed with heather, broom, and bracken fern, provide the most important source of income from the forests of the islands. The peasant custom of collecting the fallen needles from the forest floor was recorded as early as the middle of the nineteenth century. Thus an 1858 account suggested that the forest administrators for Tenerife should prohibit the extraction of the dry needles for use as fertilizer or mulch "because it removes the precious humus cap as well as many ungerminated seeds and young shoots on which the future of the forest depends."[40] By the end of the nineteenth century the dry needles

---

[37] Nougues Secall, footnote 21 above, pp. 380 and 386.
[38] Miguel Bosch, Rapid ojeada sobre el estado de los montes de Canarias, Puerto Rico, Cuba y Filipinas, *Revista Forestal, Económica y Agrícola* (Madrid), Vol. 1, 1868, pp. 169–188, reference on p. 183.
[39] Francisco Ortuño, Aprovechamientos forestales en los montes de Canarias, *Montes* (Madrid), Vol. 16, 1960, pp. 271–275.
[40] Nougues Secall, footnote 21 above, p. 390.

also were being used as a packing material for bananas that was called "grass" by British importers.

Luis Ceballos and Francisco Ortuño wrote in 1951 that the extraction of pine needles, done with increased intensity during and immediately after World War II, had all but exhausted the existent supplies.[41] As silviculturists, Ceballos and Ortuño advised termination of "this continuing drain on the only fertilizer materials available to the forest soil, which could have disastrous long term effects of the forest." But they recognized the potential social consequences of such a move. "It is difficult to imagine," they wrote, "the number of persons who presently support themselves by these forest products. In our travels through the pinares we encounter many persons each day, with loads on their heads, descending from the heights (*cumbres*) on the rocky trails. They are mostly women and children contributing their bit to the family's subsistence." The meager remuneration of five to ten pesetas that the gatherers received from their one daily load emphasized the extent of their need. The dislocations of the war years were blamed for this unecological exploitation, and conditions were improving. But Ceballos and Ortuño warned that the forest could not wait. They urged a division of the pine forests in administrative districts so that some forest areas could be rested and the soil continue to receive "the substances necessary for its maintenance." In the next years, this suggestion was implemented by a system of auctions instituted under the supervision of the Forest Service, and later the Instituto Nacional para la Conservación de la Naturaleza (ICONA). The system continues.

In recent years auctions of pine needles produced two or three times the monetary returns received from all other forest products, including lumber. On Tenerife in 1977 rights to collect 270,000 metric tons of pine needles and other organic materials from the forest floor brought 7.9 million pesetas ($US1.1 million) at public auctions in twenty-three municipios.[42] In comparison, rights to take 4,933 trees brought only 3.1 million pesetas, while grape and tomato stakes yielded another 1.3 million pesetas.

During the late 1960s, when bananas began to be boxed, demand softened, but with the recent expansion of banana terraces (12,500 hectares in 1977), the market for organic mulch from the forest floor has been strong. In 1978 ICONA supervised the auction of rights to collect more than two million tons of pine needles at $US1.50 to $US4.50 a ton on Tenerife and La Palma. Many contractors doubtlessly took more than was authorized. Trucks loaded with pine needles, often mixed with heather or bracken fern, are a common sight on the country roads, while in the pine forests great stacks of the long, reddish-brown needles, awaiting collection, stand along the roads. In addition, subsistence farmers living on the forest edges still collect pine needles to use as bedding for their animals and manure on their potato fields.

The effect of the periodic removal of needles from the forest floor continues to be debated. *Pinus canariensis* thrives on shallow mineral soils, even on fresh lava and cinder. Yet in the long run, removal of this organic matter must reduce water-infiltration rates and must encourage runoff and sheet erosion

[41] Ceballos and Ortuño, footnote 2 above, 1st ed.
[42] Information from the files of the Instituto para la Conservación de la Naturaleza (ICONA), Santa Cruz de Tenerife, courtesy of Ing. Enrique Mira.

with a consequent decline of soil nitrogen, carbon, and humus substances that regulate soil structure. The slow rate of decomposition of the needles has led to speculation that fire may be required to recycle completely their minerals into the soils of more arid habitats.

The collection of pine needles is sometimes justified by the reduced fire hazard resulting from removal of these combustible materials. After a recent destructive burn on La Palma, a correspondent to the local press criticized ICONA for the restrictiveness of its auction policy, which was said to have made more favorable conditions for fire.

A local custom on La Palma is to leave scattered pines in the midst of fields of potatoes and cultivated forage plants, principally *tedera* (*Psorlea bituminosa*) and *tagasaste* (a cultigen of *Cytisis proliferus*). The branches of these trees are pruned every three or four years. After drying on the ground, the pruned branches and any available crop residue are burned, the ashes being used to fertilize the fields in preparation for the next planting. Although the trunks sprout freely after such treatment, the branches never attain their normal horizontal spread because of the repeated pruning. Like stark skeletons with a tuft of greenery at the top, these trees suggest something "created by the genius of El Greco."

Reforestation

After four centuries of continued abuse, the pine forests of the Canary Islands have been reestablished through most of their former extent under a vigorous and well-managed reforestation program administered initially by the Patrimonio Forestal of the Spanish Forest Service and now by ICONA.[43] Almost 25,000 hectares have been planted, principally with *Pinus canariensis*, but this total includes approximately 3,400 hectares of *Pinus radiata* from California and a small area of *Pinus halepensis*. Most of this reforestation has been accomplished through partnership or *consorcio* agreements with local governments and less commonly with private landowners. On Gran Canaria, however, the Crown has retained control of most publicly held land, although the Cabildo Insular, an intermediate jurisdiction unique to the archipelago, purchased approximately 4,100 hectares for replanting with conifers under agreements with ICONA. Under the partnership agreements, the government conducts and pays for the planting, the protection, and the maintenance of the reforested tracts and supervises their exploitation. Proceeds from auctions, permits, and licenses are shared between ICONA and the landowners until the investment by the former has been recovered.

In the first years of the reforestation program there was much to be learned. Direct seeding of *Pinus canariensis* proved impractical, and there were problems with seedling transplants caused by the unusually long taproot of the tree. A unique system of transplanting from nurseries that used open-ended plastic sleeves was eventually developed. To promote the development of a secondary root system, the radical was pruned when it emerged from the bottom of the sleeve. Drought brought extensive early losses, but by 1950 some 4,200 hectares were reforested, mostly on the once well wooded northern slopes of Tenerife.

---

[43] Ceballos and Ortuño, footnote 2 above, 1st ed., pp. 187–188.

Planting continued at an accelerated pace and extended to Gran Canaria, La Palma and Hierro. Piles of rocks placed around each seedling protected it from rabbits and other animals as well as from severe storms. Stand densities were high, averaging approximately 2,500 trees per hectare. By the end of 1978, total plantings had reached 18,972 hectares in the province of Santa Cruz de Tenerife (13,000 hectares on Tenerife, 3,000 hectares on La Palma, and approximately 1,500 hectares each on Hierro and Gomera). Gran Canaria accounted for another 8,867 hectares.[44]

The tradition and the infrastructure for a lumber or wood-products industry are lacking on the islands. Most necessities are imported, the trade being facilitated by the free-port status of the archipelago. The small-scale logging operations in old-growth *Pinus canariensis* stands above Icod and Garachico on Tenerife, where clear-cutting practices have stirred up an ecological controversy, provide most of the logs processed by the three sawmills on the island. La Palma has an even more modest production, while on Gomera a small planting of the exotic *Pinus radiata* was being harvested for box wood in 1979 as part of a program to reestablish the endemic monte verde that originally covered the high slopes of the island. As kerosene has replaced charcoal and firewood for domestic use, as fire-prevention programs have become effective, and as livestock numbers have declined, conditions in the forests have greatly improved. A prolific seeder, the native pine is naturally reestablishing itself with aggressiveness in many situations, especially on bare mineral surfaces.

The program for the reforestation of the Canary Islands was a part of the grand design for the *repoblación forestal* of Spain initiated in the first years of the Franco regime. It was at a military encampment in the pine forest of La Esperanza on Tenerife that the Caudillo had first raised the cry of rebellion against the republican government in 1936. With the final victory of the Falange three years later, reforestation was given the highest priority. On the peninsula, devastated by civil war, it appears to have been strongly supported by Keynesian considerations—the creation of jobs and the stimulation of investment in depressed rural areas. In the Canary Islands, which represented only a minor part of the program, the maximizing of horizontal precipitation and water yield was the important guiding rationale. Recently aesthetic and recreational values have been given comparable emphasis.

The gradual decline of the rural population of subsistence farmers and graziers on the Canary Islands, coupled with an ambitious reforestation program, largely has checked the destructive landuse patterns of the past. Today the sunny coasts, with their water-demanding export agriculture and their swarmings of visitors on "package tours," are especially under siege. In reaction to past excesses there is currently a substantial opinion among islanders against almost all types of forest exploitation. For professional foresters this disinclination to "harvest" the standing timber of the pine lands, in particular, is seen as "clearly exaggerated," though based on the best of intentions.[45] The monte verde in most areas is only a memory, long since converted to cropland or secondary scrub. But extensive tracts on the Canary Islands today support a

[44] ICONA Memoria 1978 (Madrid: Instituto para la Conservación de la Naturaleza, 1979).

[45] Ceballos and Ortuño, footnote 2 above, 1st ed., p. 182.

vigorously growing forest of endemic pines of perhaps greater density and luxuriance than has been known at any time since the arrival of Europeans five centuries ago. The rehabilitation of these upland forests stands as an example of the positive potential of human intervention in ecologically sensitive island ecosystems.

11
From Hands to Tutors: African Expertise in the South Carolina Rice Economy

Judith A. Carney

By the end of the eighteenth century, South Carolina had emerged as one of the world's wealthiest agricultural areas. The basis for its economic prominence was rice cultivation, which had been developing rapidly from the 1690s.[1] The knowledge and expertise required for rice cultivation, however, did not exist among Barbadian planters who transferred a slave-based economy to South Carolina, nor did it come with the successive waves of British and French colonists who expanded the plantation system. The techniques of rice production were vested in the knowledge carried by many African peoples to the Americas, particularly those enslaved from Senegal to the Ivory Coast, which is home to the indigenous West African rice, *Oryza glaberrima*.[2]

Until Wood's pathbreaking research in the early 1970s, historical accounts of the development of South Carolina's rice economy credited "imaginative" planters for discovering the crop's potential and adapting

1. See Lewis Cecil Gray, *History of Agriculture in the Southern United States to 1860* (Gloucester, Mass: Peter Smith, vol.1, [1933] 1958); A.S. Salley, "The Introduction of Rice Culture into South Carolina," *Bulletins of the Historical Commission of South Carolina* 6 (Columbia: The State Company, 1919), and Wesley F. Craven, *The Southern Colonies in the Seventeenth Century, 1607–1689* (Baton Rouge: Louisiana State University, 1949); and John Otto, *The Southern Frontiers, 1607–1860* (New York: Greenwood, 1989), 33.

2. Roland Portères, "Primary Cradles of Agriculture in the African Continent," in *Papers in African Prehistory,* ed. by J. Fage and R. Oliver (Cambridge: Cambridge University Press, 1970), 43–58; J. Harlan, J. DeWet, and A. Stemler, "Plant Domestication and Indigenous West African Agriculture," in *Origins of African Plant Domestication,* ed. by Harlan, DeWet and Stemler (Chicago: Aldine, 1976), 3–19. A. Carpenter, "The History of Rice in Africa," in *Rice in Africa,* ed. by I. Buddenhagen and J. Persely (London: Academic Press, 1978), 3–10.

its cultivation to diverse environmental settings.[3] Wood challenged this perspective by arguing that Africans contributed more than mere muscle to South Carolina's rice economy: the agronomic knowledge critical to the crop's development in South Carolina originated with slaves.[4]

This paper lends further support to Wood's argument that African slaves were more than "hands" in the evolution of the South Carolina rice economy. A cross-cultural perspective is employed to show that the micro-environments planted to rice in West Africa on the eve of the Atlantic Slave Trade were identical with those that emerged in South Carolina in the period prior to American independence. Duplicate techniques for water control also appeared in the South Carolina systems. The historical reconstruction, based on empirical research as well as archival and secondary sources, demonstrates the remarkable correspondence between the rice cultivation systems established in colonial South Carolina and those indigenously developed in West Africa. It further substantiates the position that African slaves tutored planters in the requisite skills and technologies to create one of the New World's most lucrative plantation economies.

Prior to Wood, the African role in the South Carolina rice economy had been downplayed in several key respects. Educated descendants of planters dominated writing on the South Carolina rice economy into the twentieth century and provided glowing accounts of the accomplishments of their forebears. The difficulties involved in converting insalubrious swamps to productive lands as well as the sophisticated techniques of water management and tidal irrigation were unquestioningly assigned to planter ingenuity and imagination. This interpretation was bolstered by documentation that credited a ship captain from Madagascar for introducing rice to planters who then successfully seized the opportunity to experiment with its cultivation.[5] The Madagascar hypothesis assumed that knowledge of rice cultivation lay in Asia, and the South Asian origins

3. See for example, David Ramsay, *History of South Carolina* (Charleston: Walker, Evans & Co, 1858); David Doar, *Rice and Rice Planting in the South Carolina Low Country* (Charleston, S.C.: Charleston Museum, [1936] 1970); Herbert R. Sass, *A Carolina Rice Plantation of the Fifties* (New York: Wm. Morrow, 1936); Duncan C. Heyward, *Seed from Madagascar* (Chapel Hill: University of North Carolina Press, 1937); Norman Hawley, "The Old Rice Plantations in and around the Santee Experimental Forest," *Agricultural History* 23 (1949):86–91; and Converse Clowse, *Economic Beginnings in Colonial South Carolina, 1670–1730* (Columbia: University of South Carolina, 1971).

4. Peter H. Wood, *Black Majority* (New York: W.W. Norton, 1974a); and Peter H. Wood, "More Like a Negro Country: Demographic Patterns in Colonial South Carolina, 1700–1740," in *Race and Slavery in the Western Hemisphere: Quantitative Studies,* ed. by S. Engerman and E. Genovese (Princeton: Princeton University Press, 1975).

5. For the introduction of rice into South Carolina see Ramsay, *History of South Carolina*; A.S. Salley, "The Introduction of Rice Culture"; and Heyward, *Seed from Madagascar.* On planter historians diminishing the importance of skilled slaves in favor of the role of the

of the Malagasy people provided logical validity for the argument.[6] The possibility that rice could have been introduced to South Carolina from West Africa was ignored even though the Upper Guinea Coast: (1) was the source of massive numbers of slaves with experience growing rice; (2) is home to an indigenous rice species and sophisticated cultivation systems; (3) routinely provisioned rice to European slave ships for the Middle Passage; and (4) is located nearer the Americas than Madagascar, half a continent away.[7]

Clowse, writing in 1971, provided a critical shift in perspective by revealing the importance of African labor in ranching and forest extractive activities for the capital accumulation that financed the growing importation of slaves from 1700 onward. The emergence of a black majority in the colony from 1708 facilitated the economic orientation towards rice.[8] Wood (1974) built upon Clowse's research by arguing that the rice plantation economy was nurtured by the agronomic expertise as well as the labor of West African slaves. Support for an African, rather than planter, basis for the knowledge of rice growing in South Carolina increased over the following years with research that drew attention to the extent of rice

---

masters, see Leland Ferguson, *Uncommon Ground: Archeology and Early African America, 1650–1800* (Washington, D.C.: Smithsonian Institution, 1992), 61.

6. Doar's overview of South Carolina rice culture and his ancestors' roles provides an illustration: "A short recapitulation will show what has been accomplished by the enterprise of our planters in the last seventy years . . . It is one hundred and fifty-seven years since the introduction of rice into Carolina, and there are grounds for supposing that our people have accomplished more during that period, in the cultivation and preparation of this grain, than has been done by any Asiatic nations, who have been conversant with its growth for many centuries . . . in our method of irrigation we were their equal—while in economy of cultivation . . . we are greatly their superiors" (Doar, *Rice and Rice Planting,* 20). Support for diffusion of rice from Madagascar to South Carolina is additionally weakened by the year of introduction, 1685. Rice was cultivated in the Madagascar highlands, not along the lowland coasts within easy reach of Europeans (Curtin, pers. com.). Unfortunately, the debate over the year of rice introduction to South Carolina has shed little information on whether the indigenous West African rice species, *O. glaberrima,* (which have a red hue) were introduced.

7. On rice sales, purchases, and provisioning of slave ships with foods captives were accustomed to, see: John Atkins, *Voyage to Guinea, Brasil, and the West Indies* (UCLA microfilm: London, 1735); Francis Moore, *Travels into the Inland Parts of Africa* (London: Edward Cave, 1738); Theodore Canot, *Adventures of an African Slaver* (New York: Albert & Charles Boni, 1928); Mungo Park, *Travels into the Interior of Africa* (London: Eland, 1799 [1954 ed.]), 4; Paul Edwards, *Equiano's Travels* (Exeter, New Hampshire: Heinemann, 1967), 33; Philip Curtin, *The Atlantic Slave Trade* (Madison: University of Wisconsin, 1969); K.G. Davis, *The Royal African Company* (New York: Atheneum, 1970); Walter Rodney, *A History of the Upper Guinea Coast, 1545 to 1800* (New York: Monthly Review, 1970); and Paul Lovejoy, *Transformations in Slavery* (Cambridge: Cambridge University, 1983).

8. Clowse, *Economic Beginnings*; Wood, *Black Majority,* also shows that in certain South Carolina rice producing parishes blacks outnumbered whites ten to one by the first decades of the eighteenth century.

cultivation in West Africa during the Atlantic Slave Trade as well as the deliberate selection of bondsmen with the requisite expertise.[9]

This paper builds upon the cross-cultural emphasis developed by Wood (1974), Littlefield (1981), and Otto (1989) by considering the form and the process by which rice cultivation was transferred to the Americas. A detailed examination of the micro-environments planted to rice in West Africa and South Carolina as well as of the underlying knowledge systems that reappeared with the transfer of slaves to the Carolina low-country, underscores the magnitude of the African contribution. The emphasis on slave expertise in rice cultivation suggests, moreover, a different approach to understanding the process of technology transfer. Even though historical recovery of the precise manner in which West Africans tutored planters in rice growing seems unlikely, the process accompanying this knowledge exchange can perhaps be illuminated by examining the task labor system that characterized the rice plantation economy. This unusual system of plantation labor may represent the residue of a complex pattern of negotiation between slave and master in which agronomic knowledge and cooperation in developing rice plantations provided slaves leverage for limiting planter demands on their field labor. Expected work burdens on rice plantations were already established in customary practice during the colonial period. The process of technology transfer consequently may have provided slaves the opportunity to negotiate patterns of labor in rice fields to either duplicate those they knew in Africa or to redefine the nature of New World servitude so as to improve their conditions of existence.

The discussion begins with an overview of the major rice production systems of the West African rice coast, which supplied more than 40 percent of South Carolina's slaves.[10] These production types, whose

9. See Daniel Littlefield *Rice and Slaves: Ethnicity and the Slave Trade in Colonial South Carolina* (Baton Rouge: Louisiana State Univ. Press, 1981) as well as "The Slave Trade to Colonial South Carolina: A Profile," *South Carolina Historical Magazine* 91 (1990):68–99; the Littlefield book review by Philip D. Morgan, *William & Mary Quarterly* 39 (1982):709–12; David Richardson, "The British Slave Trade to Colonial South Carolina," *Slavery and Abolition* 12 (1991): 125–72. Clarence Ver Steeg, *Origins of a Southern Mosaic* (Athens: University of Georgia, 1975); James Clifton, "The Rice Industry in Colonial America," *Agricultural History* (1981):266–83; Charles Joyner, *Down by the Riverside* (Urbana: University of Illinois, 1984); Julia F. Smith, *Slavery and Rice Culture in Low Country Georgia, 1750–1860* (Knoxville: University of Tennessee, 1985); and Otto, *Southern Frontiers*.

10. Curtin, *Atlantic Slave Trade,* 103–04, 128; Littlefield, *Rice and Slaves,* 113; Otto, *Southern Frontiers,* 35; and Joseph E. Holloway, *Africanisms in American Culture* (Bloomington, Ind: Indiana University Press, 1990), 2–7. The rice classification system is based on fieldwork in Senegal, Gambia and Guinea Bissau by Carney during 1984, 1987 and 1990 as well as a review of the secondary and historical literature for the West African rice coast. Senegambia provided a disproportionate number of bondsmen to South Carolina during the first 50 years of the colony's settlement and the critical period for rice experimentation and development;

5 African Expertise

MAP 1 West African Rice Production Zone

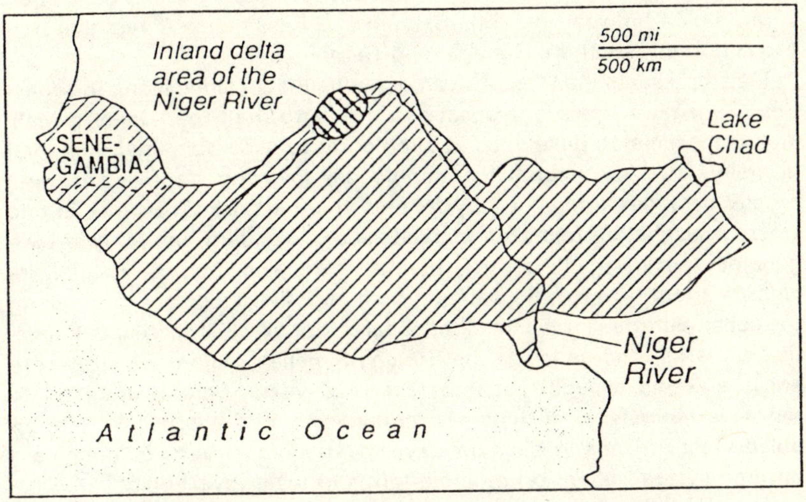

Source: Adapted from Paul Richards, "Upland and Swamp Rice Farming Systems in Sierra Leore: An Evolutionary Transition?," Comparative Farming Systems eds B.L. Turner and Stephen Brush. (New York: Guilford, 1987), pg. 156.

antiquity Islamic and European commentaries affirm, form the basis for the following section that traces the emergence of analogous rice systems in South Carolina by the time of the American Revolution. The last section builds on a comparative analysis of environmental history, labor regimes and the technological expertise of enslaved Africans to recast the traditional interpretation of both the regional economy of the rice-growing South and the trajectory of U.S. agricultural development. By highlighting the active role of African slaves in creating the foundations of a prosperous plantation economy through their technological expertise, the paper attempts to stimulate new research directions for scholars engaged in recovering the histories of subaltern groups.

Rice is an indigenous West African cultivar, domesticated during the

slaves of Senegambian provenance formed nearly 20 percent of South Carolina's bondsmen (Joyner, *Down by the Riverside,* 13).

second millennium B.C. independently of *Oryza sativa,* Asian rice.[11] The center of domestication of native West African rice, *Oryza glaberrima,* is the middle Niger River delta, but important secondary centers developed in the region between the Sine-Saloum and Casamance Rivers that meander through southern Senegal and Gambia.[12]

Map 1 presents the West African rice zone that extends along the coast from Senegal to the Ivory Coast and inland to the Niger Delta in Mali. Rice domestication developed in a region of considerable environmental diversity: riverine floodplains, highlands, lowlying depressions, and coastal estuaries influenced by forceful marine tides. This area is one where precipitation follows a marked seasonal pattern, the rainy season generally occurring during the months of June to October. Rainfall increases south, toward the equator, with slightly higher averages along the coast. But the significant climatic factor for West African rice domestication is its emergence in the Sudano-Sahelian region, which experiences episodic droughts at least every 20 years. Careful attention to moisture regimes, tidal dynamics, topography and edaphic factors enabled West Africans to overcome precipitation constraints by creating a farming system reliant on multiple forms of water availability.[13]

The West African rice production systems that predate the Atlantic Slave Trade are still cultivated in the region today. Three major water regimes regulate rice production: the upland system—originally developed in the Guinea highlands—that relies solely on rainfall; inland swamps, which draw upon supplemental water from moisture-holding clay soils and groundwater reserves for cultivation; and tidal irrigation, which allows rice to be grown along flood plains and estuaries. Each of these rice production systems were established by the 1730s in South Carolina.

Reliant solely on rainfall for moisture reserves, upland rice is generally found in areas receiving more than 1000 mm. It is commonly cultivated on soils supporting mixed woodland vegetation that is partially cleared

11. Carpenter, "History of Rice;" Philip Curtin, *Economic Change in PreColonial Africa* (Madison: University of Wisconsin, 1975); R. Charbolin, "Rice in West Africa," in *Food Crops of the Lowland Tropics,* ed. by C.L.A. Leakey and J.B. Wills (Oxford: Oxford University, 1977), 7–25; and Olga Linares, "From Tidal Swamp to Inland Valley: On the Social Organization of Wet Rice Cultivation among the Diola of Senegal," *Africa* 5 (1981):557–94.

12. Roland Portères, "Primary Cradles;" J. Harlan et al., *Origins of African Plant Domestication*; and T.T. Chang, "Domestication and spread of the cultivated rices," in *Foraging and Farming,* ed. by D. Harris and G. Hillman (London: Unwin Hyman, 1989), 408–17.

13. Paul Richards, *Coping with Hunger* (London: Allen and Unwin, 1986), 5; and Paul Richards, "Upland and Swamp Rice Farming Systems in Sierra Leone: An Evolutionary Transition?," in *Comparative Farming Systems,* ed. by B.L. Turner and S.B. Brush (New York: Guilford, 1987); 156–87. Judith Carney, "Indigenous Soil and Water Management in Senegambian Rice Farming Systems," *Agriculture and Human Values* 8 (1991):37–48.

7 African Expertise

Figure 1. Farmer in Guinea Bissau illustrating use of hollowed palms for water control. Source: Carney fieldwork.

and burned of surface debris. West African farmers have adapted seed varieties to the duration of the rainy season in areas of upland cultivation. Fertility declines, pest build-up and weeding influence the number of years an upland field remains in production, but generally fallows follow three to five years cropping. Cattle grazing is an important rotational sequence in upland rice cultivation. Following the rice harvest cattle are herded into the field to graze the stubble, their manure fertilizing the soil. Abandoned rice fields are frequently converted to cattle pastures during the fallow cycle.

The low-lying Senegambian region is dotted with numerous inland swamps, of limited areal extent, which form another important microenvironment for rice cultivation. This second type of rice production zone frees cultivation from complete dependence on rainfall by trapping supplemental moisture reserves through clay soils, and utilizing groundwater availability from perched water tables, artesian springs or catchment run-off. The major advantage of planting inland swamps is that groundwater resources can be manipulated to irrigate and drain the rice paddy.

8 agricultural history

While the accumulation of water in the root zone extends the cropping period of inland swamps, rice cultivation depends on careful attention to drainage and water control since rice seedlings are as vulnerable to excessive flooding as to drought.[14] As a consequence, inland swamps are made into water reservoirs by the construction of bunds and ridges. Bunding a plot enables the capture of rainfall for soil saturation during dry cycles of the cropping season while planting rice on ridges within the enclosed area improves soil aeration. An additional device for water control enables controlled field flooding, an important technique for weed control. Plot bunds are pierced with hollowed palms (*Borassus aethiopum*) and plugged with thatch, which permits rudimentary control over water inflow and outflow (Figure 1). Depending on type and duration of groundwater availability, inland swamp cultivation generally extends the cropping cycle beyond the rainy season. Cattle graze the harvested plot until the cultivation cycle once again resumes.

The agronomic knowledge and engineering expertise of West Africans revealed their sophistication in tidal rice cultivation, which occurs along riverine floodplains and coastal estuaries. Tidal rice relies upon the diurnal variation in river level to flood and drain the rice field and evolved in three distinct hydrological regimes: (1) freshwater rivers; (2) seasonally saline rivers; and (3) coastal estuaries menaced by marine intrusion.

Tidewater production systems are a function of coastal geomorphology, hydrology, and rainfall patterns. A low gradient characterizes the numerous rivers that discharge into the West African littoral, which enable ocean currents to push saline water considerable distances upstream. The Gambia River, for instance, is saline 70 kilometers upstream from the coast, seasonally brackish up to 250 kilometers, and permanently fresh above. Tidal rice cultivation is consequently a freshwater system above 250 kilometers and in the stretch from 250 to 70 kilometers, planting is adjusted to the downstream displacement of the saline interface. Along coastal estuaries south of the Gambia River where annual precipitation exceeds 1500 mm, rice cultivation occurs in environments swept by ocean tides.

The freshwater tidal rice system involves minimal landscape modification. Essentially, farmers let river tides irrigate alluvial floodplains. As tides sometimes reach one or two feet amplitude, the swamps are transplanted to rice varieties that elongate rapidly. The downstream seasonally saline zone is managed in a similar fashion, although sometimes a low embankment is placed between the rice field and the river to reduce tidal force. More often the task is left to nature by leaving a band of

14. D.H. Grist, *Rice* (London: Longmans, Green & Co., Ltd, 4th ed., 1968).

mangroves between the river and rice perimeter. Cultivation proceeds about one month after the rains restore freshwater reserves in river headwaters. Gradually the saltwater interface moves downstream, releasing floodplains to rice cultivation along riverine stretches with a least three months freshwater.

A variation of tidewater production in seasonally saline zones exists south of Gambia along rivers with limited freshwater but with higher levels of rainfall (>1500 mm). This system involves considerable landscape modification for water retention, drainage and controlled flooding.

The West African littoral is mantled by vast stretches of mangroves (*Rhizophera* spp) whose aerial roots trap alluvium and detritus swept into their reach by strong ocean tides. Such sediments form some of the richest soils in the West African rice zone, but require careful management since they contain quantities of iron sulfates (pyrite) that oxidize upon aeration and render the soil too acidic for cultivation.[15] Farmers prevent acidification by keeping the soil saturated during part of the year, and thus reap the higher yields from the rich organic matter retained by this method. Soil saturation is achieved by capturing rainfall and river tides.

Preparation of a tidal rice field in seasonally saline areas begins with land reclamation. A site is selected and enclosed by an earthen embankment that runs parallel to the riverine arm of the sea (Figure 2). The embankment frequently reaches more than a meter's height and span and often stretches over several kilometers. It is constructed of ample height and width to block unwelcome tidal surges and avoid the loss of rainwater impounded within the perimeter. Soil excavation for embankment construction establishes the location of the principal drainage canal. A series of lower embankments or dikes are then formed inland and perpendicular to the main one, which divide the perimeter into individual rice parcels.

Once tidal flows are barred, the system of water control is completed. A series of sluices are constructed to move water through the drainage canals for irrigation or evacuation. The sluices are fitted with valves constructed from hollowed tree trunks (frequently silk cotton species, *Bombax costatum* or *Ceiba pentandra*). Under pressure of accumulated rainfall and sea tides, the sluice gate opens and closes automatically, the valves being sealed at low tides by the pressure of rainwater retained in the plots, and being opened by the pressure of river water during high

15. F.R. Moorman and W.J. Veldkamp, "Land and Rice in Africa: Constraints and Potentials," in *Rice in Africa,* ed. by I. Buddenhagen and G. Persley (New York: Academic Press, 1978), 29–43; West African Rice Development Association (WARDA), *Types of Rice Cultivation in West Africa,* Occasional Paper #2 (Monrovia: WARDA, 1980).

10 agricultural history

Figure 2 Three examples tidewater rice system, Casamance, Senegal

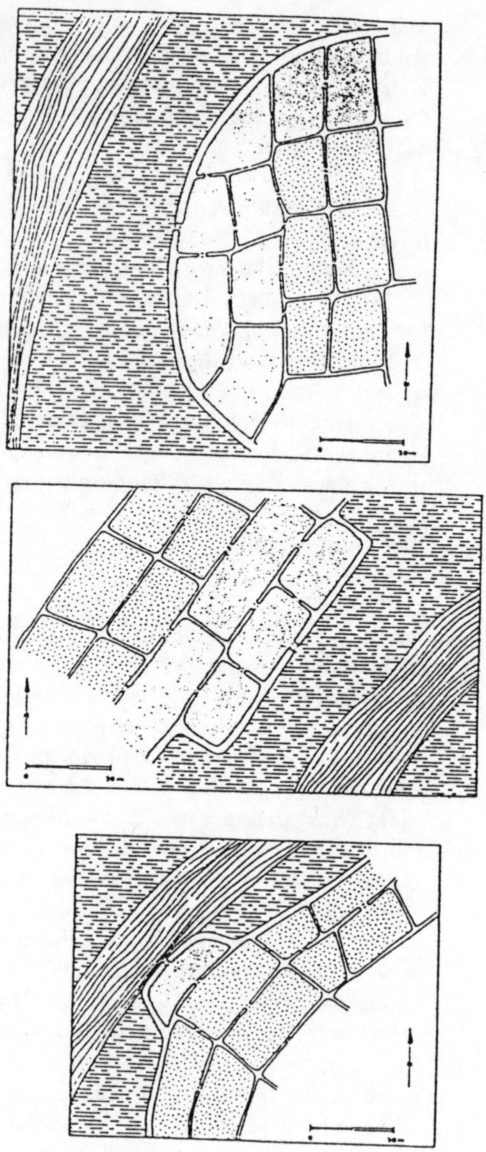

Source: Pelissier 1966, pg. 723a.

tide. An automatic valve system is not always desired, especially at critical periods in the cultivation cycle when river tides are used for field flooding and weed control, or in the pre-harvest period if brackish water returns to the estuary. Under these circumstances, farmers plug the trunk with vegetal matter or planks so water flows in one direction, from paddy to river.

South along the West African Rice Coast, increased rainfall permits yet a final variation on this system. Rice is grown along or near the coast where marine tides prevail year round. While this sub-type of tidal rice cultivation enables cultivation in areas threatened by ocean tides, greater effort is expended on plot desalination and weed control. Desalination of the marine-influenced soil is accomplished by rainfall, which over the span of several years, flushes the salts from the soil out to sea through the trunk valves. Rainfall that accumulates in the rice plot is used to drown noxious weeds with unwanted water evacuated through the valve system. A slight difference over the preceding example occurs in the valve design for water flow within this system. The tree trunks are narrowed into a cone at one end only, which keeps the valve closed against sea water intrusion but open for draining excess rainwater. Occasionally small canoes serve the same purpose and a plank is placed in position to regulate water outflow.[16] Once brought into production, these tidal fields never need fallowing.

Each of the indigenous rice systems present differing labor demands, harvest dates, weeding and fertility practices—factors that have minimized regional subsistence shortfalls for millennia. A sophisticated understanding of tidal flows, moisture regimes, soil properties, saltwater movements as well as topographic diversity created an agrarian system that maintained high levels of productivity to support some of West Africa's densest populations.[17] This agronomic knowledge and sophisticated technology was transferred by the men and women enslaved to work on rice plantations in South Carolina.

When Portuguese arrived along the West African coast in the mid-fifteenth century, they encountered rice cultivation over a broad area from Gambia to Sierra Leone. But rice cultivation was also widespread along the inland Niger Delta (Map 1), as noted by Islamic scholars who journeyed overland to the Mali Empire during the fourteenth and six-

16. Rodney, *History of Upper Guinea Coast,* 22.

17. On population densities in West African rice farming systems see Linares, "From Tidal Swamp;" D. Paulme, *Les gens du riz* (Paris: Librairie Plon, 1954); and Rodney, *History of Upper Guinea Coast.*

teenth centuries.[18] While medieval Islamic sources provide little insight into rice farming systems except to note the frequency of flood recession agriculture, early European commentaries provide more detail. By these accounts it is clear that African rice culture antedated the arrival of Europeans and was highly sophisticated.

By the 1450s Portuguese accounts begin to mention rice cultivation in the region from Senegal to Gambia, the initial area of rice culture encountered as ships inched southwards. The first commentary on rice dates to 1453 when Portuguese mariners disembarked along one Senegambian river to find a broad river plain sown to rice.[19] Two years later Alvise da Cadamosto, the first European to describe the River Gambia, commented on rice cultivation in the region as well as the diversity of varieties planted.[20] Rice proved so abundant along the West African coast that by 1480 Portuguese ships were purchasing the cereal for provisions, often from female traders.[21]

With the European presence largely confined to the Atlantic coast and river estuaries, the rice systems adjusted to marine and riverine tides received early comment. Valentim Fernandes, who journeyed along the Senegambian coast between 1506–1510, was probably referring to tidal and flood recession agriculture when he mentioned Senegambian farm-

18. For Arabic sources on West African food systems from the tenth through sixteenth centuries, see Tadeusz Lewicki, *West African Food in the Middle Ages* (Cambridge: Cambridge University Press, 1974).

19. ". . . they arrived sixty leagues beyond Cape Verde, where they met with a river which was of good width, and into it they entered with their caravels . . . they found much of the land sown, and many . . . fields sown with rice . . . And . . . all that land seemed . . . like marshes." Gomes Eannes de Azurara, *The Chronicle of the Discovery and Conquest of Guinea,* vol.II (London: Hakluyt Society, 1899), 263–64.

20. "In this way of life they conduct themselves in almost all respects similarly to the negroes of the kingdom of Senega [Senegal]; they eat the same foods except they have more varieties of rice than grow in the country of Senega . . ." G.R. Crone, *The Voyages of Cadamosto* (London: The Hakluyt Society, 1937), 70.

21. See Paul Pélissier, *Les Paysans du Senegal* (St. Yrieix: Impr. Fabregue, 1966), 711–12 for review of rice commentaries during voyages of Eustache de la Fosse (1479–80), Duarte Pacheco Pereira (1505–06), and Valentim Fernandes (1506–07). On rice sales during the fifteenth century see Crone, *Voyages of Cadamosto,* 48–49. Women's involvement in the rice trade was first noted by Ibn Battuta who travelled to Mali in the mid-fourteenth century; see H.A.R. Gibb, *Ibn Battuta: Travels in Asia and Africa, 1325–1354* (London: Routledge & Kegan Paul, 1969), 322. The earliest European account (1580) also draws attention to women's role in the rice trade: ". . . and here the black women hold a market when ships are in port; they bring for sale rice . . ." in A. Donelha, *An Account of Sierra Leone and the Rivers of Guinea and Cape Verde* (Lisboa; Junta de Investigações Cientificados do Ultramar, 1977), 149. References to female traders and the rice trade increase during the seventeenth and eighteenth centuries with the augmented demand for cereals to provision slave ships. See Judith Carney, *The Social History of Gambian Rice Production: An Analysis of Food Security Strategies* (Ph.D. dissertation, Geography, University of California Berkeley, 1986).

13 African Expertise

ers harvesting two rice crops a year.[22] De Almada, describing the same area in 1594, noted the use of river tides for floodplain rice cultivation while Jobson, the English explorer who journeyed up the Gambia River in 1620–1621, observed rice being transplanted to riverine swamps.[23] Caillié left a clear description of tidal rice cultivation from his overland trip through Senegambia during the early nineteenth century: ". . . rice grows to the height of four feet. The soil, which is composed of a very hard grey sand, is fertilized by the inundations of the Tankisso."[24] Widespread on floodplains from the Gambia River northwards where precipitation falls below 1000 mm, tidal rice was likely the initial production system encountered by Europeans.

As the Atlantic slave trade expanded along the Upper Guinea coast, commentaries proliferated on the tidal saline system that sustained high population densities. Alvares De Almada in 1594 described this system south of the Gambia River, noting the use of dikes to impound rainwater for desalination and seedling submersion as well as the technique of transplanting.[25] When Sieur de la Courbe journeyed overland from Gambia to Guinea Bissau in 1685 he marvelled at the extensive system of dikes and rice polders developed along river estuaries.[26] One hundred years later Samuel Gamble, a slave captain who worked the Upper Guinea coast, described one ethnic group—the Bagas—as expert in rice cultivation. His diagram of tidal rice and accompanying account provide information on the water management techniques employed in the tidal saline system, the gender division of labor, and use of the specialized long handled shovel (*kayendo*) for turning heavy clay soils (Figure 3).[27]

22. "Du Cap Vert jusqu'ici il y a deux hivernages et deux hivers chaque année, deux fois ils sèment et deux fois ils récoltent le riz et le mil etc, à savoir une fois ils récoltent en avril et une en septembre et quand ils moissonent le riz alors ils sèment les ignames et ainsi ils cultivent toute l'année." In Valentim Fernandes, *Description de la Côte Occidentale d'Afrique. Memoire No.11* (Bissau: Centro de Estudos de Guiné Portuguesa, 1951), translated from Portuguese by A. Teixeira da Mota and R. Mauny.

23. Rodney, *History of Upper Guinea Coast,* 20–21; and Richard Jobson, *The Golden Trade* (Teignmouth, Devonshire: Speight and Walpole, 1904 edition), 59.

24. Rene Caillié, *Travels through Central Africa to Timbuctoo, and across the Great Desert, to Morocco, performed in the Years 1824–1828* (London: Henry Colburn and Richard Bentley, 1830), 210.

25. "Fazem os negros as searas dos arrozes naquelas lalas, e fezem valados de terra por amor da venida do rio, mas nem por isso deixa o rio muitas vezes de os romper e alargar as searas. Depois deste arroz, nado, o arrrancam e transpõem em outras lalas, mais enxutas, onde de logo mantimento." In Pélissier, *Les Paysans du Senegal,* 714.

26. ". . . il avait déjà commencé a pleuvoir . . . je vis des lougans de riz qui sont tout le long du bord de la rivière; ils sont traversés de petites chausées [dikes], d'espace en espace, pour empêcher que l'eau ne s'écoule . . ." Ibid., 714.

27. "The Bagos are very expert in Cultivating rice and in quite a Different manner to any of the Nations on the Windward Coast. The country they inhabit is chiefly loam and swampy. The rice they first sew [sic] on their dunghills and rising spots about their towns; when 8 or

14 agricultural history

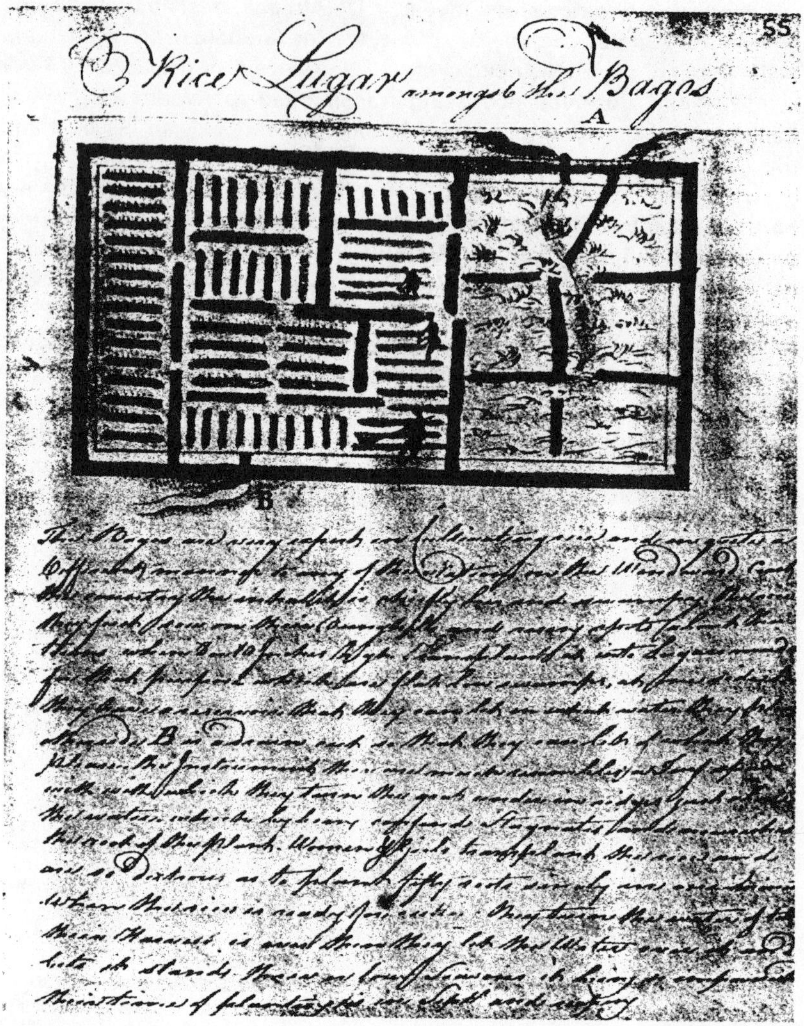

Figure 3. Illustration and description of Baga rice cultivation, in Captain Samuel Gamble's journal, *ca.* 1793. Courtesy of National Maritime Museum, London. Source: Daniel Littlefield, 94. Reprinted with permission of the author.

15 African Expertise

Thus, from the sixteenth century the techniques of diking, desalinating, ridging, transplanting, and drainage that continue to characterize tidal saline rice cultivation south of The Gambia received repeated mention by European traders along the West African littoral.[28]

Commentaries on rice systems located away from the coast and riverine arteries used by European traders, however, were generally absent before the eighteenth century, until European overland expeditions set forth to discover the sources of major West African rivers. From this period onwards the inland swamp cultivation system receives frequent mention as well as the techniques of plot diking, ridging, and water impoundment.[29] Frances Moore, a Gambia-based factor for a European trading company at the height of the Atlantic Slave Trade during the 1730s, commented on the seasonal rotation of inner tidal floodplains as cattle pasture: "The sides of the River are for the most part flat and woody, for about a quarter of a Mile inland, in some Places not so much, and within that are pleasant open Grounds, which they use for their Rice, and in the dry Season it serves the cattle for Pasture."[30] With the dispersal of European explorers further inland throughout the nineteenth century, commentaries on rainfed cultivation and particularly, the burn-

---

10 Inches high [they] transplant it into Lugars made for that purpose which are flat low swamps, at one side . . . they have a reservoir that they can let in what water they please, other side . . . is a drain out so they can let off what they please. The Instrument they use much resembles a Turf spade with which they turn the grass under in ridges just above the water which by being confined Stagnates and nourishes the root of the plant. Women & Girls transplant the rice and are so dextrous as to plant fifty roots singly in one minute. When the rice is ready for cutting they turn the water off till their Harvest is over then they let the Water over it and lets it stands three or four Seasons it being so impoverished." In Daniel Littlefield, *Rice and Slaves,* 93–95.

28. Pélissier, *Les Paysans du Senegal,* 711–16; and Rodney, *History of Upper Guinea Coast,* 20–23.

29. On planting on ridges: "The rice, which is . . . esteem'd their choicest Food, they set in Rills . . .; it grows in wet grounds, the Ears like Oats, in Moore, *Travels into the Inland Parts of Africa,* 31. Descriptions of the phreatic rice system include: (1) "Rice is almost the only grain sown at Gambia in the lands overflown by the rains of the high season. The negroes cut all these lands with small causeys which with-hold the waters in such a manner, that their rice is always moistened," in M. Adanson, *A Voyage to Senegal, The Isle of Goree and the River Gambia* (London: J. Nourse and W. Johnston, 1759), 166; and (2) "As the country is flat, they take care to form channels to drain off the water. When the inundation is very great, they take advantage of it and fill their little reservoirs, that they may provide against the drought and supply the rice with the moisture which it requires," in Caillié, *Travels through Central Africa,* 162. See also Park, *Travels into the Interior of Africa;* and G. Mollien, *Travels in Africa* (London: Sir Richard Phillips & Co, 1820).

30. Moore, *Travels into the Inland Parts of Africa,* 37.

ing of forested tracts for agro-pastoral rotations become more numerous.[31]

Several conclusions can therefore be drawn from these fragmentary historical records. Rice cultivation was widespread throughout West Africa prior to contact with Europeans. Numerous seed varieties were known and developed with differing tolerances to drought and salinity as well as maturation dates. The major production zones utilized for rice cultivation were skillfully husbanded by manipulating moisture regimes, topography and soil type with techniques that eased labor demands and simultaneously enhanced subsistence security. And finally, it was this sophisticated technology that provided the impetus for slave traders to procure their quarry from the Rice Coast and transport them across the Atlantic to the South Carolina plantations.

Although South Carolina remained a rice producer through the early 1900s, the diverse micro-environments were already sown to rice when the Stono slave rebellion erupted in 1739. While cultivation systems analogous to those of the West African rice belt developed in the colony well before the rebellion, the reliance on each type of cultivation described above varied with the evolution and commoditization of the rice economy. The early-eighteenth century serves as a point of departure for this section, which presents an historical overview of the changing emphasis on rice systems shaped in part by the technical expertise that slaves carried with them from West Africa to South Carolina.

Initial efforts in rice cultivation were mainly carried out in upland areas of South Carolina and complemented the colony's early economic focus on stock raising.[32] Broadleaf hardwood forests were girdled, felled and

31. Thomas Winterbottom, *An Account of the Native Africans in the Neighbourhood of Sierra Leone* (London: C. Whittingham, 1803), 49. For early archival reference (1616) on forest clearing for agriculture in Sierra Leone, see Rodney, *History of Upper Guinea Coast,* 22–23.

32. For accounts of upland rice as the first system implemented in South Carolina see: Thomas Nairne, "A Letter from South Carolina," [1710] and John Norris, "Profitable Advice for Rich and Poor," [1712] in *Selling a New World: Two Colonial South Carolina Promotional Pamphlets,* ed. by Jack P. Greene (Columbia, S.C.: University of South Carolina Press, 1989); John Drayton, *View of South Carolina* (Charleston, S.C.: W.P. Young, [1802] 1972 reproduction of original by The Reprint Company, Spartanburg, S.C.); R.F.W. Allston, "Essay on Sea Coast Crops," *De Bow's Review* 16/1 (1854):589–615; T.A. Richards, "The Rice Lands of the South," *Harper's New Monthly Magazine* 114/19 (1859):721–38; Gray, *History of Agriculture,* 46; Clowse, *Economic Beginnings;* Ver Steeg, *Origins of a Southern Mosaic;* Sam B. Hilliard, "Antebellum Tidewater Rice Culture in South Carolina and Georgia," in *European Settlement and Development in North America: Essays on geographical change in honour and memory of Andrew Hill Clark,* ed. by James R. Gibson (Toronto: University of Toronto Press, 1978), 91–115; Clifton, "Rice Industry in Colonial America;" Richard D. Porcher "Rice Culture in South Carolina: A Brief History, The Role of the Huguenots, and Preservation of Its Legacy," *Transactions of the Huguenot Society of South Carolina* 92 (1987):1–22; and Otto, *Southern Frontiers.* For a detailed description of rice cultivation systems in South Carolina, see Judith

17 African Expertise

burned for agro-pastoral production.[33] A land use system prevailed in which cereals—increasingly rice cultivation—rotated with cattle pasture.[34] Livestock figured importantly in export earnings along with naval stores based on the extraction of tar, pitch, and turpentine from less fertile, pine-forested tracts.[35] Rice began to surpass livestock exports in value during the first decades of the eighteenth century in response to West Indian demand for the crop as the slave food staple on sugar cane plantations.[36]

South Carolina's agro-pastoral economy rested on slave labor. Slaves were charged with clearing, planting, extracting marketable forest produce such as resin, and herding cattle—to prohibit stock damage to cultivated fields during the day, to corral the animals at night for safety, and to move herds to other areas for seasonal grazing and forage.[37] Even though metropolitan advisors had recommended rice as a potential cash crop for the region prior to white settlement, the techniques of its production were unknown to European planters of English and French Huguenot descent. An English observer writing in 1704 recalled Europeans ". . . being unacquainted with the manner of cultivating rice" or with the methods for processing the crop.[38] Although rice was a valued food staple during Lent in Catholic Europe by the sixteenth century and had been cultivated in Italy since 1475, the crop was not widely consumed or

---

Carney and Richard Porcher, "Geographies of the Past: Rice, Slaves and Technological Transfer in South Carolina," *Southeastern Geographer* (forthcoming).

33. Ver Steeg, *Origins of a Southern Mosaic,* 115; Clifton, "Rice Industry in Colonial America;" Peter Coclanis, *The Shadow of a Dream* (New York: Oxford University Press, 1989), 36; and Otto, *Southern Frontiers,* 30–35.

34. Norris, "Profitable Advice," 37–42.

35. On the livestock trade Nairne ("Letter from South Carolina," 41) wrote: ". . . South Carolina abounds with black Cattle to a Degree much beyond any other English colony . . . These creatures have mightily increas'd since the first settling of the colony, about 40 years ago. It was then reckon'd a great deal to have three or four Cows, but now some People have 1000 Head, but for one Man to have 200 is very common." Otto, *Southern Frontiers,* 30, reports South Carolina exports of salted meat to the English West Indies by 1682; and Ver Steeg, *Origins of a Southern Mosaic,* 120–21 refers to communications by Surveyor General Edward Randolf during his 1699 visit to the colony that mention the significance of pitch, tar and turpentine to the Carolina economy.

36. On the changing basis of exports see, Ver Steeg, *Origins of a Southern Mosaic,* 117; Clowse, *Economic Beginnings;* and Stephanie McCurry, "Defense of their World: Gender, Class, and the Yeomanry of the South Carolina Low Country, 1820–1860" (Ph.D. dissertation, History, State University of New York at Binghamton, 1988). For the role of rice on sugar plantations: Marietta Morrissey, *Slave Women in the New World* (Lawrence, Kansas: University Press of Kansas, 1989), 24.

37. See Peter H. Wood, " 'It was a Negro Taught Them,' a New Look at African labor in Early South Carolina," *Journal of Asian and African Studies* 9 (1974b):160–69; and Otto, *Southern Frontiers,* 30–33.

38. Quoted in Wood, *Black Majority,* 56 who first drew attention to European ignorance of rice cultivation and suggested an African contribution to its development in South Carolina.

Figure 4 Inland and Tidal Rice Systems on the Bluff Plantation*

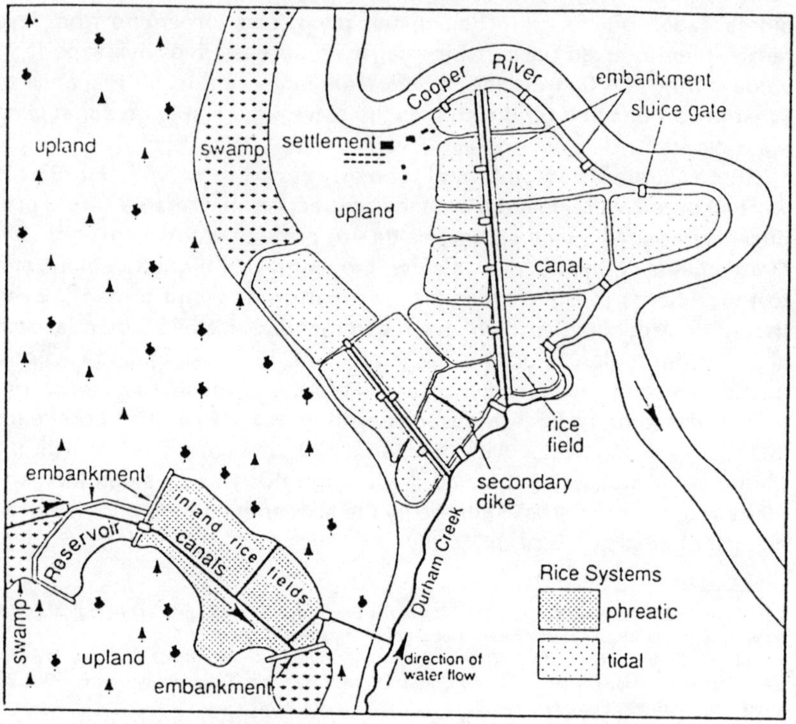

* Located at western branch of the Cooper River, Berkeley County, South Carolina.

Note: This map is a modification of the Bluff Plantation map in Richard D. Porcher, A Field Guide to the Bluff Plantation (New Orleans: The Kathleen O'Brien Foundation, 1985) pg. 266. The modification shows both phreatic and tidal fields on one map.

19 African Expertise

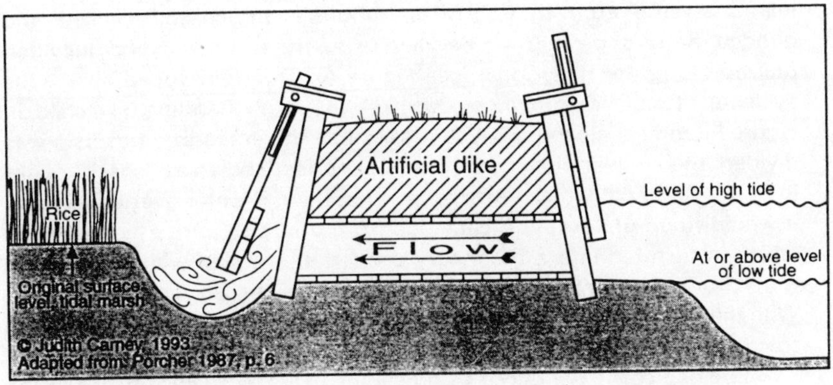

Figure 5. Cross Section of Tidewater rice trunk

planted in Protestant Europe.[39] Rice cultivation as well as its rotation with cattle pasture was, however—as historical records clearly indicate—a widespread land use pattern along the Upper Guinea coast, frequented by seventeenth- and early-eighteenth century slavers.[40]

The capital accumulated from ranching and forest extractive activities by the beginning of the eighteenth century was used to finance the swelling importation of slaves from West Africa for accelerated development of the rice export economy.[41] The mounting economic reliance on rice exports in the early period of the eighteenth century was based on expansion of cultivation: production spread from the uplands to inland swamps.[42] Rice yields were better in these forested swampy soils, rich in organic matter and of high moisture retention.[43] But the arduous process of first clearing the cypress and gum swamps was only achieved by escalating the import of slaves during the early decades of the century. Slaves converted the swamps to rice fields and frequently constructed reservoirs to supplement rainfall for improved yields.

After surveying topographic features and groundwater availability for

39. Grist, *Rice*, 7.

40. See K.G. Davis, *The Royal African Company* (New York: Atheneum, 1970); Rodney, *History of Upper Guinea Coast;* Curtin, *Economic Change,* 100–01, 144–45, 157; Wood, *Black Majority,* 45, 131; Otto, *Southern Frontiers,* 35–37.

41. Clowse, *Economic Beginnings;* Wood, *Black Majority;* Ver Steeg, *Origins of a Southern Mosaic,* 114, 129, 131; Littlefield, *Rice and Slaves;* and Otto, *Southern Frontiers,* 33.

42. See Norris, "Profitable Advice," 97–98; Nairne, "Letter from South Carolina," 40; Whitten, "American Rice Cultivation," 6 and Otto, *Southern Frontiers,* 34.

43. Heyward, *Seed from Madagascar,* 11–16; Clowse, *Economic Beginnings,* 125–26; Hilliard, "Antebellum Tidewater Rice," 94–100; Whitten, "American Rice Cultivation," 6–7; Porcher, *Field Guide to The Bluff Plantation,* 26–28; and Otto, *Southern Frontiers,* 34–35.

inland swamp cultivation, African laborers embanked, levelled and bunded marshlands in South Carolina using the technology that they had utilized along the Rice Coast (see Figure 4). They developed analogous systems of sluices (known as "trunks" in South Carolina, sketched in Figure 5) and canals to replenish reservoirs. The embanked rice field was divided into smaller plots by a system of bunds (known as "check" banks in South Carolina) which ensured an even flow of water over the crop so it would ripen uniformly. A canal led from the reservoir to the rice field, which in turn fed into a secondary canal that emptied with gravity flow from the lower bank through a trunk into an adjacent creek (Figure 4). With the system thus in place, water could be drawn or drained according to crop requirements.[44]

The close correspondence in principles of water control between the West African and South Carolina inland swamp systems, as well as the emergence of identical mechanisms for regulating water flow indicates that both brain and brawn of slave labor was being put to diligent use by planters. One pertinent example is the use of the term "trunk" by South Carolina planters for sluices. Technology transfer from West Africa is revealed, albeit unwittingly, by Doar (writing in 1936) whose curiosity over the origin of the term led to an unexpected discovery: " 'For years the origin of this name bothered me. I asked every old planter I knew, but no one could enlighten me. One day a friend of mine who planted on one of the lowest places . . . said to me with a smiling face: 'I have solved that little trunk question. In putting down another one, I unearthed the granddaddy of plug trunks made long before I was born.' It was simply a hollow cypress log with a large hole from top to bottom. When it was to be stopped up a large plug was put in tightly and it acted on the same principle as a wooden spigot to a beer keg."[45] South Carolina rice fields, like their West African counterparts, consequently used hollowed logs as sluice valves before the wooden gate system developed (compare Figures 1 and 5). The hollowed logs were plugged to regulate water entry and exit in the same manner as in West Africa. It is likely that the initial use of tree trunks for regulating water flow led to the later sluice system retaining the name, "trunk."

The inland swamp cultivation method, like its West African counterpart, presented numerous advantages over upland rice.[46] The rice field

44. This account draws on references from footnotes 49 and 50.

45. Doar, *Rice and Rice Planting,* 12.

46. Two sub-types of the West African system also were planted in South Carolina: inland swamps fed by underground springs and those carved from saline marshes. The latter was found along the Cooper River where "The rich marshes tempted planters as far down the river as Marshlands [Plantation], nearly within sight of the ocean. Here they had to depend entirely

no longer needed fallowing to sustain fertility and thus could be cultivated continuously. Besides providing an auxiliary water reserve for mid-season droughts and a means to drain fields in excessively wet years, the reservoir system could also be used for field flooding and thus, weed suppression—a major labor-demanding activity.[47] Reviewing early colonial references to rice cultivation, Gray underscored the significance of weeding and its associated labor demands for inland and subsequently, tidal cultivation: "Only two flowings were employed [inland] as contrasted with the later period when systematic flowings [tidal] came to be largely employed for destroying weeds, a process which is said to have doubled the average area cultivated per laborer . . . The later introduction of water culture [tidal] consisted in the development of methods making possible a greater degree of reliance than formerly on systematic raising and lowering of the water."[48]

Despite the advantages over the upland system that preceded it, inland rice production faced three problems: (1) the limited availability of suitable swamps; (2) an uncertain water supply in the rainfed reservoirs during dry years; and (3) in some years, severe end-of-season rainstorms (freshets) that caused creeks and rivers to overflow and destroy the rice crop.[49] Even though these drawbacks existed, the inland system remained the dominant method of cultivation until the 1750s when the techniques of harnessing river tides for rice production had diffused throughout the colony.

Tidal, or tidewater, rice dominated South Carolina production from the Antebellum period until the demise of cultivation in the twentieth century (Map 2). The earliest reference dates to 1738 with a notice of land sale by William Swinton of Winyah Bay: ". . . that each contains as much River Swamp, as will make two Fields for 20 Negroes, which is overflow'd with fresh Water, every high Tide, and of Consequence not subject to the Droughts."[50] The statement clearly establishes that: (1) the principles of tidewater cultivation were already known in the colony; (2) reliance on tidal flow reduced drought stress in rice production; and (3) the emphasis on tidewater rice depended on large numbers of slaves for clearing the

---

on 'reserve' waters formed by damming up local streams." John B. Irving, *A Day on Cooper River* (Charleston, S.C.: The R.L. Bryan Co., 1969), 154. See also N.R. Hawley, "The Old Plantations in and around the Santee Experimental Forest," *Agricultural History* 23 (1949):86–91.

47. Hilliard, "Antebellum Tidewater Rice;" and Heyward, *Seed from Madagascar,* 14.

48. Gray, *History of Agriculture,* 281.

49. Letters of George Ogilvie, 25 June 1774, South Carolina Historical Society; Heyward, *Seed from Madagascar;* Hilliard, "Antebellum Tidewater Rice,"; and Porcher, "Rice Culture in South Carolina.

50. The land sale was advertised on 19 January 1738 in the *South Carolina Gazette.*

Map 2

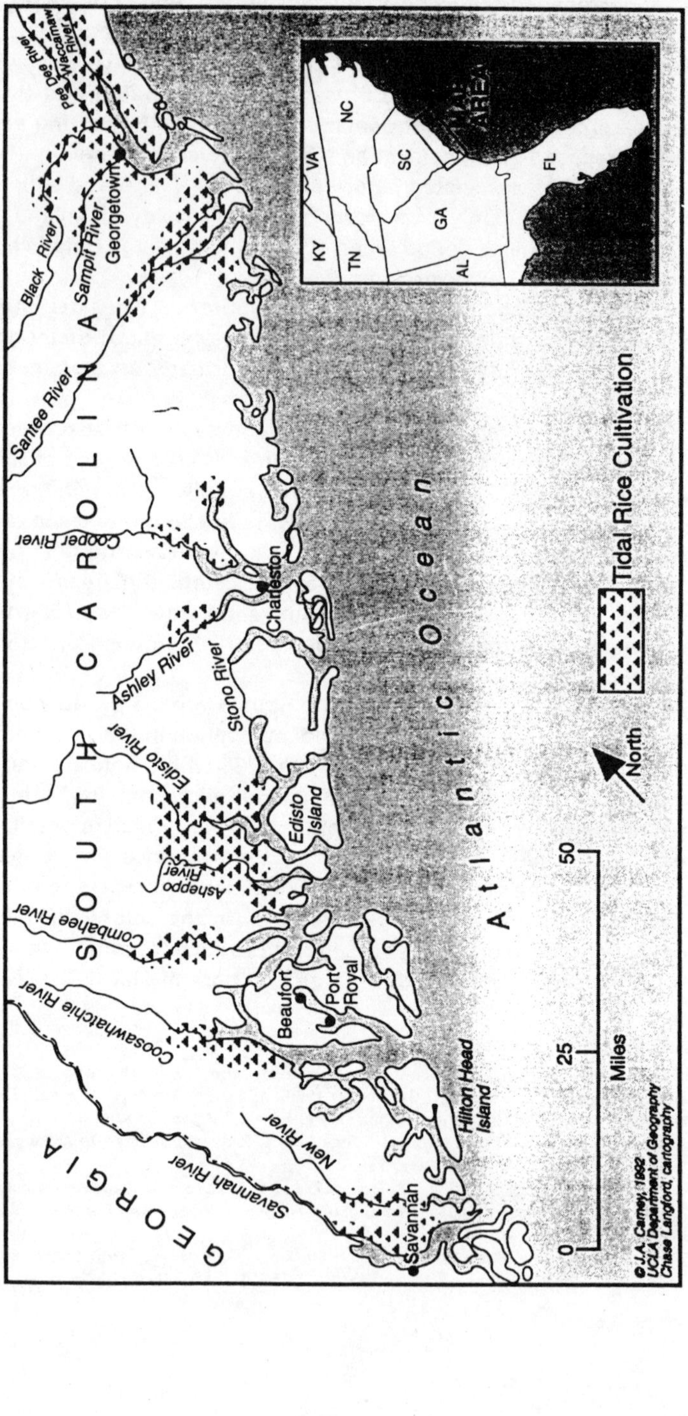
GEORGIA
SOUTH CAROLINA
Savannah River
New River
Savannah
Hilton Head Island
Coosawhatchie River
Beaufort
Port Royal
Combahee River
Ashepoo River
Edisto River
Edisto Island
Stono River
Ashley River
Charleston
Cooper River
Santee River
Black River
Sampit River
Georgetown
Pee Dee River
Waccamaw River
Atlantic Ocean
Tidal Rice Cultivation
North
0 25 50
Miles
© J.A. Carney, 1992
UCLA Department of Geography
Chase Langford, cartography
KY
VA
NC
TN
SC
GA
AL
FL

cypress and gum swamps as well as constructing and maintaining the elaborate infrastructure.[51] The agronomic principles involved in field flooding appeared in South Carolina during the first half of the eighteenth century. The growing emphasis on tidewater rice production became evident in the dramatic increase in slave imports, which nearly doubled from 39,000 to 75,000 between 1750–1770.[52]

The tidewater system, which one South Carolina planter termed a "huge hydraulic machine," rested on an "apparatus of levels, floodgates, trunks, canals, banks, and ditches . . . requiring skill and unity of purpose to keep in order."[53] Slaves from the Upper Guinea coast possessed the engineering know-how to transform tidal swamps into a rice-producing "hydraulic machine" as well as the technical expertise to apply the "apparatus of production" for swamp reclamation. Controlled flooding not only depressed weeding but also greatly reduced labor demands over previous systems. A slave was consequently able to manage five acres instead of the two typically planted to inland rice cultivation.[54] In short, tidewater cultivation led to improved labor output and increased yields, giving planters a "unity of purpose" for obtaining West African slaves with the requisite knowledge and skills.

Preparation of a tidewater plantation for rice cultivation followed principles evident in West African tidal production (compare Figure 4 with 2). An embankment of sufficient height to prevent tidal spillover was constructed along the river as well as through the swamp area for additional water control (Figure 4). The earth removed in the process resulted in an adjoining canal, used to receive or drain water from the fields. A series

51. For discussions on the evolution and operation of the tidewater rice cultivation system, see: Edmund Ruffin, "General Description of the Tide Water Swamps in their Natural State," *Commercial Review* 4 (1847):505–11; Allston, "Sea Coast Crops;" Richards, "Rice Lands;" A.M. Ferster, "Rice Culture in South Carolina," *Proceedings of the Elliott Society of Natural History* (December 1860):40–46; Doar, *Rice and Rice Planting;* Heyward, *Seed from Madagascar;* Douglas C. Wilms, "The Development of Rice Culture in 18th Century Georgia," *Southeastern Geographer* 12 (1972):45–57; Sam B. Hilliard, "The Tidewater Rice Plantation: An Ingenious Adaptation to Nature," *Geoscience and Man* 12 (1975):57–66; James Clifton, "Rice Industry"; David Whitten, "American Rice Cultivation, 1680–1980: A Tercentenary Critique," *Southern Studies* 21 (1982):5–26; James Clifton, "Jehossee Island: The Antebellum South's Largest Rice Plantation," *Agricultural History* 59/1 (1985):56–65; Timothy Silver, *A New Face on the Countryside. Indians, Colonists and Slaves in South Atlantic Forests, 1500–1800* (New York: Cambridge, 1990) and Margaret Washington Creel, *"A Peculiar People," Slave Religion and Community-Culture Among the Gullahs* (New York: New York University, 1988), 34.

52. Otto, *Southern Frontiers,* 41.

53. Doar, *Rice and Rice Planting,* 8.

54. Allston, "Sea Coast Crops;" Clifton, "Rice Industry," 275; Whitten, "American Rice Cultivation," 9–15.

of trunks with vertically positioned flaps operated as valves to control water flow between the cultivated area and the river.[55]

Tidewater plots were indeed highly productive as a result of the sophisticated knowledge and agronomic skills brought to South Carolina plantations by West African slaves. Two factors determined where tidewater fields could be constructed: tidal amplitude and saltwater encroachment.[56] A location too near the ocean faced saltwater incursion while one too far upstream removed a plantation from tidal influence.[57] As in West Africa, a three- to five-foot tidal pitch was necessary to flood and drain the rice fields—a condition matched in South Carolina only along riverine stretches within 20 miles of the coast. Whether or not tidal cultivation reached to the coast, however, depended on even more complex hydrological conditions: namely, differences in freshwater dynamics between rivers draining the uplands and those flowing inland from the sea. Since rivers of piedmont origin deliver freshwater nearly to the coast, tidal cultivation can occur within a short distance from the ocean. But rivers that are arms of the sea must reach further inland for freshwater flows. Along these rivers the freshwater stream flow forms a pronounced layer on top of the heavier saltwater, the former being tapped for tidal irrigation.[58] The sites suitable for tidal cultivation in South Carolina consequently required skilled manipulation of tidal flows and saline–freshwater interactions to attain high levels of productivity.

Like the West African tidal saline system, South Carolina tidewater production depended upon diurnal variation in marine tides and its effect on estuary and river freshwater levels. When the tide raised freshwater in the river above the field level, flooding was possible and trunks opened to allow water onto the fields. With successive high tides, fields could be flooded to the required height. Field drainage occurred through a reversal of this process—opening the trunks at low tide so water could run off the fields.

By the time of the American Revolution, the colony's economic prominence was anchored to tidewater cultivation. The appearance and rapid diffusion of the tidewater system in South Carolina during a century of

55. Doar, *Rice and Rice Planting,* 7–15; Clifton, 7; Whitten, "American Rice Cultivation," 9–10; and Hilliard, "Tidewater Rice Plantation," 58–60 and "Antebellum Tidewater Rice Culture," 98–110.

56. Ibid.

57. Ibid., 101.

58. Reliance on the sharp horizontal discontinuity between fresh and saltwater in tidewater cultivation, however, was not without risk. Climatic factors were also important: periodic hurricanes drove saltwater upriver, which destroyed the rice crop; and unusually severe rainstorms at the end of the cultivation season (freshets) sometimes drowned the plants. See Allston, "Sea Coast Crops"; Doar, *Rice and Rice Planting;* Heyward, *Seed from Madagascar;* and Elizabeth Pringle, *A Woman Rice Planter* (Cambridge: Harvard University, 1961).

25 African Expertise

escalating slave imports from Africa's Rice Coast suggests that South Carolina reaped the transfer of an elaborate rice cultivation system that had been perfected by West Africans over centuries, if not millennia. While the inland swamp system proved pivotal in orienting South Carolina to an export economy based on rice, tidewater cultivation made it one of the most prosperous economic enclaves in the South.

The evidence presented in this paper further substantiates Wood's original argument for recognizing the importance of African tutelage in developing South Carolina's rice economy, yet poses some questions: why would West Africans transfer a sophisticated technology to develop a plantation economy that, in turn, imposed on them onerous labor in insalubrious swamps? Why would they painstakingly develop an elaborate tidewater system they rejected as "mud work" following slave emancipation?[59]

The answer to these questions can perhaps be sought by reexamining the definition of slavery, as well as the changing lineaments of the social relations of production in South Carolina's rice economy. It is possible that the attempt to convey the meaning of slavery and bondage in an essential, clear and forceful manner often obscures significant and complex details—details that reveal how subaltern societies engage in dialectical struggles between active and passive, acceptance and resistance, coercion and consent.[60] By recognizing the complex ways in which social relations are challenged and transformed we may develop an alternative perspective toward how slaves may have seized rare opportunities to negotiate and improve the conditions of their unfree labor.

During the first half of the eighteenth century, all the major rice typologies were present in South Carolina. By 1720 slaves outnumbered whites in the colony—ten to one in some areas—their population burgeoning rapidly with new arrivals until the 1739 slave rebellion known as the Stono uprising.[61] A bias for slaves from rice-growing Senegambia and Sierra Leone characterized the period prior to Stono, which was also marked by the appearance of an innovative form of labor organization, the "task" system, quite different from the more pervasive "gang" form.[62] Gray succinctly captured the essence of this "new" organizational

59. James Scott Strickland, " 'No More Mud Work': The Struggle for the Control of Labor and Production in Low Country South Carolina, 1863–1880," in *The Southern Enigma* ed. by Walter J. Fraser and Winfred B. Moore (Westport, Conn: Greenwood, 1983), 43–62.

60. Judith Carney and Michael Watts, "Manufacturing Dissent: Work, Gender and the Politics of Meaning in a Peasant Society," *Africa* 60 (1990):207–241.

61. Wood, *Black Majority,* 131; Curtin, *Atlantic Slave Trade,* 127–62.

62. Philip Morgan, "Work and Culture: The Task System and the World of Low Country Blacks, 1700 to 1880," *William and Mary Quarterly,* 3rd series, 39 (1982):563–99; and Julia F. Smith, *Slavery and Rice Culture in Low Country Georgia*, 61.

form: "Under the task system the slave was assigned a certain amount of work for the day, and after completing the task he could use his time as he pleased." This was clearly different from the more pervasive gang system where "the laborer was compelled to work the entire day."[63] The task system represented a significant improvement in that it provided slaves the opportunity to allocate more time to cultivate their garden plots, hunt, fish, and extract natural resources that could either go toward improving their nutritional status or be sold in local markets for small profits.[64]

The origin of the task labor system is probably African as it was already a pervasive feature of African slavery along the Upper Guinea coast and its hinterland during the Atlantic Slave Trade. Task labor guaranteed those in servitude partial control over their labor power through a moral economy that regulated the terms under which slavery could be exercised. European comments on African slavery from the earliest period of contact mention the use of slaves for cultivation but also refer to slave rights, both of which are critical aspects of a task labor system. Cadamosto, a crew member on a Portuguese caravel that travelled to the Senegal River between 1455–1456, reported a Wolof king turning war captives to agricultural production; a half century later Valentim Fernandes mentioned Wolof domestic slaves having one week day to work for themselves; while De Almada, writing in 1594, referred to "Fula slaves ruling the Wolofs."[65] The earliest European references to indigenous slavery consequently indicate that in Africa those enslaved were not considered merely as animate commodities; they retained rights within

63. Gray, quoted in Morgan, "Work and Culture," 564; Gray, *History of Agriculture,* 550–51.

64. While Doar, *Rice and Rice Planting,* 32 presents a nostalgic view of the paternalism of the South Carolina rice planter, he does present an account of the activities which a task labor system enabled hard-working slaves: "They used to get other fresh meat by fishing and some of the thrifty raised poultry which they could sell or eat. The family was allowed land for a garden, and they also could plant rice, if they wished, on outside margins of the river, a privilege which a great many availed themselves of, judging by the little fields, which could be seen on the plantations."

65. De Almada's familiarity with chattel slavery appears to have distorted his views on African indigenous slavery. 'Fula slaves ruling the Wolofs' likely is little more than an observation of Wolof dependence on slave labor and an exaggeration of slave rights. Cadamosto quoted in Paul Lovejoy, *Transformations in Slavery,* 27 with Fernandes and De Almada cited in Rodney, *History of the Upper Guinea Coast,* 236. Many African societies, like their European counterparts, were stratified prior to the arrival of the Portuguese. The nomenclature, "slaves," designated war captives and those pawned in famine, but was frequently a temporary social position from which one might obtain freedom. See John Thornton, *Africa and Africans in the Making of the Atlantic World, 1400–1680* (New York: Cambridge University Press, 1992), esp. 80–91.

the social structure and there were established limits on the hours or days their labor could be appropriated.[66]

The Atlantic Slave Trade intensified indigenous slavery from the seventeenth century as nobles increasingly relied on servile agricultural labor to produce the food surpluses necessary to support the incessant wars that provided the trade's quarry. By the early-eighteenth century, slave villages were in evidence along coasts, rivers, and major trade routes throughout the Guinea region, producing the food surpluses for provisioning armies, caravans, aristocracies and slave ships.[67] But indigenous slaves retained rights that specified days of the week or hours of the day for working on their own fields, a system more akin to European serfdom than what was to be found on New World sugar plantations.[68] Caillié clearly described this prevailing task labor system, remarking that slaves in Guinea farmed five days a week from early morning until early afternoon in specific agricultural tasks but "are allowed two days in the week to work in their own fields . . ."[69] A similar task system was already

66. A more appropriate term to describe African slaves would be "serfs." The labor obligations resembled those of serfs, with individuals providing service within limits mediated by social consensus. By 1600 the function of indigenous slaves in West African societies had shifted from an earlier emphasis on reproduction and domestic retainers to production and agriculture. (Lovejoy, *Transformations in Slavery,* 31.) It is in this period that the African task system likely took shape, reaffirming a preexisting moral economy.

67. Walter Rodney, *History of the Upper Guinea Coast,* 253–70; Lovejoy, *Transformations in Slavery,* 108–34.

68. The number of European traders and explorers to the Upper Guinea coast and its hinterland increased during the eighteenth century, which also saw the appearance of narratives by manumitted slaves. These documents provide more detailed observations of social structure and agricultural production. Oral histories collected in the twentieth century for Senegambia and Guinea attest to the continuity of the task system after abolition of the international trade. See Terry Alford, *Prince Among Slaves* (New York: Oxford University Press, 1977), 7; Douglas Grant, *The Fortunate Slave* (New York: Oxford, 1968); Paul Edwards, *Equiano's Travels* (London: Heinemann, 1967), 10; William Derman, *Serfs, Peasants, and Socialists* (Berkeley: University of California, 1973); and Peter Weil, "Agrarian Production, Intensification and Underdevelopment: Mandinka Women of the Gambia in Time Perspective," proceedings of a Title XII Conference, Women in Development, held by the University of Delaware in Newark, Delaware on May 7, 1981. A remnant of the task system survives in present day Senegambia as "strange farming" [French: navetanne]—based on seasonal migrants providing agricultural labor on fixed days (usually 3–4 half days/week) in exchange for lodging, food and access to farm land; see Ken Swindell, "Serawoolles, Tillibunkas and Strange Farmers: The Development of Migrant Groundnut Farming along the Gambia River, 1848–95," *Journal of African History* 21 (1980):93–104; and Philippe David, *Les navetannes: histoire des migrants saisonniers de l'arachide en Senegambie des origines a nos jours* (Dakar: Nouvelles Editions africaines, 1980).

69. Quoted from Derman, *Serfs, Peasants, and Socialists,* 36. While Caillié's observation is from the 1820s, it details the labor rights of those in agricultural servitude mentioned more generally but a century earlier by Equiano and the Guinean Prince, Abd Rahman Ibrahima. See Edwards, *Equiano's Travels,* and Alford, *Prince Among Slaves.* Another commentary on the task system by Hugh Clapperton (1824), British envoy to the Sokoto Caliphate in northern Nigeria, mentions slave responsibility in agriculture fixed between sunrise and 2 p.m., quoted

in place in the South Carolina rice fields by 1751 when Johan Bolzius remarked, "If the Negroes are Skilful and industrious, they plant something for themselves after the day's work."[70] Whether the New World task labor system represents the transfer of an entire labor process from West Africa or is the hybrid result of negotiation and struggle between master and slave over agronomic knowledge and the labor process, awaits additional historical inquiry.

Ver Steeg, who has questioned conventional representations of white-black relations in colonial South Carolina, suggests that prior to the Stono uprising slaves were not considered incompetent and inferior; they were, in fact, entrusted with considerable responsibility to the extent of bearing arms for the colony's defense.[71] Perhaps the basis for this initial esteem derived from their vital role in transforming marshlands into rice swamps and the creativity they exercised in tailoring crop production to South Carolina's diverse lowland environments. From this perspective the innovative task system – documented to have evolved in the New World first in the United States and regionally in South Carolina – may well have been the result of a dialectical struggle between slaves and planters in which slaves provided critical expertise necessary for the plantation to succeed in exchange for improving the conditions of their bondage. If this were so, West African bondsmen may have attempted to ameliorate European chattel slavery to the less brutal form they knew in their own territories.[72] The historical juncture marking the convergence of distinct rice production systems with the appearance of the task labor organization in South Carolina prior to the Stono rebellion invites speculation and certainly deserves further study in colonial planter–slave relations.

While South Carolina and Georgia provide the only documented examples of tidewater cultivation outside West Africa, the African role in

---

in Lovejoy, *Transformations in Slavery,* 206–07. In addition to Lovejoy the partial autonomy of slave villages is mentioned by Mollien, *Travels in Africa;* Rodney, *History of the Southern Guinea Coast.*

70. Bolzius quoted in Philip Morgan, "Work and Culture," 565. The entire document appears: "Johan Bolzius Answers A Questionnaire on Carolina and Georgia," (translated and edited by Klaus G. Loewald, Beverly Starika and Paul Taylor) *William and Mary Quarterly,* 3rd series, 14 (1957):218–261.

71. Ver Steeg, *Origins of a Southern Mosaic,* 105–06.

72. Ira Berlin attributed the emergence of the task system to slave isolation from the Anglo-American world, which enabled the retention of an African cultural heritage under similar environmental conditions. While this view suggests an African origin for the task system, it did not develop on many other slave plantation systems which were characterized by absentee owners. Ira Berlin, "Time, Space, and the Evolution of Afro-American Society on British Mainland North America," *American Historical Review* 85 (1980):44–78. For yet another view, see Philip Morgan, "Work and Culture," 567–69.

transferring rice growing to the New World extended far beyond the plantation system of the U.S. South. Slaves in the French West Indies frequently cultivated rice in provision gardens attached to coffee and sugar plantations.[73] As early as 1579, a Spanish land grantee in Tabasco, Mexico commented on rice cultivation in an area where Africans were enslaved for tobacco production.[74] Fugitive slaves in Surinam sustained their maroon communities by growing rice in inland swamps established outside the sugar plantation zone—a region similar to the low-lying depressions along Mexico's Gulf Coast that were planted to rice by runaways.[75] Under slavery Africans brought their culture of rice cultivation to New World swamps. Even after release from bondage, rice remained a favorite food staple cultivated by maroons in Surinam and Mexico and later, by emancipated slaves in South Carolina and Georgia.

Long before millions were transported across the Middle Passage, West Africans had refined an elaborate food production system that displayed acute knowledge of moisture regimes, soil principles, swamp farming, and hydrology. The result was an array of rice production zones, more diverse than those in Asia and attuned to different water resources, which maximized subsistence security and thus, human survival over millennia. This knowledge and the expertise to adapt rice planting to temperate zone conditions were among the scant "possessions" remaining to slaves pressed into the Atlantic Slave Trade. For those arriving in South Carolina, the lowland swamps presented precisely the environmental diversity to enable cultivation of their preferred food staple. The view of slaves as mere hands in the extractive-pastoral economy took on a new light with the quickening of planter interest and demand for rice within the larger New World plantation sector. Slave knowledge of rice cultivation offered the means to transform lowcountry environments into a lucrative economic system and to establish a plantation sector that profited from a startlingly unusual form of sophisticated technology transfer through their own toil. Yet in teaching their masters the needed skills, slaves were able to use their command of knowledge over rice

73. Morrissey, *Slave Women,* 51.

74. "Relaciones de Yucatan," I (*Coleccion de Documentos Ineditos de Ultramar,* 2a serie, tomo num.11, Madrid: Impresores de la Real Casa, 1898). The relevant quote for coastal Mexico is: ". . . en esta tierra a senbrado el arroz e millo [millet, another crop of African origin] y se da muy bien en ella . . .," 368.

75. Richard and Sally Price, eds., *Stedman's Surinam,* (Baltimore, MD: Johns Hopkins Press, 1992), 208–19. Rebel settlements in eighteenth-century Surinam widely cultivated rice, experienced abundant harvests, transferred the crop between communities, and even took their names from rice, such as Reisse Condre (translation: from the quantity of rice it afforded). Information from the states of Veracruz and Tabasco derived from oral histories recorded by Carney in 1992.

cultivation to alter, albeit slightly, the social conditions of work through the task labor system.

The agronomic contribution of South Carolina's black population remained without commentary in the written record until the 1970s when some researchers began to examine the role of slaves in the rice economy. Although considerable headway has been made to dispel the traditional view of slaves as mere hands, to recognizing their role as tutors, our society has yet to accord full respect to the men and women who toiled and endured enormous tribulations only to create wealth for those few people who treated them as animate commodities. The persistent view of Africans as mere hands and bodies, with nothing more to contribute than brute physical labor, has kept alive the rather unsubstantiated and ill-founded myth that Africans are incapable of comprehending the sophisticated concepts embodied in science and technology.[76] This cross-cultural perspective therefore must be seen as yet another attempt to whittle away at the foundations of this myth by bringing historical evidence to illuminate the far-reaching contributions of Africans to agrarian development in the Americas.

76. Paul Richards, *Indigenous Agricultural Revolution* (London: Hutchinson, 1985); and Michael Adas, *Machines as the Measure of Men. Science, Technology, and Ideologies of Western Dominance* (Ithaca, NY: Cornell, 1989).

12
Environmental Change in Colonial New Mexico

Robert MacCameron

In recent years scholars of North American history have paid increasing attention to the interrelationship between human societies and their physical and natural worlds. Their studies analyze the various ways in which people interact with their surrounding environment and how their choices affect not only the human community but the larger ecosystem as well. Just as nature shapes human society, humans, in significant and far-reaching ways, shape nature.

Exemplary works by William Cronon and Richard White have focused, from an ecological perspective, upon the English frontier experience in North America. Cronon, in his seminal book, *Changes in the Land: Indians, Colonists and the Ecology of New England* (1982) demonstrates that the English colonization of New England produced a number of "fundamental reorganizations...in the region's plant and animal communities";[1] a result, fundamentally, of the "colonists' more exclusive sense of property and their involvement in a capitalist economy."[2] Similarly, White, in his study entitled *Land Use, Environment and Social Change: the Shaping of Island County, Washington* (1980), describes how the introduction of European technologies along with the "Columbian Exchange" of plants, animals, and diseases dramatically altered the operation of natural systems, producing in turn what one botanist has described as the "most cataclysmic series of events in the natural history of the area since the Ice Age."[3]

This present study focuses upon environmental change in another area of North America: the upper Rio Grande valley which came under Spanish rule between 1598 and 1821. This largely semiarid ecosystem encompasses today the area of north central New Mexico, from approximately Belen, just below Albuquerque in the south, to Taos in the north, and from the Sangre de Cristo and Sandia mountain ranges in the east, to the Chama, Jemez and Rio Puerco river valleys in the west. The study examines just how Spanish culture and society brought about new relationships between human societies, including Pueblo Indians[4] and the land, and how changes in the land occurred as a result of those relationships. The English experience in North America serves as a principal point of reference.

The kind of physical and natural world in which the Spanish settled provides an important context for any discussion of environmental change. Scholars today describe north-central New Mexico as possessing essentially five different life zones: the Upper Sonoran, Transitional, Canadian, Hudsonian, and Arctic-Alpine. The Upper Sonoran zone includes mostly valleys, foothills, and plains, and extends from approximately 4,000 to 7,000 or 8,000 feet in elevation. Average annual rainfall is approximately ten inches. The Transition zone encompasses the middle slopes of the higher mountains and begins at approximately 7,000 to 8,500 feet above sea level on northeast slopes and 8,000 to 9,500 feet on southwest slopes. The Canadian zone covers most of the higher peaks of the Sangre de Cristo range and extends from approximately 8,500 to 12,000 feet depending upon cold or warm slopes. The Hudsonian and Arctic-Alpine zones are found on peaks that are around or above the timberline. Each of these zones is characterized by distinct flora and fauna, as they were at the time of Spanish arrival, although the boundaries between them are sharply marked only on steep-sided slopes. The zones themselves and their individual characteristics result from differences in altitude, temperature, precipitation, and barometric pressure. Significantly, human settlement has occurred almost exclusively in the first two zones. The Upper Sonoran has encompassed, past and present, most of the agricultural and grazing lands of the region, and here farming, except at the highest elevations, requires irrigation or some other water control system. The Transition zone is a bit wetter and therefore allows for some dry farming as well as seasonal grazing. But in fact, both zones constitute a semiarid environment in which access to water has been crucial for human survival.

Within this environmental context, Pueblo-Spanish contact produced its own particular form of the Columbian Exchange, that is, the reciprocal introduction of Old and New World plants, animals, and disease. More broadly construed, the exchange also included forms of material culture and even distinct values relating to economic, social, political, and religious organization.[5] Students of New Mexico's environmental history differ in their emphasis on the effects that this exchange has had on the land and the human societies occupying it. An anthropological view emphasizes the front end of the exchange, the Spanish introduction of new seeds, wheat, vegetables, and the tools for environmental destruction, including the iron axe and livestock, which changed the ecology of the upper Rio Grande valley forever.[6] A historical view more often stresses the rather remarkable adjustment achieved by Hispanic settlers in a very rugged frontier environment. A community-based agro-pastoral system of subsistence evolved that successfully sustained, both environmentally and culturally, many small communities for generations.[7] It was only after the arrival of the Anglo-Americans into the upper Rio Grande valley and the introduction of commercial agriculture that changes in the land occurred in any dramatic fashion.[8]

This second view, in essence, subscribes to a traditional interpretation of the relationship between economics and ecology, as outlined by Donald Worster and others, that subsistence agriculture, such as that employed by the Spanish in New Mexico, tends to preserve much of an environment's diversity and complexity, producing in turn an ecological and social stability. On the other hand, capitalism, involving specialization in production and competition in markets for profit leads to "a radical simplification of the natural ecological order, number of species found in an area, and in the intricacy of their interconnections."[9]

Interpreted in the broadest sense, this essentially linear model for environmental change in colonial New Mexico is basically correct. There is no question that human impact on the land, from Pueblo to Spanish to Anglo-American, was increasingly severe. But this model also contains certain limitations when applied to the case of colonial New Mexico. A closer look at the Spanish period reveals a far more complex process at work. In fact, the evidence indicates that a number of factors or determinants worked toward or mitigated changes in the land, producing in turn, in a largely nonlinear fashion, different kinds and rates of change.

Factors include demographic features of both Pueblo and Spanish, the Spanish introduction of grazing animals into the upper Rio Grande valley, colonial New Mexico's relative social-economic isolation, Spanish material culture, dimensions of Spanish land institutions, systems of political control, the varying nature of an inclusive frontier, changes in climate, and the vulnerability and resilience of a semiarid ecosystem. Many of these factors were paradoxical in their effect. They, at once, simplified the ecosystem and thereby accelerated change, and sustained ecosystem diversity and complexity, and produced little or no change.

Demography, especially numbers of people and where they relocated, is an important starting point for understanding the factors affecting environmental change. Throughout the Spanish colonial period approximately 99 percent of the population of the Province of New Mexico, including the general categories of Indians, Spanish, and *castas* (people of mixed blood), occupied about only one percent of the land. Access to water, in the form of rivers, creeks, and streams, was one obvious reason. But so also was the intermittent danger posed by the nomadic tribes—Apaches, Comanches, Navajos, and Utes—surrounding the upper Rio Grande Valley. Attempts to expand beyond the Rio Grande itself out to its principal tributaries and elsewhere were often discouraged by the presence of such nomads. Need for water and the presence of unfriendly Indians acted as a centripetal force upon both Pueblo and Spanish settlements. But at the same time, as Spanish population increased, especially during the eighteenth century, overcrowding along the Rio Grande occurred and access to resources, especially water and pasturage, diminished. These factors served as a countervailing or centrifugal force to Spanish settlement.[10]

The Spanish colonial population of the upper Rio Grande valley was centered principally on three villas: Santa Fe, founded in 1610; Santa Cruz de la Canada, founded in 1695; and Albuquerque, founded in 1706. The first two are located in the area known as the Rio Arriba, or the "upper river," while Albuquerque is located in the Rio Abajo, or "lower river." La Bajada mesa, 19 miles south of Santa Fe, served as the dividing line; there the Santa Fe Plateau drops about 1500 feet. Over the course of the colonial period these three centers and their environs, consisting of many smaller communities, contained the bulk of the Spanish population. Colonists also tended to radiate out from the three villas and establish small farms and ranches (ranchos), and hamlets (plazas). Yet frequently these more isolated settlements were abandoned and sometimes returned to again as shifting migration became an important demographic feature of colonial New Mexico.

The Spanish population, (including castas), in the upper province grew very gradually at first, set back by the Pueblo Revolt of 1680 and the economic unattractiveness of the region, and then increased more rapidly by the end of the 18th century. It was less than 1,000 in 1600, approximately 2,900 in 1680, 2,000 in 1700, 3,400 in 1752, 5,800 in 1776, 19,000 in 1800 and 28,000 in 1821.[11] More specifically, figures for the eighteenth and nineteenth centuries indicate shifts in population between and among the three villas: enumerations indicate that Spanish concentration was highest in Santa Fe in 1752, in Albuquerque in 1760, and 1776, and in Santa Cruz in 1790, 1800 and 1817. By 1817 Santa Cruz had a Spanish population of 12,903; Albuquerque, 8,160; and Santa Fe, 6,728.[12]

More Spanish and *castas* eventually came to live in the heart of the Rio Arriba, around the Villa of Santa Cruz, than in the vicinities of either Santa Fe or Albuquerque, reflecting both environmental realities and Spanish policies based upon them. In this protective terrain, with perennial supplies of water for irrigation, Spanish governors awarded communal land grants to groups of people, believing that they could protect themselves from attack by nomadic Indians. These grants, with their concentrated agricultural communities, served the purposes of the government by establishing controls over the citizens themselves and by creating an effective defensive frontier. In contrast, the settlement pattern in the Rio Abajo, along the broad, fertile plain of the Rio Grande came to be characterized by a more dispersed population occupying private land grants.[13] Meanwhile, the Pueblo, dramatically reduced in numbers, clung tenaciously to some twenty-one discrete settlements, ranging from Taos in the north to Isleta in the south, and from Pecos in the east to Acoma, the sky Pueblo, in the west. Paradoxically, this reduction in Pueblo population meant that there were more human beings in the upper Rio Grande valley at the beginning of the Spanish period in 1598 than at its close in 1821.

The Spanish introduction of grazing animals, termed "ganado mayor," cattle and horses, and "ganado menor," sheep and goats, into colonial New Mexico altered the face of the land in dramatic ways. In contrast to most other areas of Spanish America where cattle constituted the principal livestock, sheep came to dominate the landscape of the upper Rio Grande valley. There were several principal reasons for this development. Immense herds of buffalo on the plains to the east provided a source for hides, jerky, salted tongues, and tallow, products of a quality equal to those from domestic cattle. The hostile Indians surrounding the province, with the exception of the Navajo, usually coveted cattle and horses because they could be driven easily away. Sheep, on the other hand, could be scattered and then recovered after a raid. Throughout the course of the colonial period, the mining settlements of Nueva Vizcaya in Chihuahua and Durango provided a strong market for New Mexico sheep. During the eighteenth and early nineteenth centuries the sheep trade became the primary export industry.[14]

Unlike the human population of the region, sheep came to be concentrated principally on the broad plains of the Rio Abajo, in and around the Villa of Albuquerque. A livestock enumeration in 1827, six years after Mexican independence, reveals nearly 250,000 sheep and goats in the province, with Albuquerque possessing 155,000, Santa Fe, 62,000, and Santa Cruz de la Canada, 23,000. At the same time, there were only 5,000 cattle, 2,150 mules, and 850 horses.[15]

These grazing animals, both sheep and cattle, had an important impact on the ecology of the area. Various descriptions of the land by early visitors provide a benchmark from which to view later changes. Two pre-settlement observers of New Mexico, Hernan Gallegos and Diego Perez de Luxan, representing the Chamuscado-Rodriquez (1581) and Espejo (1582) expeditions respectively, noted lush grasslands, untouched pastures, highly suitable for both sheep and cattle.[16] After Spanish settlement, evidence indicates that portions of these grasslands were, over time, dramatically overgrazed. While many fewer in number than sheep, cattle also effected changes in the land in several principal ways. Whereas sheep were often grazed on distant pastures and required intensive labor, cattle were turned loose, unattended, on commons close by agricultural plots for safety from hostile Indians. As a result, the commons were frequently overgrazed. Marc Simmons says that Albuquerque's east mesa, for example, was virtually denuded of grass by 1750, forcing cattle to invade the fenceless crop fields of both the Spanish and the Pueblo.[17]

Over the course of the colonial period, the Spanish Archives of New Mexico contain numerous cases of New Mexico governors warning Spanish settlers to control their livestock, particularly from damaging Pueblo fields and irrigation ditches, or face severe penalties. As the number of sheep dramatically increased over the eighteenth and early nineteenth centuries Spanish settlers frequently petitioned governors for fresh pasturage away from the core settlements of the Rio Abajo and Rio Arriba, looking particularly to the south of Albuquerque, to the Rio Puerco, and to the Chama valley. Petitioners in the late 1730s, for example, complained about the inadequacy of land and water, especially the shortage of grass, near Albuquerque.[18]

Yet the needs of the settlers and goals of the state were often in conflict. Governors, representing the interests of the defensive function of the frontier, wanted to control and regulate shifts in settlement, especially on frontiers where conflicts with Indians were likely to occur. As a result, settlers' requests for fresh Crown land were often denied, and even in cases where their petitions were approved, they frequently were forced to return to the safe harbor of the three villas to escape Indian depredations. These barriers to expansion merely exacerbated livestock pressures on already settled land. By the 1820s, New Mexico sheep growers still sought fresh grass, this time to the east, onto the plains beyond the Sandia and Manzano mountains.[19]

In contrast, the isolation of colonial New Mexico from the rest of New Spain and English North America clearly moderated the process of environmental change. In 1803 Governor Fernando de Chacon, in a report to officials in New Spain, vividly portrayed the region's general economy by noting that New Mexico's "natural decadence and backwardness is traceable to the lack of development and want of formal knowledge in agriculture, commerce and the manual arts."[20] Chacon thought that Spaniards and castas were little dedicated to farming and contented themselves with sowing and cultivating only what was necessary for their sustenance. While remarking on the abundance of sheep in the province, he indicated that no great number of swine existed. The continual raids of the nomadic Indians discouraged raising horses and mules. Deposits of minerals such as lead, tin, and copper were located in various parts of the province but they were virtually untapped, while large scale smelting or amalgamation operations were nonexistent. Throughout New Mexico, high quality mica or gypsum (yeso) covered windows in place of glass panes. Formal apprenticeships, examinations for the office of master, and organized guilds, customary elsewhere in New Spain, did not exist. But out of necessity and the natural ingenuity of the people, according to Chacon, some trades were practiced skillfully, including weaving, shoemaking, carpentry, tailoring, smithing, and masonry. The exports of the colony, transported by the annual mule trains to Sonora, Vizcaya, Coahuila, and points south, consisted of oxen, sheep, woolen textiles, some raw cotton, hides, and piñon nuts.

The total value of these products, including wine from the El Paso district, was estimated at only 140,000 pesos annually. The products brought back into the colony included linen goods, chocolate, sugar loaves, soap, rice, iron, leather goods of all sorts, pelts, paper, drugs, and some money. Because hard currency was in chronic short supply throughout the colonial period, barter was the principal mode of exchange.

Trade also existed between the Spanish, the Pueblo, and nomadic tribes. Taos became a principal center for exchange during times of peace, where, for horses, an array of metal tools, corn, and trinkets, the "uncivilized" Indians traded pelts, buffalo skins, and Indians whom they had captured from other tribes.[21] The fact that pelts were obtained either through import or at Taos indicates that little hunting of small or large game went on in the upper Rio Grande valley itself.

In essence, colonial New Mexico demonstrated a low level of technological development, little or no occupational specialization, generally self-reliant local production for local consumption, and a broad utilization of the entire environment—all characteristics of a subsistence economy. It would be a mistake to assume that such characteristics did not lead to the overuse of resources or environmental degradation, but surely the degree of change was less than what might have occurred in a society distinguished by specialization in production and labor, and by the export and import of products to and from similarly specialized producers.

As an example, if large deposits of silver had been discovered in New Mexico during the Spanish period, as they were in other semiarid regions of central and northern New Spain, the ecology of the area would have been transformed to a significantly greater degree. Silver mining required the special input of both raw materials and labor: a large supply of wood for fuel, shoring in mines, construction in buildings, and machinery; water for power and washing; and thousands of grazing animals to produce hide, leather, meat, and energy for transport and powering machines. Shifts in population occurred as well to meet the intensive labor demands in the mines, and the human wastes from such concentrations were merely dumped into the local waterways. The amalgamation of silver also required the use of such products as copper sulfate, common salt, and in large quantities, mercury. Metallic mercury, mercury vapor, and lead, the residues of this process, poisoned both plants and animals as they readily invaded the air, water, and ground.

After only a few decades, the environmental effects of this economic activity were devastating. Entire areas in and around mining communities were denuded of grass, deforested, and eroded by wind and water. Vegetation loss also reduced transpiration, leading in turn to a decline in local rainfall. As food chains were destroyed, animals and fish disappeared, and it can be assumed that the results of mercury and lead poisoning in human beings, mostly Indian labor, were frequently fatal.[22] Nothing on this ecological scale occurred in the Province of New Mexico during the Spanish period.

Yet the Spanish introduction of metal tools into the upper Rio Grande valley did allow both the Spanish and Pueblo to manipulate that environment in entirely new ways. The introduction of the axe alone enabled bench lands, bounded above and below by steeper slopes, to be cleared of dense vegetation, and woodland along the rivers and in the higher elevations to be cut. Wood became an important source for construction, tools, and fuel. The piñon, a scrub pine, was used to make plowshares and the legs of spinning wheels; the cottonwood to make wine barrels and *carreta* (cart) wheels; the oak to make stirrups and the *carreta* frame; and the Douglas fir to make the shafts of plows and provide large timbers for bridges and vigas (beams) in roofs.[23]

The degree to which the upper Rio Grande valley was deforested as a result can be inferred from several commentaries made at the beginning and end of the Spanish period. In 1582, Luxan wrote that "this Province [New Mexico] boasts of many pine forests..." and that "there are also fine wooded mountains with trees of all kinds...."[24] In 1839, in one of the first Anglo accounts of the region, Josiah Gregg, explorer and trader, wrote that "on the water-courses there is little timber to be found except cottonwoods, scantily scattered along their banks. Those of the Rio del Norte [Rio Grande] are now nearly bare throughout the whole range of the settlements, and the inhabitants are forced to resort to the distant mountains for most of their fuel."[25] Another Anglo account sixteen years later corroborates this description. W. W. H. Davis observed that "wood is exceedingly scarce all over the country. The valleys are generally bare of it, and that found upon the mountains consists of a growth of scrub pine called the piñon. The country is said to have been well wooded when the Spanish first settled it, but in many parts it has been entirely cut off, and in some instances without even leaving a tree for shade."[26]

The slow development of Spanish material culture is most apparent when viewed in relation to that of English North America. As a result of the colony's isolation from European influence, the use of agricultural tools remained largely unchanged over the period of Spanish rule. The scratch plow, essentially as described in the Bible, was equipped with an iron, steel, or wooden share and cut a shallow furrow about six inches deep instead of turning the soil. It was not until the end of the Mexican period in 1846 that two-handle steel plows with a moldboard for turning the soil reached New Mexico.[27]

The ecological implication of this low-level technology is significant. Whereas in English North America the use of deep cutting plows, in the absence of contour plowing, crop rotation and manuring, caused soil erosion on a massive scale, the scratch plow of New Mexico generated comparatively little soil loss. In fact, the farmlands of the upper Rio Grande valley were auto-replenishing through the agency of silt-laden irrigation and flood waters and appeared, over time, to a number of colonial visitors as particularly abundant and fertile.

Another side of material culture was the Pueblo and Spanish use of soil as a building material. While the Spanish introduced the formed standard adobe brick to the Pueblo, the techniques and materials for construction remained essentially unchanged from what the Indians had used before.[28] Adobe construction tapped several available and replenishing natural resources: loamy soil, sand, water, and straw. Wood was used only for ceiling beams or *vigas*. For the Pueblo, the adobe was one of the most prized Indian crafts as it represented the sacredness of the land itself. In contrast, the English use of wood for both construction and sale, while entailing a sense of craft, was predicated largely on an ethos of function and profit. As a result, Pueblo and Spanish buildings, decrepit when seen through the eyes of some nineteenth century Anglo-American visitors, arose out of and were an integral part of the physical landscape, and unlike the acute deforestation which occurred in English North America, caused few changes in the land.

After the Pueblo Revolt the Spanish imposed a pattern of land use and settlement on the upper Rio Grande valley in marked contrast to the widely scattered large estates, worked primarily by Indians through the *encomienda*, of the early seventeenth century. On the basis of individual and communal land grants the Spanish came to live in smaller units strung out along the Rio Grande and its tributaries. In the Rio Arriba, where the colony's Spanish population became most concentrated, the communal land grant dominated. In this case individuals in a group of settlers would each receive an allotment of land for a house, a plot for irrigation, and the right to use the unallotted land on the grant in common with the other settlers for pastures, watering places, firewood, logs for building, and rock quarrying, among other activities.[29] The intensity of land use under these circumstances certainly differed by degree between and among the various grants. But the very configuration of these settlements, coupled with barriers to out-migration, helped to produce such changes in the land as overgrazing and deforestation, the subsequent loss of topsoil, and the silting of irrigation ditches and streams.

A pattern of land-holding developed in the region, both on and outside of the community grants, of long-lots or narrow strips of land emanating from the water ways. They were a necessary response to the physical environment and represented a practical and equitable method of partitioning irrigable land. Yet over generations, through the institution of partible inheritance, long-lots were divided and redivided lengthwise, leading to more densely populated communities living on increasingly smaller parcels of land.[30] As a result settlers had to move their grazing and cutting operations ever higher up the mountain sides to accommodate the loss of grass and trees on the original commons. At the same time over-used irrigation plots for growing wheat and garden vegetables became increasingly less productive.

On the other hand, metes and bounds, a legal system by which property boundaries were determined, was a feature of Spanish land institutions that favored environmental conservation. Boundaries were not established with any prescribed shape in mind, nor were they necessarily contiguous with any others. Instead, they were drawn to follow the natural contours of the land and to include the most valuable resources: soil, woodlands, and access to water. It was a system highly adaptive to the local environment,[31] and predicated, at least in the case of colonial New Mexico, on the needs of subsistence agriculture.

A system of grids, based upon the rectangular survey, came to prevail in many parts of Anglo-America with far more severe environmental consequences. There little allowance was made for local topography, hydrology, or climate. Fields were arranged according to a rigid north-south, east-west alignment often resulting in enhanced soil erosion. The distribution of surface water, arable land, grass suitable for grazing, and timber for securing wood was often unequal between and among individual holdings. As a result, property owners tended to make extreme demands upon natural resources that were frequently in short supply. In contrast to the communal restraints evident in parts of colonial New Mexico, in order to maximize profit in a market economy, owners were free to exploit their local environment as they saw fit.[32]

With roots deep in Iberia, Spanish irrigation was at once vital to agricultural productivity and a source of land change and deterioration. The pre-contact Pueblo use of water control devices, the complexity of which is still debated, led to the accumulation of salts and other mineral deposits in the soil and the consequent need to seek new growing areas in the face of an expanding population and shrinking bottomland. The Spanish system, borrowed by the Pueblos, placed a more complex pattern of ditches on the face of the land with even greater natural and physical effects. In both the Rio Arriba and Rio Abajo, the Spanish system of irrigation consisted primarily of a main ditch, the *acequia madre*, receiving water from a river or stream at a higher elevation than the lands being irrigated, and then relying on gravity to carry the flow. Secondary ditches (*sangrias*) branched off the main ditch and directed water to individual fields. Gates were made of earth and boulders to regulate the flow, and flumes (hollowed out logs), provided an elevated channel for water to cross gullies and ravines.[33]

Over the years riparian lines of trees and shrubs developed alongside well-maintained ditches, replicating the biotic environment found along unimpeded streams. In other instances, grazing animals invaded both Spanish and Pueblo fields and trampled the sandy banks along the water courses, filling them or causing breaks which allowed water to escape.[34] Abandoned ditches were often transformed into gullies or small arroyos as the result of runoff and soil erosion.

The problem of salinization that confronted pre-contact Pueblo farmers was intensified under the Spanish. Alkali compounds, consisting of various salts, are characteristic components of arid and semiarid soils; they are also highly soluble in water. When dissolved by irrigation, the water and alkali enter the soil together, then return to the surface by capillary action, much the same way that oil flows up a lamp wick. The sun then evaporates the water and leaves behind the salts to act as corrosive agents on the stems of plants.[35]

The effects of such phenomena as salinization, deforestation, overgrazing, and population increase reveal that selected areas in both the Rio Arriba and Rio Abajo were exceeding their carrying capacity by the end of the eighteenth century. While remaining relatively small in numbers and employing a subsistence-based economy, the Spanish in the upper Rio Grande valley had not yet achieved a totally sustainable society in relation to their environment.

While issues of power and control, particularly during the seventeenth century, were often contested at the local level between the Franciscan missionaries and the governor, the latter official influenced the course of environmental change more strongly. Directly responsible to the Viceroy, and after 1776, the Commandante General of the Provincias Internas, the governor appointed or approved lesser officials, enforced royal decrees, and ordered the formation of militia from settlements. More importantly from an environmental perspective, he made land grants which were intended primarily to manage the settlement of the Spanish and casta population within the colony. For in theory at least, people were simply not allowed to live where they liked or to move at their own discretion to another location. In the process, Spanish law strictly outlawed such hallmarks of the Anglo-American frontier as land speculation and absentee ownership.

This kind of social control had several ecological effects. In some cases, a governor's decision to award, or even revoke, a land grant led to further deterioration of already occupied land. One such instance occurred in 1735 when Governor Cruzat y Gongora nullified a number of grants made by the acting governor, Juan Paez Hurtado, in the area of the Chama valley. Facing overcrowded living conditions in the vicinity of the lower valley and Santa Cruz, settlers had sought fresh land for grazing. Gongora's decision to rescind the grants was based in part on his personal need to control and regulate the advancement of settlements, but it was also based on his belief that the upper Chama was a place where colonists and Utes in close contact might precipitate a war.[36] Here environmental concerns took a back seat to issues of colonial security.

Built into the very land grant process was a step tantamount to the environmental impact study of today. The alcalde mayor, a lower-level but important official appointed by the governor, had to determine whether the proposed land grant in his jurisdiction would adversely affect any Pueblo settlement or other third party, and whether there was sufficient water for irrigation and livestock and enough cultivable, grazing, and wood-producing land to support the proposed number of settlers.[37] While this system was hardly perfect, the land grant papers of New Mexico are replete with examples of the alcalde mayor addressing these issues in writing to the governor, and sometimes recommending to him that a grant be denied because criteria were not met. While the system was intended to preserve the economic survival of the colony and not the environment, it nonetheless had the long term effect of conserving the land and water of colonial New Mexico.

The degree to which Pueblo and Spanish society created an inclusive frontier also directly affected issues of land use and change. While Pueblo people embraced Old World plants, animals, and material culture introduced by the Spanish, they did not accept, in any profound sense, Western belief systems, particularly those centering upon religion or property. In anthropological terms, the Pueblo became acculturated, but not assimilated into Spanish society. Although greatly reduced in numbers, they lived under Spanish rule in their own compact, autonomous communities, as they do in the modern era. Through intermarriage and other forms of contact, the Spanish borrowed culturally from the Pueblo. This occurred most frequently in outlying communities rather than the missions or villas. The long-term effect of this two-way process was to produce both a hard and soft impact upon the land. The soft impact emphasizes the influence of Pueblo beliefs and customs upon the Spanish settlers. The hard impact focuses principally upon Pueblo acculturation.

Richard Ford has offered a succinct view of the existence of one Pueblo group, the Tewa, prior to contact and the subsequent effects of the Columbian Exchange upon them and their relationship to the land. The Tewa lived on maize supplemented with squash and beans, and gathered plant products. Rabbits, hares, deer, and other game provided a source for meat. Firewood came from deadwood, and construction timbers were usually recycled from older structures. In essence, according to Ford, "this was an ecosystem that could rapidly recover when fields were abandoned or when the human population founded a new village elsewhere."[38]

With the arrival of don Juan de Oñate in 1598, the Tewa economy changed dramatically. Five Spanish contributions had particular impact: the introduction of spring wheat; kitchen gardens grown with irrigation water; orchards of peach, apricot, plum, and cherry trees; grazing animals; and the metal axe. The result for the Pueblo was a strange admixture of environmental degradation to their land and "a beneficial and more secure subsistence base."[39] Deforestation and overgrazing occurred on Pueblo land just as it did on that of the Spanish. But the new food sources only reinforced the adaptive capacity of the Tewa. Wheat, in particular, while not displacing corn, came to serve as "a high yield caloric safety valve" for the Pueblo. It is likely that the Pueblo partially compensated for the local overgrazing and loss of cool-season grasses such as mutton grass, Indian rice grass, and June grass, by raising wheat varieties that matured about the same time.[40] These additions to the Pueblo economy also tempered their need to migrate as they might have in precontact times owing to political disputes, scarcity of wood, or unproductive fields. Even as their landscape was simplified, the Tewa expanded their control over the productivity of their land and created a measure of economic security.[41] At least in a material sense, the Pueblo came to more closely resemble their European neighbors.

The inclusive nature of Pueblo-Spanish society entailed accommodative elements as well as more disruptive ones. Whereas students of Pueblo-Spanish relations can determine with some certainty the far-reaching ecological effects of the Spanish donations to the Pueblo in the form of plants, animals, and material culture, it is much more difficult to assess how Pueblo belief systems, and their practice of a particular subsistence agriculture, affected Spanish settlers and the environment of the upper Rio Grande valley. As Christopher Vecsey has noted, all Native American groups "established a religious association with nature that transcended but did not nullify their effective exploitation of the environment. They achieved an integration of subsistence and religious activities."[42] He includes the Pueblo among those American Indians whose religious core was molded by environmental relations: their fertilization ritual, for example, was about maintaining life in humans, plants, animals, and the world at large, benefiting, in turn, nature and culture alike.

To what degree Spanish settlers adopted these attitudes from the Pueblo is impossible to determine, but more cultural borrowing likely went on between the two groups than in almost any other region of Spanish America. Frances Swadesh has suggested that what evolved in the Province of New Mexico after the seventeenth century was a non-dominant frontier community. Relations between settlers and their Indian neighbors, especially in outlying areas, were far more egalitarian than the colonial model for social relations intended.[43] The traditional Spanish barriers to social mobility, including caste, class, and ethnic identity, were largely absent. Archival records also reveal that a not insignificant number of Spanish settlers moved into Pueblo communities; both men and women married Pueblo Indians and raised their children in the spouse's pueblo. As a result, according to Swadesh, Spanish settlers often times "found themselves more at odds with their own colonial authorities than with their Indian trading partners, compadres, and friends."[44] As settlers acquired knowledge of herbs and wild plant foods from the Pueblo, and as they witnessed or participated in an agricultural cycle both religious and secular in meaning, they may also have internalized a view of the land predicated on principles other than the usual European ones of strict utility. And while this process hardly mitigated the harsher effects of Spanish culture on Pueblo land use, it may have changed or softened the attitude of some Spanish toward exactly how the physical and natural world should be exploited.

Any analysis of land change must also address natural forces, especially variations in climate, and their impact upon biophysical processes. Prolonged drought, heavy rains, and extremes of heat and cold may effect environmental changes entirely independent of human agency. They may also interact with human activities and speed along the process of change. In the case of New Mexico, there are essentially three types of climate: arid, semiarid and subhumid/humid. Areas of arid climate have scrubby, heat- and drought-resistant desert plant cover; semiarid areas have a vegetation of short, bunchy grasses; and the subhumid/humid areas, in the hilly or mountainous regions, have woodland or forest cover.[45] Latitude, elevation, and location with regard to moisture-laden winds primarily determine these variations in climate and consequent vegetation. Albuquerque, at the northern point of the arid range, and Santa Fe, in a semiarid range closely bounded by the mountains, although only sixty miles apart, often exhibit marked differences in local climate. Such differences mean that it is very hard to associate general climate patterns with ecological change in any specific locale.

From a broader perspective, evidence from tree ring studies (dendroclimatology) indicates that the climate of the upper Rio Grande valley between 1598 and 1821 fluctuated fairly regularly between wet and cold and warm and dry periods. For Western North America, periods of widespread drought occurred between 1626 and 1635, and 1776 and 1785, while periods of above average moisture occurred from 1611 to 1625, 1641 to 1650, and 1741 to 1755. Chronologies from the upper Rio Grande valley (lat. 35-43; long. 106-06; based on rings from Douglas-fir, ponderosa pine, and piñon pine) reveal dryer than normal periods between 1661 and 1675, 1726 and 1765, and 1796 and 1830, and wetter than normal periods between 1601 and 1645, 1706 and 1730, and 1781 and 1800.[46] These changes in climate clearly had the potential to effect changes in the land, both on their own and in concert with human activity. Drought or extended periods of deficient precipitation may produce, among other effects, a marked decline in plant cover, even in the absence of grazing animals. Particularly vulnerable to drought are perennial grasses like the gramas, which may die off and then be replaced by invaders such as sagebrush, snakewood, and rabbit brush.

But several variables bear strongly on the amount of plant life lost during any particular drought. They include the time of year in which the drought occurs, winter and spring having more adverse effects than summer and fall, and the texture of the soil. More heavily textured soils may retain more moisture, release it to plant use more slowly, and thus allow perennial grasses to survive. Likewise, periods that are colder and wetter than normal can produce environmental change. An excess of rain and snow, and their runoff, have the potential to cause soil erosion, arroyo formation, changes in stream flow, and flooding. Such effects are particularly acute when the climate changes quickly and dramatically from an excessively dry to wet period.[47]

Further inferences can be drawn from the relationship between fluctuations in climate, human activity—including that of Pueblo and Spanish—and ecological change. While both climate and humans, acting independently of one another, may simplify an ecosystem, together their effects may be far greater and more long lasting. The formation of arroyos, valley bottom gullies characterized by steeply sloping or vertical walls, offers a case in point. Both human land use and climatic changes are among the complex causes leading to their creation. On the human side, logging, fire, grazing, cultivation, and the existence of roads, trails, and irrigation ditches can lead to the local removal of vegetation on valley floors and alter valley bottom soils. This increases the erodibility of valley floor material. Both an increase and decrease in humidity can contribute to arroyo formation. Increase in precipitation and decrease in temperature may increase runoff from slopes, while a decrease in precipitation and increase in temperature may reduce the vegetation cover over drainage basins. Both phenomena increase the velocity of flows through valley bottoms. Individually, but especially together, human and climatic factors lead to localized erosion and arroyo initiation.[48]

Finally, this discussion of environmental factors must be viewed in the context of the inherent vulnerability or resilience of a semiarid ecosystem. Ecologists no longer assume that the degree of diversity and complexity of an ecosystem determines stability or change. Other factors such as elasticity (ability to recover from damage), and inertia (ability to resist displacement), are now considered more important. Thus desert or semidesert regions may be more resilient and less vulnerable to change than other, more complex ecosystems such as woodlands or rain forests. The flora and fauna of arid regions evolved in an environment where the normal pattern is more or less random alteration of short favorable periods and long stress periods. Ecologists posit that plants and animals have preadapted resilience. Applying this idea to the upper Rio Grande valley leads to several conclusions. On the one hand, grassland and woodland, for example, might well have survived in some instances the invasion of sheep, cattle, and axe. In fact, some argue that limited or optimal grazing may even enhance the growth of certain grasslands.[49] In contrast to the total absence of grazing or excessive grazing, light grazing favors production of some grasses by restoring nutrients through feces and urine. Even if damaged, grasslands have the potential often to undergo a regeneration of growth. The carrying capacity of any specific ecosystem might be sustained on account of plant and animal resilience, or even through the "benefits" of limited human economic activities upon them. Yet it is clear that human agency, both Pueblo and Spanish, in this area contributed to fundamental changes in the land: the carrying capacity of selected areas was exceeded as the land proved to be more fragile than resilient. It can even be argued that a process of desertification began in selected areas where grazing was particularly heavy. That Anglo-Americans, upon their arrival after 1821, exacerbated those changes manifoldly does not negate this fact.

It is tempting to assess the long run effects of the Spanish colonial presence in the upper Rio Grande valley on environmental change as constituting a middle ground between those changes which occurred first under the Pueblo and then later under the Anglo-Americans. The interaction of various factors, in effect, produced a rate and kind of change certainly exceeding that experienced under the Pueblo but falling far short of that produced by Anglos. The acceleration in deforestation and overgrazing in New Mexico after the Anglo-American occupation in 1846, and especially after the introduction of the railroad in 1880, demonstrates that this assessment is essentially correct. But the kind of change that occurred between 1598 and 1821 is more complex than merely a linear one. Change was also cyclical and layered or superimposed in nature. The rate of change was at once fast, slow, intermittent and inexorable, and only rarely constant.

Several primary factors account for these variations in change. People did not settle evenly or randomly on the land. Instead they concentrated themselves in some areas, settled sparsely in others, on some occasions moved for environmental or other reasons, and on other occasions abandoned their land and moved back to where they had come from. Nor were these same people engaged in uniform economic activities having equal impact upon the land. Natural forces acted both alone and in concert with human activities on the land in an unpredictable and capricious manner. For instance, the frequently changing course of the Rio Grande, resulting from its own natural meander, precipitated the cyclic destruction and regeneration of the woodland or bosque habitat of the river.[50] Overgrazing and drought together, followed by above normal precipitation, enhanced the opportunity for arroyo formation. Finally, the resilience or fragility of the land itself depended upon variations in the adaptability of local flora and fauna, and the degree to which both humans and natural forces could disturb or modify them. How changes occurred in the land was integrally tied to all of these factors.

Change was most linear in the heavier population centers of Santa Fe, Santa Cruz de la Canada, and Albuquerque. Descriptions of these areas from the sixteenth century *entradas*, when juxtaposed to Anglo-American accounts of the early to middle-nineteenth century, indicate the sort of dramatic shift akin to descriptions of environmental change in colonial New England and the Pacific Northwest. The archival record for Santa Fe over the colonial period notes steady loss of natural resources in the form of water, grass, and wood to the point that a governor in the late eighteenth century recommended that the capital be moved to the confluence of the Santa Fe and Rio Grande Rivers.[51] Similarly, the Albuquerque area underwent a gradual yet inexorable process of desertification that resulted from overgrazing by sheep.

In contrast, cyclical change took place most frequently when settlers, usually on account of land pressure in and around the three villas, occupied land at the edge of the New Mexico frontier. Settlements in the north around Taos, in the northwest on the Chama River valley, in the west out to the Puerco River, in the south beyond Isleta, and in the east to the Pecos River, throughout most of the colonial period, were often in flux, sometimes abandoned outright on account of Indian hostilities, and sometimes even resettled after a several-year hiatus. As a result, changes in the land that would have occurred from permanent settlement might have commenced only to be interrupted by abandonment, and then recommenced. Deforestation and overgrazing in these areas rarely took place in a sustained way.

In this context, the year 1790 is notable in colonial New Mexico history in that it marked the beginning of relatively peaceful times between the Spanish, the Pueblo, and the nomadic tribes, especially the Apache and Comanche. The outward expansion that followed came to resemble more closely the Anglo-American frontier of an ever-receding line. It also meant that the cyclical nature of ecological change came largely to an end, as places such as the Puerco River valley began to fill with settlers so steadily that by the middle to late-nineteenth century it had become as severely overgrazed and eroded as any place in New Mexico.

Environmental change might also be characterized as layered or superimposed, as in a kind of palimpsest. This occurred most often where Pueblo communities adopted the plants, animals and material culture of the Spanish, moving from a system of extensive to intensive agriculture. In effect, Pueblo land use changed by degree rather than by radical transformation. The growing of wheat required the expansion of Spanish-style irrigation, with concomitant environmental effects: and as the Pueblo came to rely on new domesticated plants and animals, their hunting and gathering of wild plants and animals declined. So the precontact Pueblo practices of land use and their effects on land change intensified as Spanish agricultural practices became superimposed on Pueblo practices through the process of acculturation. The dramatic decline in Pueblo population prevented their communities from exceeding carrying capacity as, in selected cases, they were in danger of on the eve of contact.

Environmental change in colonial New Mexico was sui generis, predicated on circumstances and conditions reflective of the particular climate, geography, and cultures of the region. The dialectical relationship between different kinds of factors produced varied kinds and rates of change unlike that experienced on any other North American frontier. This study shows the limitation of relying on modes of production or economic systems as a single or even primary explanation for change. They are clearly more useful as a means of analysis from a macro perspective such as representing, in broad strokes, the environmental change which occurred under preindustrial Pueblo and Spanish society through Anglo occupation. A look at the micro level reveals the difficulty in assigning any such clear causality for change. Perhaps the challenge before us is to mediate between the two approaches.

Acknowledgements

Research for this article was partially funded by a grant from the National Endowment for the Humanities. I would especially like to thank Dan Scurlock, Rose Diaz, Bill Tydeman, Don Dreesen, and Bob Delaney, in Albuquerque, and Phyllis Moseley MacCameron, Tom Rocco, Peter Murphy, Anne Bertholf and Candy Carroll, in Buffalo, for their encouragement and assistance. I would also like to thank Hal Rothman of Environmental History Review *and two anonymous reviewers for helpful comments on the manuscript.*

---

1 William Cronon, *Changes in the Land: Indians, Colonists and the Ecology of New England* (New York: Hill and Wang, 1983), vii.

2 Cronon, viii.

3 Richard White, *Land Use, Environment, and Social Change: The Shaping of Island County, Washington* (Seattle: University of Washington Press, 1980), 36. Neither Cronon nor White subscribes to the notion that Native Americans lived in any sort of perfect harmony or static relationship with nature prior to contact with Europeans. Cronon notes that the Indian practice in New England of burning forests to clear land for agriculture and to improve hunting "could sometimes go so far as to remove the forest altogether, with deleterious effects for trees and Indians alike." And White points out that the natives of Island County did not hesitate to alter natural systems when it was to their advantage to do so. For example, through the manipulation of the environment the Salish increased such desired plant species as bracken, nettles and camas for use as food crops. Rather, most remarkable was the degree of environmental change that occurred in these far corners of English North America after contact. For both Cronon and White such change was both rapid and essentially linear.

4 Scholars generally agree, on the basis of archaeology and Spanish accounts, that precontact Pueblo settlements used some form of irrigation farming to grow maize, beans, squash, cotton and tobacco. While access to surface water removed certain restrictions and risks to agriculture, and, indeed, created surpluses allowing for greater social elaboration, studies now also indicate that environmental factors, some the result of using water control devices, constrained Pueblo farming and in the process effected changes in the land.

Irrigation farming, even in its crudest forms, likely sets in motion ecological chain reactions. Plants are introduced into habitats where they could not have sustained themselves previously; terracing changes the natural flow of streams; and man-made water diversions modify the natural vegetation, change the organic matter in the soil, and perhaps alter the migrating pattern of birds and animals. See Michael C. Meyer, *Water in the Hispanic Southwest: A Social and Legal History, 1550-1850* (Tucson: The University of Arizona Press, 1984), 19.

In addition, poor drainage and high evaporation can lead irrigation waters to deposit salts and other minerals that inhibit crop production. And as crop irrigation brings both fields and plants closer together the risk of crop loss due to diseases and insect pests is increased. See Linda S. Cordell, *Prehistory of the Southwest* (Orlando: Academic Press, Inc.,1984), 203-204.

Due to these factors, Cordell believes that good bottomland in the upper Rio Grande valley was very likely in increasingly short supply over the two centuries prior to the arrival of the Spanish. Studies of Pueblo land use and evidence of expansion, away from the river onto the plains, indicate that intensive efforts were being made to support a large population increase as Pueblos were abandoned and new communities founded. See Cordell "Prehistory: Eastern Anasazi," in *Handbook of North American Indians*, v.9, ed. Alfonso Ortiz (Washington: Smithsonian Institution, 1979),151. If this interpretation is correct, the carrying capacity (the maximum population that a particular environment can support indefinitely without leading to environmental degradation) along the Rio Grande itself may well have been reaching its limit at the time of Spanish contact.

5 A useful point of departure for understanding similarities and differences in frontier societies, and their relationship to the land, is the notion of an inclusive versus exclusive frontier. Spanish colonists frequently settled in areas of sedentary Indians, seeking Indian labor at the same time that they strove for Indian souls and mated with their women. Where there were no Indians there were no Spanish. Without a sharply defined racial barrier, this was a frontier of inclusion. In contrast, English colonists, while integrating economically with certain native populations to a degree, did not intermarry. Nor on any significant scale did they attempt to convert Indians to Protestantism. This then was a frontier of exclusion. See Alistair Hennessy, *The Frontier in Latin American History* (Albuquerque: University of New Mexico Press, 1978), 146-147; and Alfred Crosby, Jr., *The Columbian Exchange: The Biological and Cultural Consequences of 1492* (Westport, CT: Greenwood Press, 1972) and *Ecological Imperialism: The Biological Expansion of Europe 900-1900* (Cambridge: Cambridge University Press, 1986).

Add to this broad characterization the fact that the Province of New Mexico served as a defensive and missionary outpost over most of its colonial history. It never was an economic frontier on the order of New England, for example, a region which quickly became integrated into an international commercial trade network. In fact, Spain's loss of portions of its North American empire can be attributed to the advantages of England's expanding economic frontier, over Spain's defensive frontier, operating at long distance from centers of resources and population.

[6] Richard I. Ford, "The New Pueblo Economy," in *When Cultures Meet: Remembering San Gabrial Del Yunge Oweenge* (Santa Fe: Sunstone Press, 1987), 73.

[7] John R. Van Ness, "Hispanic Land Grants: Ecology and Subsistence in the Uplands of Northern New Mexico and Southern Colorado," in *Land, Water and Culture: New Perspectives on Hispanic Land Grants*, eds. Charles L. Briggs and John R. Van Ness (Albuquerque: University of New Mexico Press, 1987), 204.

[8] Hal Rothman, "Cultural and Environmental Change on the Pajarito Plateau," *New Mexico Historical Review* 64 (April 1989), 186. Rothman agrees fundamentally with Van Ness. He states that "American influence telescoped into a few years much more environmental and cultural change than Spanish practices had produced in nearly three hundred years." There were two reasons for these varying rates of change, according to Rothman. First, a marginal area such as New Mexico did not attract sufficient numbers of Europeans to effect dramatic environmental change; and, second, the "un-European," semiarid climate of New Mexico protected it from the "full brunt of the portmanteau biota of the Spanish."(188) Because many Old World plants, such as fruit trees, melons, and wheat could exist only in proximity to water, more was needed than merely the presence of the Spanish and their descendents to "Europeanize" the plants and animals of New Mexico. What was required was the transformation of New Mexico into an economic frontier, creating opportunities to produce and transport commodities to market on a large scale.

[9] Donald Worster, "Toward an Agroecological Perspective in History," *The Journal of American History* 76 (March 1990), 1101.

[10] Over the period of colonial rule, the Spanish conducted regular, and, by most accounts, fairly accurate censuses of the Province of New Mexico. (Because this study focuses upon the upper Rio Grande valley, the following population figures exclude the areas of El Paso and Zuni.) Only the estimates of Pueblo population at the time of the first Spanish settlement in 1598 and the Benavides counts of 1630 and 1635 have been the subject of any real debate. The figure for 1598 is generally agreed to be around 38,000, while Benavides estimated some 26,000 Pueblo Indians in 1630. The sharp decline was attributable to disease, starvation owing to Spanish tribute and labor institutions, and flight to western Pueblos. Throughout the remainder of the 17th century and for most of the 18th century Pueblo population continued to decline. It was approximately 23,600 in 1680, the year of the Pueblo Revolt, 7,200 in 1706, 5,200 in 1752, 6,500 in 1776, 6,400 in 1805 and 1821. See Marc Simmons, "History of Pueblo-Spanish Relations to 1821," in *Handbook of North American Indians*, vol.19, ed. Alfonso Ortiz (Washington: Smithsonian Institution,1979),185.

[11] Oakah L. Jones, Jr., *Los Paisanos: Spanish Settlers on the Northern Frontier of New Spain* (Norman: University of Oklahoma Press, 1979)117-129.

[12] Jones, 117-129.

[13] Alvar W. Carlson, *The Spanish-American Homeland: Four Centuries in New Mexico's Rio Arriba* (Baltimore: The Johns Hopkins Press,1990), 9.

[14] Marc Simmons, "The Rise of New Mexico Cattle Ranching," *El Palacio* 93 (Spring 1988), 7.

[15] John O. Baxter, *Las Carneradas: Sheep Trade in New Mexico,1700-1860* (Albuquerque: University of New Mexico Press, 1987), 90.

[16] George P. Hammond and Agapito Rey, eds., *The Rediscovery of New Mexico, 1580-1594* (Albuquerque: University of New Mexico Press,1966) 89, 230.

[17] Simmons, "New Mexico Cattle Ranching," 7.

[18] Baxter, 24.

[19] Baxter, 92.

[20] Marc Simmons, "The Chacon Economic Report of 1803," *New Mexico Historical Review* 60 (January 1985), 87.

[21] David J. Weber, *The Taos Trappers: The Fur Trade in the Far Southwest, 1540-1846* (Norman: University of Oklahoma Press,1971),10.

[22] Peter Bakewell, "Ecological Effects of Silver Mining in Colonial Spanish America," Paper presented at the American Historical Assoc. annual meeting, December, 1985.

[23] Hester Jones, "Uses of Wood by the Spanish Colonists in New Mexico," *New Mexico Historical Review* 7 (July 1932).

[24] Hammond and Rey, 221, 230.

25 Josiah Gregg, *Commerce of the Prairies* (Norman, University of Oklahoma Press, 1954),113.

26 W. W. H. Davis, *El Gringo: New Mexico and Her People* (Lincoln: University of Nebraska Press, 1982), 356.

27 Simmons, "New Mexico's Colonial Agriculture," *El Palacio* 89 (Spring 1983), 9.

28 Edward W. Smith, *Adobe Bricks in New Mexico* (Socorro: New Mexico Bureau of Mines and Mineral Resources, 1982), 1113.

29 Malcolm Ebright, "New Mexican Land Grants: The Legal Background," in *Land, Water, and Culture: New Perspectives on Hispanic Land Grants*, 23.

30 Carlson, 31, 69-70.

31 D.W. Meinig, *The Shaping of America: A Geographical Perspective on 500 Years of History*, v.1 (New Haven: Yale University Press, 1986), 240.

32 John Van Ness, "Hispanic Land Grants," 193-194.

33 Nancy Hunter Warren, "The Irrigation Ditch (Photo Essay)," *El Palicio* 86 (Spring 1980), 28. See also Daniel Tyler, "Dating the Caño Ditch: Detective Work in the Pojoaque Valley," *New Mexico Historical Review* 61 (January 1986) and Stanley Crawford, *Mayordomo: Chronicle of an Acequia in Northern New Mexico* (Albuquerque: Univ. of New Mexico Press, 1988).

34 Marc Simmons, "Spanish Irrigation Practices in New Mexico," *New Mexico Historical Review* 47 (April 1972), 145.

35 Arthur Goss, "The Value of Rio Grande Water for the Purpose of Irrigation," New Mexico College of Agriculture and the Mechanic Arts, Agricultural Experiment Station, *Bulletin* (November 1893), 34.

36 Frank E. Wozniak, "Irrigation in the Rio Grande Valley, New Mexico: A Study of the Development of Irrigation Systems Before 1945" (Santa Fe: The New Mexico Historic Preservation Division, 1987), 38, photocopied.

37 Wozniak, 25.

38 Ford, "The New Pueblo Economy," 74.

39 Ford, 86.

40 Vorsila L. Bohrer, "The Prehistoric and Historic Role of the Cool-Season Grasses in the Southwest," *Economic Botany* 29 (July-Sept. 1975), 203.

41 Ford, 87.

42 Christopher Vecsey, "American Indian Environmental Religions," in *American Indian Environments: Ecological Issues in Native American History*, eds. Christopher Vecsey and Robert W. Venables (Syracuse: Syracuse University Press, 1980), 10.

43 Frances Leon Swadesh, "Structure of Spanish-Indian Relations in New Mexico," in *The Survival of Spanish American Villages*, ed. Paul Kutsche (Colorado Springs: Research Committee, Colorado College, 1979), 53-61.

44 Swadesh, 61.

45 Yi-Fu Tuan, Cyril E. Everard, Jerold G. Widdison, *The Climate of New Mexico* (Santa Fe: State Planning Office, 1969), 158.

46 Harold C. Fritts, "Tree-Ring Evidence for Climatic Changes in Western North America," *Monthly Weather Review* 93 (July 1965), 421, 430-431.

47 Fluctuations in the climate of Colonial New Mexico can be viewed as well in the broader context of the climatic phenomenon known as the Little Ice Age. Between 1430 and 1850, the Northern Hemisphere's climate was allegedly cooler than periods either before or after. Therefore, the upper Rio Grande valley, over this time, may have experienced, on average, larger snow cover, enhanced freezing, and, therefore, shorter growing seasons. On the other hand, cooler and more moist summers may have produced excellent harvests. More importantly, however, climate conditions during the Little Ice Age were far from stable, and there were complex spatial patterns of warming and cooling throughout the period. To draw causal relations between the LIA and long-term environmental change in the upper Rio Grande valley is therefore risky at best. See T.M.L. Wrigley et alia, eds., *Climate and History: Studies in Past Climates and Their Impact on Man*. (Cambridge Univ. Press, 1981), 17.

48 Ronald U. Cooke and Richard W. Reeves, *Arroyos and Environmental Change in the American South-West* (Oxford: Clarendon Press, 1976), 16.

49 John R. Kummel and Melvin I. Dyer, "Consumers in Agroecosystems: A Landscape Perspective," in *Agricultural Ecosystems: Unifying Concepts*, eds. Richard Lowrance, Benjamin R. Stinner, and Garfield J. Hause (New York: John Wiley & Sons, 1984), 65.

50 Dan Scurlock, "The Rio Grande Bosque: Ever Changing," *New Mexico Historical Review* 63 (April 1988), 135.

51 Spanish Archives of New Mexico I, 1118.

13

"Saw Several Finners But No Whales":

The Greenland Right Whale (Bowhead) – An Assessment of the Biological Basis of the Northern Whale Fishery During the Seventeenth, Eighteenth and Nineteenth Centuries

Chesley W. Sanger

The nature of the Northern whale fishcry, the success or failure of the efforts of masters, seamen and companies, and the overall development and subsequent decline of the industry, were all influenced by the habits and character of the resource. Yet, compared with later whale fisheries the biological basis of commercial Arctic whaling is still poorly understood. This paper considers the characteristics of the Greenland Right whale or Bowhead (*B. mysticetus*) which made it the target species of Northern whalers during the seventeenth, eighteenth and nineteenth centuries.

The Greenland Right was the second cetacean species--after the Black Right whale (*B. glacialis*) in the Bay of Biscay and off southern Labrador--to become the object of an organized, large-scale

---

I would like to thank A. Small, Department of Geography, University of Dundee, and W.G. Handcock, Department of Geography, Memorial University of Newfoundland, for advice and assistance. The comments and suggestions by Valerie Burton have also helped in the revisions of this paper. Particular thanks are due to C. Conway and G. McManus, Memorial University of Newfoundland Cartographic Laboratory, Department of Geography, for drawing the figures, and Carole-Anne Coffee for typing the manuscript. This research was funded by the Social Sciences and Humanities Research Council of Canada.

whale fishery.[2] Given the relatively primitive open-boat and hand-thrown harpoon technology available to the early whalers, the Greenland Right whale was easy to pursue. It was slow moving, gentle, did not sink when killed, and from an economic point of view, produced excellent yields of oil and bone. As Scoresby pointed out in 1820, "this valuable and interesting animal...is the object of our most important commerce to the Polar Seas--is productive of more oil than any other of the Cetacea, and, being less active, slower in its motion, and more timid than any other of the kind, of similar or nearly similar magnitude, is more easily captured."[3] Most Right whale stocks had, however, been seriously depleted by this date. Other cetaceans referred to by Scoresby were more numerous: rorquals, for example, collectively referred to as "finners" because of their distinguishing dorsal fin, not found in either the Greenland or Black Right whale.

While the logbooks and journals of British Northern whalers contain numerous references to "finners" being sighted, it was not until the latter half of the nineteenth century that the development of new technology and techniques permitted the successful harvesting of whale species which previously had been immune to attack.[4] The master of the *North Pole*, whaling out of Leith in 1837, recorded in

---

[2]For a full discussion of Basque whaling operations within the Bay of Biscay and at Labrador, see C.W. Sanger, "The Origins of the Scottish Northern Whale Fishery" (Unpublished Ph.D. thesis, University of Dundee, 1985), 55-72.

[3]W. Scoresby, Jr., *An Account of the Arctic Regions with a History and Description of the Northern Whale-Fishery* (Edinburgh, 1820), I, 449-450.

[4]The "modern" era of commercial whaling dates from 1868, when the Norwegian, Svend Foyn, successfully used a harpoon-cannon mounted on the bow of a steam-powered catcher to kill whales in coastal waters off Norway, the carcasses being towed back to shore for subsequent processing. Modern whaling also required the development of exploding harpoons and winching apparatus capable of retrieving the bodies of whale species which usually sank when dead. Thus, the stocks of great whales which had previously been immune to attack by whalers using traditional equipment and hunting methods could now be harvested.

his logbook: "Saw several finners but no whales."[5] Similarly, one entry of a journal maintained throughout the 1822 whaling voyage of a London whaler, the *Leviathan*, noted that several rorquals had been sighted, "but they are too formidable to attack with safety, and the chance of killing them when struck is very slight."[6]

Any attempt to capture other than the "Right" whale, besides jeopardizing life and limb, would also have increased the likelihood of incurring financial loss. The rewards simply did not justify the risks. Following the sighting of several finners on the southwest fishing grounds (Figure 1) in 1831, the surgeon of the Aberdeen whaler, *Hercules*, explained that "they are seldom or never taken as they are so swift and strong that they would soon run out the lines in a ship."[7] Nevertheless, the temptation of having these gargantuan creatures breach close by, especially as the Greenland Right whale became scarcer, could not always be resisted. The results, however, were invariably the same. The first mate of the Kirkcaldy whaler *Chieftain* in 1841 recorded in the vessel's log that one of the boats had struck a "Finner Fish," but the whale "taking the lines out with such rapidity the Harpooner thought proper to cut them after there was three lines out with the intention to save the rest of the lines."[8] Still another illustration of the strength and speed which placed the Humpback, Sei, Fin, and Blue whales beyond the catching capability of the early hunters is provided by Captain Adams, master of the Dundee whaling

---

[5]Logbook of the *North Pole*, 21 June 1837, Dundee Public Library. See, as well, W. Elking, "A View of the Greenland Trade and Whale Fishery, with the National and Private Advantages Thereof," in J.R. McCulloch (ed.), *A Select Collection of Scarce and Valuable Tracts on Commerce* (London, 1859), 79.

[6]R.P. Gillies, *Tales of a Voyager to the Arctic Oceans* (London, 1826), II, 15.

[7]Journal kept by medical officer of the *Hercules*, 30 April 1831, University Library, King's College, Aberdeen.

[8]Logbook of the *Chieftain* kept by D. Kerr, first mate, 31 May 1841, Kirkcaldy Public Library.

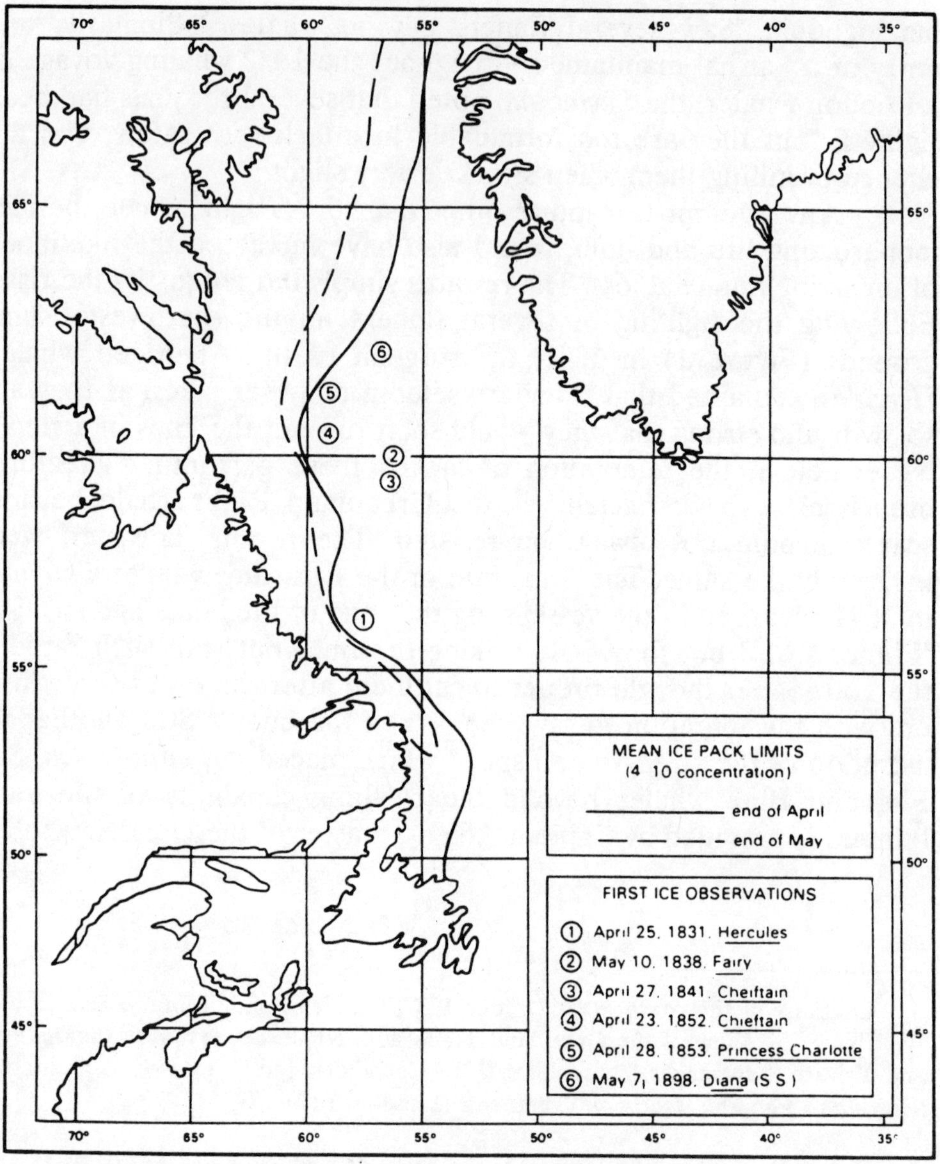

Figure 1: Southwest Whaling Grounds

Sources: Logbooks; and Monthly Ice Charts, Meterological Office, Bracknell, Berkshire.

steamer *Esquimaux* in 1895. His logbook recorded: "Saw one finner tonight. Tried her but she was off like lightning."[9]

Other species of whales, then, were familiar to the early whalemen. Virtually all of them, however, "were considered too fast, too small, too dangerous, or too rare to interest the whalemen."[10] Only two species, in fact, were available to the hunters--the Greenland and Black Right whales.[11] As early as 1722, Henry Elking, in a proposal to re-establish an English presence in the Northern whale fishery, noted the differences between the various species of baleen whales and the inappropriateness of attempting to capture any other than the Greenland Right whale:

> The fin-fish is as big as the [Greenland Right] whale ... and is distinguished from the whale by a large fin on his back. He is not so fat, nor hathe such fins in his mouth as a whale; so that they never are thought worth the trouble of catching, or to venture the harpoons and lines upon him; being more nimble than the whale, he might run away with the lines, or endanger the shallops.[12]

---

[9]Logbook of the *Esquimaux*, 7 May 1895, Broughty Castle Museum.

[10]E.N. Flayderman, *Scrimshaw and Scrimshandas: Whales and Whalemen* (New Milford, Conn., 1972), 18.

[11]The Greenland Right whale (Bowhead) was more timid when approached, yet more robust when struck. It was larger, produced more oil, and had longer and finer baleen. It was also restricted in its spatial distribution to the higher latitudes of the northern hemisphere, while the Black Right whale had a wider distribution throughout the middle and lower latitudes of both hemispheres. For a more detailed discussion of the major differentiating features between the two species, see R.J. Harrison and J. King, *Marine Mammals* (2nd ed., London, 1980), 53-55; and J.B. Holder, "The Atlantic Right Whales," in *Bulletin of the American Museum of Natural History*, I, No. 4 (1 May 1883), 99-137.

[12]Elking, "A View of the Greenland Trade and Whale Fishery," 79.

The nature and habits of the Bowhead, then, made it the principal target of the Northern whalers.

II

Although there has been considerable improvement in both the quantity and reliability of information on whales during the present century, many species are still poorly understood. Hunting reduced the Eastern Canadian Arctic Bowhead to such low numbers, for example, that research opportunities are infrequent.[13] There is even less information available in published form on Greenland Right whale stocks which were exploited at Spitzbergen and along the Greenland Sea ice-edge throughout the seventeenth, eighteenth and nineteenth centuries. While recent study has concentrated on the Western Arctic Bowhead,[14] Ellis' comment is still pertinent: "For a

---

[13]Nevertheless, useful work by Reeves and Mitchell in particular has drawn on historical evidence to provide information on the nature, distribution and migration of Bowheads which supported commercial whale fisheries in Hudson Bay, Davis Strait and Baffin Bay: see, for example, R.R. Reeves, E. Mitchell, A. Mansfield and M. McLaughlin, "Distribution and Migration of the Bowhead Whale, *Balaena mysticetus*, in the Eastern North American Arctic," *Arctic*, XXVI, No. 1 (1983), 5-64; E. Mitchell and R.R. Reeves, "Catch History and Cumulative Catch Estimates of Initial Population Size of Cetaceans in the Eastern Canadian Arctic," *Report of the International Whaling Commission*, XXXI (1981), 645-82. For an indication of the highly selective nature of other contemporary information available on this stock, see A. Jonsgard, "Bowhead Whales, *Balaena mysticetus*, Observed in Arctic Waters of the Eastern North Atlantic after the Second World War," *Report of the International Whaling Commission*, XXXI (1981), 511.

[14]Even in this instance, as Nerini, *et al.* pointed out as recently as 1984, "virtually no information relating to their life history was recorded," although an estimated ten thousand animals were taken between 1848 and 1915: M.K. Nerini, H.W. Braham, W.M. Marquette and D.J. Rugh, "Life History of the Bowhead Whale, *Balaena mysticetus* (Mammalia: Cetacea)," *Journal of Zoology*, CCIV (1984), 444. For evidence of the Bering Sea Stock, based largely on material provided by the aboriginal fishery, see for example R.E. Durham, *The Catch of Bowhead Whales (Balaena Mysticetus) by Eskimos, with Emphasis on the Western Arctic* (Los Angeles, 1979), *passim.*; R. Gambell, "Bowhead Whales and Alaskan Eskimos: A Problem of Survival," *Polar Record*,

huge animal that was the object of a concentrated fishery for over three centuries, surprisingly little is known of the biology of the bowhead whale."[15] It is not too surprising, however, when one considers the fact that, despite a lengthy exploitation period, the harvesting of Bowheads occurred at a time when investigative instruments and techniques were sadly deficient. By the time public opinion brought sufficient pressure to bear on the whaling trade to stimulate proper scientific research for conservation and management purposes, Greenland Right whale stocks could no longer sustain such large-scale, commercial hunting. Populations, in fact, had declined to such drastic levels that this was the first whale species to be officially protected. Opportunities to conduct research on the Bowhead, therefore, have been severely restricted. Its higher latitudinal distribution has also hindered investigative efforts. Marquette, for example, points out that "because of its isolation in the harsh arctic environment, the bowhead has been subject to little biological research."[16]

Much of what is known about the Greenland Right whale is the result of the work of William Scoresby, Jr. As Ellis explains: "Even though we now have more sophisticated methods of analysis and observation, Scoresby had more whales."[17] Almost as important, however, is the information provided by Scoresby's American counterpart, Charles Scammon, who, in 1874, also following a successful career as a Bowhead whaling master, published an account of whales and whaling in the Pacific.[18] The extent to which Scoresby and Scam-

---

XXI (1983), 467-473; and K.W. Hazard and L.F. Lowry, "Benthic Prey in a Bowhead Whale from the Northern Bering Sea," *Arctic*, XXXVII, No. 2 (1984), 166-168.

[15]R. Ellis, *The Book of Whales* (New York, 1980), 82.

[16]W.M. Marquette, "Bowhead Whale," in D. Haley (ed.), *Marine Mammals of Eastern North Pacific and Arctic Waters* (Seattle, 1978), 73.

[17]Ellis, *The Book of Whales*, 79.

[18]C.M. Scammon, *The Marine Mammals of the North-Western Coast of North America: Together with an Account of the American Whale-Fishery* (San Francisco, 1874).

mon have contributed to the present understanding of *B. mysticetus* is indicated in the comments of a prominent Canadian marine biologist:

> From the beginning of the twentieth century until quite recently, the right whale has been a rare mammal and few biologists have been fortunate enough to see one. It is not surprising, therefore, that our knowledge of the habits of this species is still based largely on the accurate observations recorded by the whaling captain William Scoresby in his history and description of the whale fishery (1820), and by Scammon in his account of the American whale fishery in the North Pacific.[19]

Although not the result of properly planned and carefully implemented scientific research, the publications of Scoresby and Scammon are excellent examples of attentive observation (often under adverse conditions), as well as detailed and accurate descriptions of whales and all activities associated with the whaling industry. This paper therefore draws heavily on both. A number of other sources, however, while not as well-known, have also been used. In addition, logbooks and journals, especially those of British whaling vessels, provide useful first-hand information.

III

The Bowhead, although not the very largest whale (Figure 2), is still huge. While Scammon noted that an individual animal "seldom attains the length of sixty-five feet," the largest of Scoresby's 322 whales only measured fifty-eight feet.[20] The most enormous Greenland Right whale taken by the eighteenth and nineteenth century

---

[19] A.W. Mansfield, "Occurrence of the Bowhead or Greenland Right Whale (*Balaena mysticetus*) in Canadian Arctic Waters," *Journal of the Fisheries Research Board of Canada*, XXVIII, No. 12 (1971), 1873.

[20] Scammon, *The Marine Mammals of the North-Western Coast of North America*, 52: Scoresby, *History and Description of the Northern Whale-Fishery*, I, 451-52.

European Northern whalers, therefore, probably would have been slightly less than sixty feet in length, while the average size, especially as resource stocks were depleted, and as fewer specimens were able to reach full maturity, would have been considerably smaller.

Because of their tremendous size, Bowheads produced large quantities of blubber which rendered excellent yields of high quality oil. Two additional features ensured that Greenland Right whales would provide proportionately higher oil yields than any other cetacean. An affinity for cold water required the development of an especially thick layer of insulating blubber and the seasonal availability of huge quantities of food enabled the Bowhead to sustain a large energy reservoir of fat. The Bowhead, therefore, is a "stout whale," with the thickness of the blubber layer being "eight or ten to twenty inches, varying in different parts as well as in different individuals."[21] Even in the average sized whale taken by the *Royal George* of Hull, on 23 July 1818, however, one which was "called middling, being neither large, nor yet small...its thickest part was a foot."[22]

The truncated body of the Greenland Right whale (Figure 2) is due mainly to "its ponderous head, forming...more than one-third of the whole creature, which is short, bulky, and bloated in its appearance."[23] The immense proportions of its head reflect the extreme to which the feeding apparatus had evolved, thus permitting the creature

---

[21]Scammon, *The Marine Mammals of the North-Western Coast of North America*, 52: Scoresby, *History and Description of the Northern Whale-Fishery*, I, 460.

[22]R. Phillips, *Journal of a Voyage of Discovery to the Arctic Regions Performed between the 4th of April and the 18th of November, 1818 in His Majesty's Ship Alexander, Wm. Edw. Parry Esq. Lieut. and Commander* (London, 1819), 44-45.

[23]Scammon, *The Marine Mammals of the North-Western Coast of North America*, 52.

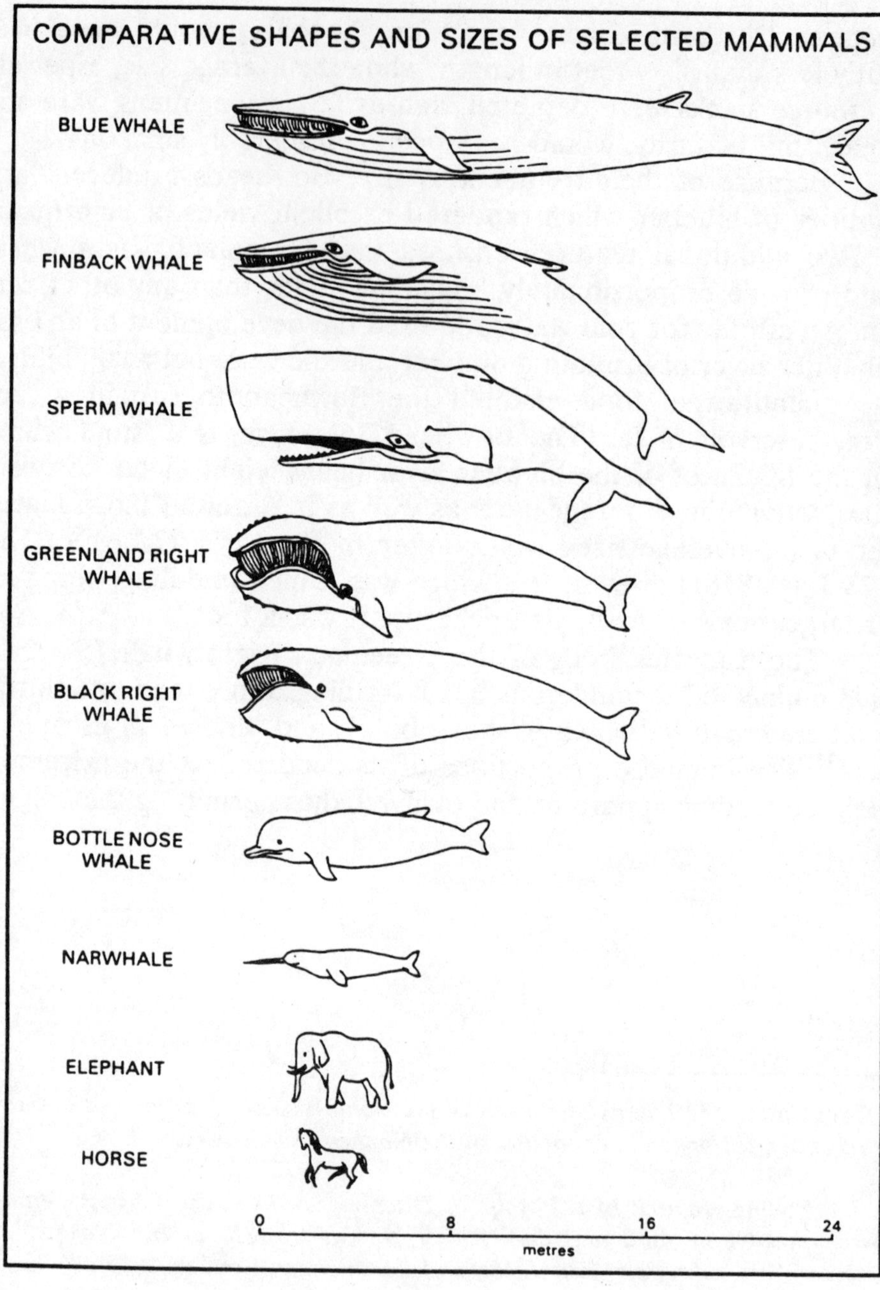

Figure 2: Comparative Shapes and Sizes of Selected Mammals

to make maximum use of its feeding opportunities during the relatively short, but extremely productive summer season in the higher latitudes. The American whalers' term for this whale, Bowhead, referred to the extreme arching of the upper jaw which permitted the growth of baleen plates that were longer, finer, and more numerous than the whalebone found in any other Mysticeti species. An indication of the massive proportions of the Greenland Right whale's jaw is provided by Scoresby: "When the mouth is open, it presents a cavity as large as a room, and capable of containing a merchant-ship's jolly-boat, full of men, being six or eight feet wide, ten or twelve feet high (in front), and fifteen or sixteen feet long."[24] Greenland Right whales, therefore, besides yielding the greatest quantities of high quality oil, also provided their pursuers with the largest amounts of very fine whalebone. As can be seen in Tables 1 and 2, the length of the baleen taken by Scottish whalers operating from both sailing vessels and steamers during the nineteenth century is just less than the figures provided by Scoresby who, although referring to an earlier period, noted that "fifteen feet is the greatest length of the whalebone; but ten or eleven feet is the average size, and thirteen feet is a magnitude seldom met with."[25] Although Greenland Right whale stocks had been seriously reduced by the early 1820s, the value of baleen continued to reward Northern whalers for their enterprise and persistence right through until the latter half of the century when, despite the adoption of more efficient technology in the form of steamers, the average number of whales taken per voyage declined dramatically.[26]

The truncated body shape, raised snout, and two large fins "of

[24] Scoresby, *History and Description of the Northern Whale-Fishery,* I, 455.

[25] *Ibid.*, I, 457.

[26] For a more detailed discussion of the influence of technological innovation on a similar marine resource-based industry, the Newfoundland seal fishery, see C.W. Sanger, "Technological and Spatial Adaptation in the Newfoundland Seal Fishery During the Nineteenth Century" (Unpublished M.A. Thesis, Memorial University of Newfoundland, 1973).

Table 1
Whaling Results of Ten Scottish Sailing Voyages between 1830 and 1853: Size of Baleen When Provided at Davis Strait and Baffin Bay

	May	July	Aug.	Sept.	Oct.
No. of catches per month	8	35	4	8	2
Average size of baleen (feet)	10.5	9.0	8.8	10.3	11.0

Note: Nursing young excluded. Average size for all whales 9.6 feet. Average number of whales per voyage 5.7.

Source: Logbooks (Appendix)

Table 2
Whaling Results of Sixteen Scottish Steamer Voyages between 1885 and 1900: Size of Baleen

	May	June	July	Aug.	Sept.	Oct.
No. of catches per month	3	3	8	2	8	5
Average size of baleen (feet)	10.4	10.2	10.3	10.4	11.2	10.3

Note: Nursing young excluded. Average size for all whales 10.5 feet. Average number of whales per voyage 1.8.

Source: Logbooks (Appendix)

an immense size,"[27] averaging "seven to nine feet in length, and four or five in breadth,"[28] make the Greenland Right whale one of the least streamlined cetaceans and a relatively slow swimmer. Its average

---

[27]Phillips, *Journal of a Voyage in His Majesty's Ship Alexander*, 45 .

[28]Scoresby, *History and Description of the Northern Whale-Fishery*, I, 455.

speed, according to Scoresby, and confirmed by a British zoologist who sailed in 1861 on the Dundee whaling steamer *Narwhal*, "seldom exceeds four miles an hour."[29] This placed it within range of the sail and oar-propelled open boats used by Northern whalers. Even the Bowhead's maximum speed posed no real problem. As Scoresby commented, "their extreme velocity may be at the rate of eight or nine miles an hour; yet we find this speed never continues longer than for a few minutes, before it relaxes almost to one-half."[30]

Moreover, the Greenland Right whale is not an exceptionally deep diver, since it does not forage on the ocean floor for food as do other species, although on occasion it may descend to substantial depths. Scoresby, for example, experienced incidents in which harpooned whales submerged "to the depth of 400 fathoms, with the average velocity of seven or eight miles per hour."[31] He also described experiences of other masters who drew up whales "by the line attached, from a depth of 700 or 800 fathoms, and...[had] been found to have broken their jaw-bones, and sometimes crown-bone, by the blow struck against the bottom."[32] Scammon also recorded occasions when a Bowhead had "been known to 'sound out' a line, in its descent and return, equal to a mile in length."[33] He added, however, that "it must not be inferred that this was done by the perpendicular course of the whale, for it is found that the line runs out with great swiftness, when the creature begins its return to the surface."[34]

---

[29]*Ibid.*, I, 467; R. Brown, "Notes on the History and Geographical Relations of Cetacea Frequenting Davis Strait and Baffin's Bay," *Proceedings of the Zoological Society of London*, Part III, (June/December 1868), 541.

[30]Scoresby, *History and Description of the Northern Whale-Fishery*, I, 467.

[31]*Ibid.*

[32]*Ibid.*, I, 468.

[33]Scammon, *The Marine Mammals of the North-Western Coast of North America*, 57.

[34]*Ibid.*

The breathing method of the Greenland Right whale also made it susceptible to attack. Although there is a slight disagreement, Scoresby and Scammon agree that Bodfish was essentially correct when, following thirty-one years as an Arctic whaler, he wrote that "the bowhead, if of average size, will remain underwater from eighteen to twenty-two minutes, rising about three times an hour, when he will remain on the surface from one and a half to two minutes. Very large whales will remain underwater for half an hour."[35]

The fact that Bowheads surface to breathe at short and regular intervals made them extremely vulnerable. As Bodfish explained: "Now the thing is to work the boat to the spot where the whale will come up, or as close as may be judged, so that he can be approached and struck during the minute or two that he is on the surface."[36] Even the act of exhaling assisted their pursuers, especially when visibility was curtailed by fog, since the sound could be heard over considerable distances. The surgeon of the *Hercules* during the 1831 season noted that "the blowing of a whale is a prolonged sighing noise which can be heard in calm weather two miles off and continues about seven seconds when the fish is lying still."[37] Similarly, the spout, with its characteristic double, or "V" shape, besides signalling that this was the "right" whale to pursue, could also be detected at considerable distances.[38]

There is a strong correlation between the migration habits of the Greenland Right whale and sea-ice margins. Scammon, for ex-

---

[35]H.H. Bodfish, *Chasing the Bowhead* (Cambridge, Mass., 1936), 90. Harpooned animals could remain submerged for longer periods. Scammon, for example, reported that one "creature descended to the muddy bottom, and there remained for an hour and twenty minutes." Scammon, *The Marine Mammals of the North-Western Coast of North America*, 57.

[36]Bodfish, *Chasing the Bowhead*, 91.

[37]Logbook of the *Hercules*, 19 July 1831.

[38]As Scoresby explained: "The vapour they discharge, is ejected to the height of some yards, and appears at a distance, like a puff of smoke." Scoresby, *History and Description of the Northern Whale-Fishery*, I, 465.

ample, observed that "everything tends to prove that the *Balaena mysticetus* is truly an "ice-whale," for among the scattered floes, or about the borders of the ice-fields or barriers, is its home and feeding-ground."[39] This affinity for ice had important implications for when, where, and how the Bowhead could be captured in the Greenland Sea and at Davis Strait/Baffin Bay.

The evidence gathered from British whaling logbooks (Figures 3.1, 3.2, 4.1, 4.2) confirm in essential details the important work of Reeves, *et al.*, who used published records on commercial whaling prior to 1915 to identify the spatial and temporal dimensions of the Davis Strait/Baffin Bay fisheries.[40] While there are no comparable data for the Greenland Sea whale fishery, British logbook accounts of sightings and kills provide a picture of the nature, timing and routes of these migrations with a similar precision (see Figure 5).[41]

To find a niche on the boundaries of sea and ice, the Greenland Right whale developed several highly specialized adaptations. While the massive proportions of the head are primarily to permit the efficient capture of food, its size also assists in the clearing of breathing holes when needed. Scammon, for example, reported instances in which the Greenland Right whale had "been known to break through

---

[39]Scammon, *The Marine Mammals of the North-Western Coast of North America*, 58.

[40]They also used published and unpublished records from the post-commercial whaling period (1915-1979) to produce tables. The data were then plotted on charts to indicate the nature, timing and routes of eastern North American Arctic Bowhead migrations: Reeves, *et al.*, "Distribution and Migration of the Bowhead Whale." Eschricht and Reinhardt used Danish West Greenland settlement records to compile a comprehensive data set which, when used with whaling information, also provides a fairly reliable indication of the Greenland Right whale's range during the "non-hunting seasons" of the year: D.F. Eschricht and J. Reinhardt, "On the Greenland Right Whale (*Balaena mysticetus*)" in Eschricht, et al. (eds.), *Recent Memoirs on the Cetacea* (London, 1866).

[41]For a more detailed analysis of spatial and temporal patterns in the seventeenth, eighteenth and nineteenth century Greenland Sea and Davis Strait/Baffin Bay whale fisheries, see Sanger, "Origins of the Scottish Northern Whale Fishery," 183-357.

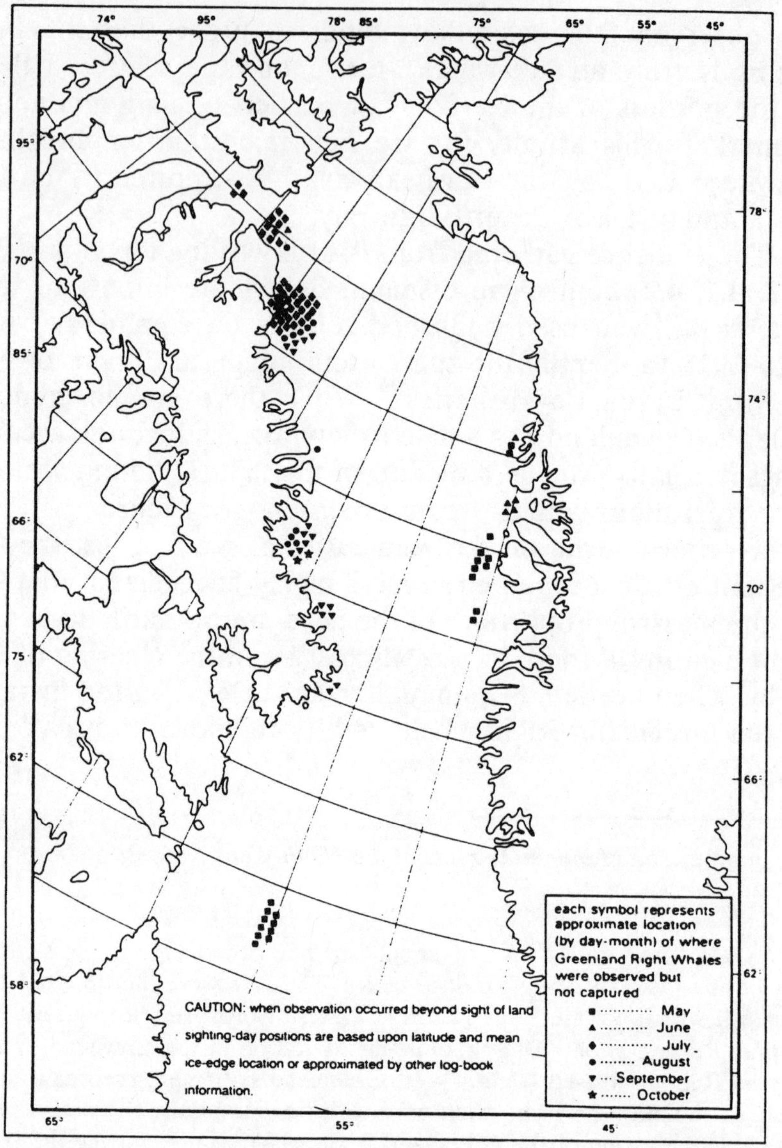

Figure 3.1: Whale Sightings: 10 Scottish Sailing Vessel Voyages on the Davis Strait - Baffin Bay Hunting Grounds (1830-1853)

Sources: Logbooks; and Monthly Ice Charts, Meterological Office, Bracknell, Berkshire.

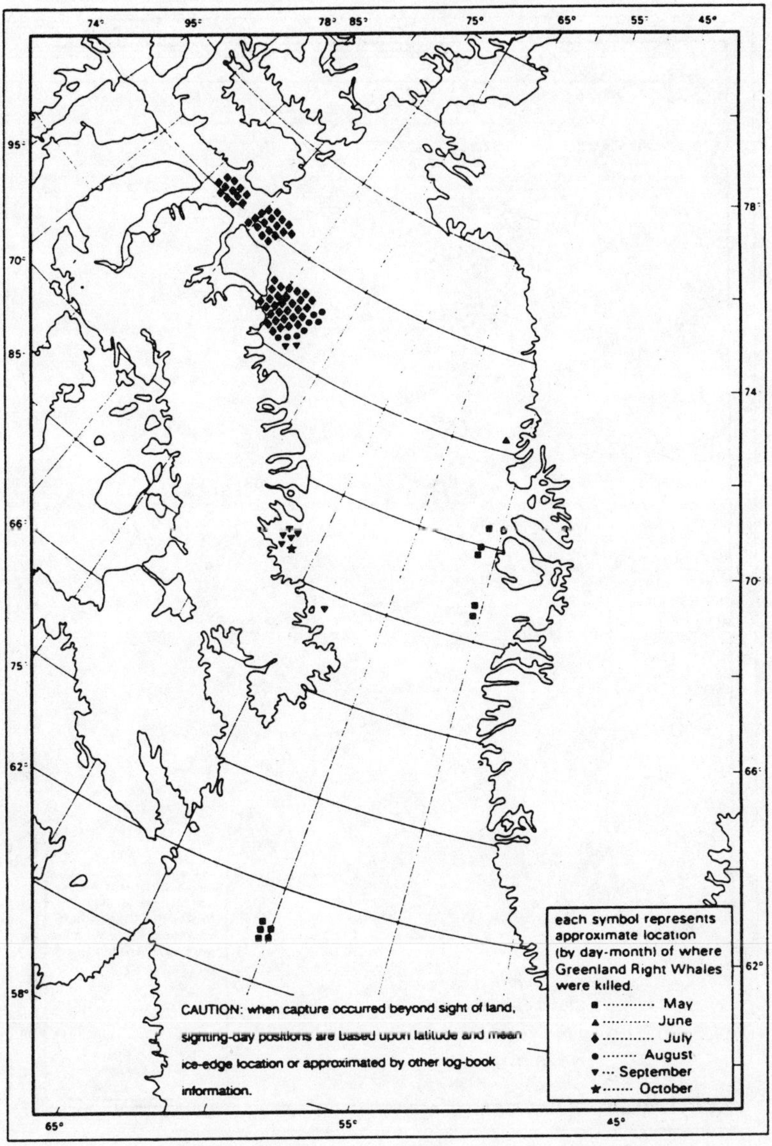

Figure 3.2: Whale Captures: 10 Scottish Sailing Vessel Voyages on the Davis Strait - Baffin Bay Hunting Grounds (1830-1853)

Sources: Logbooks; and Monthly Ice Charts, Meterological Office, Bracknell, Berkshire.

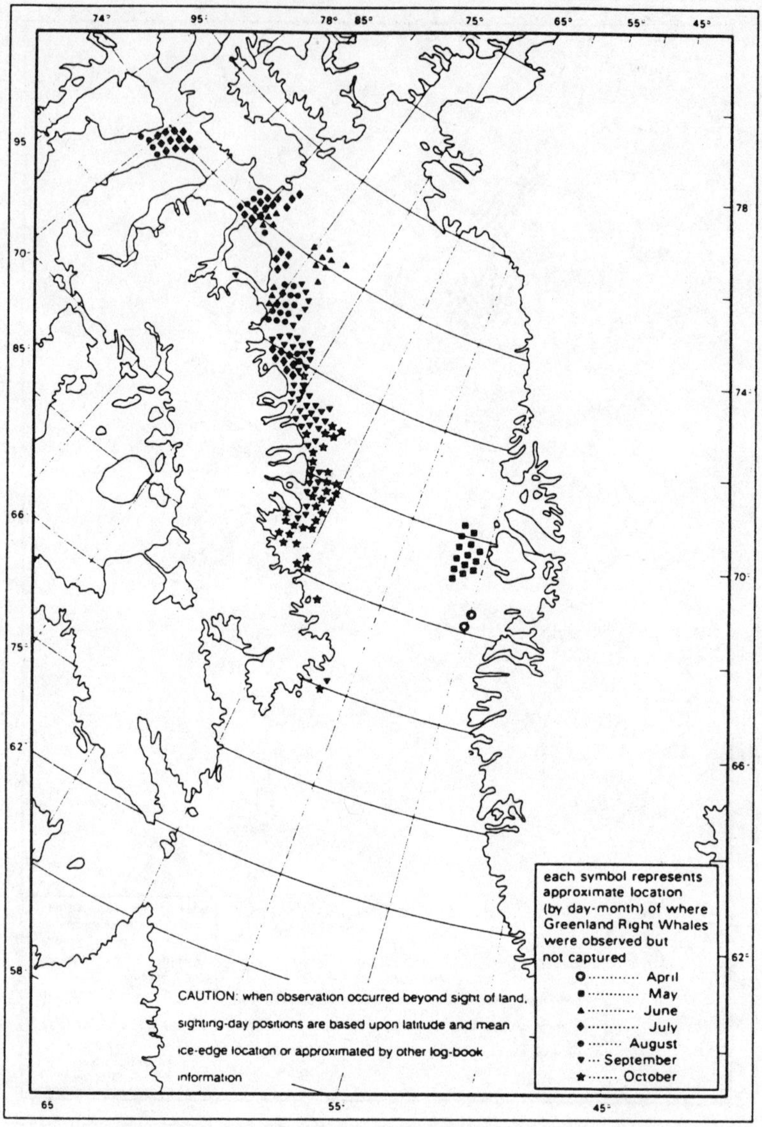

Figure 4.1: Whaling Sightings: 16 Scottish Steamer Voyages on the Davis Strait-Baffin Bay Hunting Grounds (1885-1900)

Sources: Logbooks; and Monthly Ice Charts, Meterological Office, Bracknell, Berkshire.

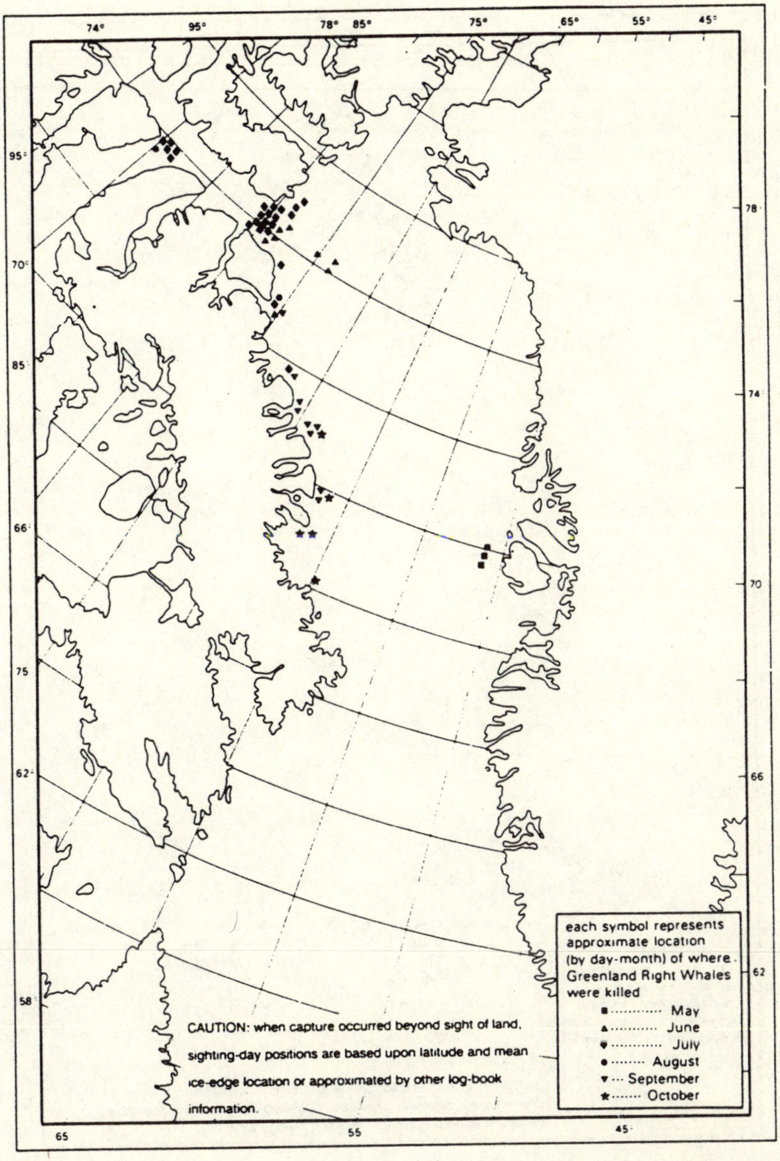

Figure 4.2: Whale Captures: 16 Scottish Steamer Voyages on the Davis Strait-Baffin Bay Hunting Grounds (1885-1900)

Sources: Logbooks; and Monthly Ice Charts, Meterological Office, Bracknell, Berkshire.

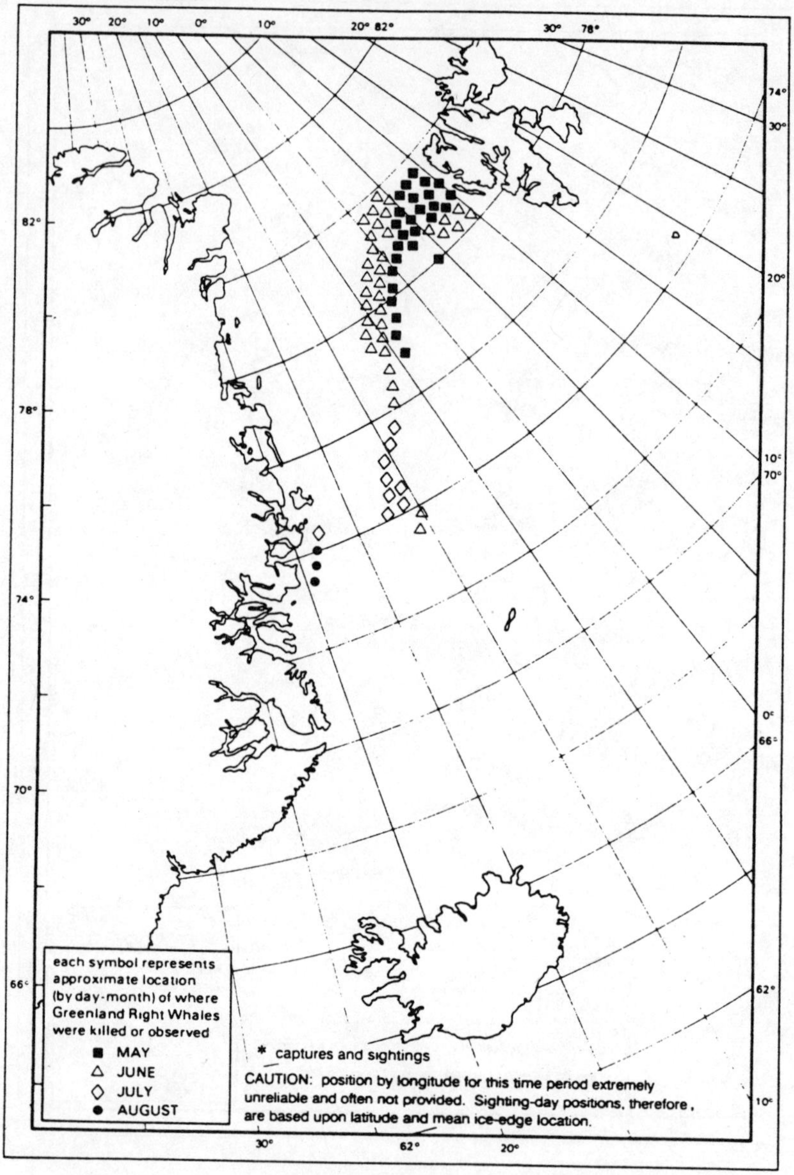

Figure 5: Whaling Results of 9 British Sailing Vessel Voyages on the East Greenland Hunting Grounds, 1822-1866

Sources: Logbooks; and Monthly Ice Charts, Meterological Office, Bracknell, Berkshire.

ice, three inches in thickness, that had been formed over water between the floes."[42] Similarly, the double blow-holes are located "on the most elevated part of the head,"[43] thus permitting it to surface for air in small openings that are only large enough to admit this protuberance, but not the entire head. One final adaptation which, as noted earlier, distinguishes the Greenland Right whale from other mysticetes, is the absence of a dorsal fin, an obvious advantage for moving between, and under, ice floes.

Not only was the Greenland Right whale, given the available technology, the only cetacean that could be successfully captured following the demise of the Black Right whale fishery during the late seventeenth century, but it also provided far greater quantities of very fine oil and bone. Although many and varied items were produced from these two commodities, they were, for the most part, essential in that they "added greatly for many centuries to the comfort and welfare of the civilized world."[44] An excerpt from a nineteenth century Scottish publication provides a good indication of the utility and value of these Greenland Right whale products:

> Man, ever searching the remotest parts of the globe for objects which might contribute to his use and accommodation, discovered, in those huge animals, a variety of substances fitted for the supply of important wants. Even after his more refined taste rejected their flesh as food, the oil was required to trim the winter lamp, and to be employed in the various branches of manufacture; while the bone, from its firm, flexible, and elastic qual-

---

[42]Scammon, *The Marine Mammals of the North-Western Coast of North America*, 58-59. See also, R.K. Nelson, *Hunters of the Northern Ice* (Chicago, 1969).

[43]Scoresby, *History and Description of the Northern Whale-Fishery*, I, 456.

[44]J.A. Allen, "The North Atlantic Right Whale and its Near Allies," *Bulletin of the American Museum of Natural History*, XXIV (1908), 278.

> ity, is peculiarly fitted for various articles of dress and ornament.[45]

Oil was initially the most important product. Towards the end of the eighteenth century, however, and periodically throughout the nineteenth, whalebone became increasingly important until, as Ommanney points out, it "was nearly fourteen times more valuable than the oil, so that the Greenland whalers often took only the baleen and threw the rest of the carcass away."[46]

From the beginning of commercial exploitation of the Black Right whale, through to the beginning of the nineteenth century, whale oil was used primarily in the manufacture of soft soaps and as lamp oil.[47] An indication of the essential nature of this commodity is provided by a local historian, who, in 1811, while lamenting Aberdeen's reluctance to expand its Northern whaling fleet, wrote:

> As oil is an indispensable necessity of life, whether for the purpose of giving light, or as an ingredient in the manufacture of soap, the whale fishing is an object of the highest natural importance...The importance of the species of fishing, both to the individuals concerned, and to the country at large, is so obvious that it is astonishing the citizens of Aberdeen, who are in general extremely enterprising, should have directed so little of

---

[45][J.] Leslie, [R.] Jameson, and H. Murray, *Narrative of Discovery and Adventure in the Polar Seas and Regions: With Illustrations of their Climate, Geology, and Natural History: and an Account of the Whale-Fishery* (2nd. ed., Edinburgh, 1831 [first published in 1830]), 361.

[46]F.D. Ommanney, *Lost Leviathan* (London, 1971), 61.

[47]See, for example, Scoresby, *History and Description of the Northern Whale-Fishery*, II, 420-421.

> their capital and attention to so profitable a business.[48]

Despite the introduction of coal gas during the first half of the nineteenth century, whale oil remained an essential commodity until the first commercial exploitation of petroleum in the 1850s reduced its importance. Whales, as Colwell points out, "were almost the only source of a pure, clean oil which could be used for illumination, instead of sputtering, smelly tallow candles or the costly beeswax candles."[49] Whale oil, however, was also used in a wide variety of other ways. Besides being a high quality lubricant, it was used in the manufacture of cordage and played an important role in the preparation of rough fibres, such as jute, and finer woollen clothes. It was an ingredient in paints and varnishes, and, when combined with tar to form a preservative, was used extensively in the construction of wooden vessels. Whale oil also functioned as a leather softener and was used in the preparation of such widely divergent products as perfume and explosives (glycerine).

An 1874 article which described Dundee's involvement in Northern whaling reported that the "whalebone or baleen is a remarkable substance, applicable to many purposes for which no other material will suit so well."[50] Until the development of steel and plastics, baleen, because it combined strength with elasticity, continued to fulfil an important if not quite so essential need as that being satisfied by whale oil. As Captain Middleton, a retired Aberdeen whaling master, observed in 1888:

> Those advanced in years will recollect the time when the streets in our native land were lit by oil lamps, and when with the exception of a dip or a mould in the

[48]W. Thom, *The History of Aberdeen: Containing an Account of the Rise, Progress, and Extension of the City* (Aberdeen, 1811), 171.

[49]M. Colwell, *Whaling Around Australia* (London, 1970), 1.

[50]"The Dundee Whaling Fleet," *The Practical Magazine* (1874), 173.

> parlour, nothing but train oil was used to light our homes. Whalebone also was a common article of commerce, and being tough and flexible, was used for a variety of purposes, to which from the want of it, steel is now applied.[51]

The value and utility of whalebone was also enhanced by the fact that it required "comparatively little preparation to fit it for use."[52] Scoresby provides the details:

> This singular substance, when softened in hot water, or simply by heating it before a fire, has the property of retaining any shape which may be given it, provided it be secured in the required form, until it becomes cold. This property, together with its great elasticity and flexibility, renders it capable of being applied to a great variety of useful purposes.[53]

Whalebone was used primarily by the fashion industry for such items as hoop skirts, stays, and hats. It was, consequently, subject to the whims of changing women's fashion styles which on occasion raised baleen prices to such high levels that the capture of but one animal could often pay all the expenses of a Northern whaling voyage. Whalebone was also used in a wide variety of other ways: as springs for chairs, sofas, and carriages; in the construction of umbrellas, whips, fishing rods, walking canes, travelling trunks, hat boxes, and covers for telescopes; for making brushes (especially chimney sweeps' brooms); as upholstery stuffing; and in the manufacture

---

[51] L. Middleton, *Whaling Recollections: 1818-1888* (Private printing, 1888 [?]), 3-4.

[52] A.H. Clark, "History and Present Condition of the Fishery," in G.B. Goode (ed.), *The Fisheries and Fishery Industries of the United States*, Section 4, *History and Methods of the Fisheries* (Washington, 1887), II, 5.

[53] Scoresby, *History and Description of the Northern Whale-Fishery*, II, 435.

of a broad range of minor items such as pen holders, shoe horns and combs.

IV

The nature and habits of the Greenland Right whale made it the only species of whale frequenting northern waters which could be successfully harvested by commercial whalers prior to the introduction of steam-powered chasers and explosive harpoons in the latter half of the nineteenth century. The physical characteristics of the Bowhead determined largely which strategies, equipment, and techniques could be used in the hunt. Its ability to adapt to harsh high-latitude environmental conditions ensured that each whale killed produced very large quantities of fine quality oil and bone. The seasonal migrations, in turn, dictated when and where vessels could operate, and severely reduced stocks during the eighteenth and nineteenth centuries made a thorough knowledge of the resource an absolute prerequisite of a successful Northern whaling master. An engineer on a Dundee whaler observed in 1861, "vessels in this trade it is well known, require for a chance of success, to reach a certain place, by a given date."[54] Thus even in the age of steam the habits and nature of the Greenland Right Whale still determined the character and organization of the industry.

[54]Journal of the *Camperdown*, kept by A. Smith, Engineer, 1861, Dundee Public Library, 90.

Appendix

The following logbooks, journals and related publications, in the archives of Broughty Castle Museum, Dundee (BCM), Dundee Public Library, Dundee (DPL), Glasgow University Library, Glasgow (GUL), Kirkcaldy Public Library, Kirkcaldy (KPL), Peterhead Public Library, Peterhead (PPL), Scottish Fisheries Museum, Anstruther (SFM), and University Library, King's College (ULKC), were used in the preparation of tables 1 and 2 and figures 1 and 3.1-4.2.

Active 1853 Peterhead. Sail; Captain David Gray; East Greenland; Journal kept by J.B. Arbuthnot, passenger; PPL.

Alexander 1825 Aberdeen. Sail; Captain Fairburn; East Greenland; Journal kept by Thomas Scoresby, second mate, son of Wm. Scoresby, Jr.; ULKC (Microfilm).

Aurora 1893 Dundee. Steam; Captain McKay; Newfoundland and Davis Strait; Journal kept by Dr. J.W. Allen; medical officer, and titled "On Board the *Aurora* in '93. The Record of a Sealing and Whaling Voyage;" GUL.

Caledonia 1834 Kirkcaldy. Sail; Captain Gray; Davis Strait; SFM.

Chieftain 1841 Kirkcaldy. Sail; Captain R. Tod; Davis Strait; Logbook kept by D. Kerr, first mate; KPL.

Chieftain 1842 Kirkcaldy. Sail; Captain R. Tod; Davis Strait; Logbook kept by D. Kerr, first mate; KPL.

Chieftain 1852 Kirkcaldy. Sail; Captain W. Archibald; Davis Strait; KPL.

Diana 1898 Dundee. Steam; Captain Adams; Davis Strait; BCM.

Dorothy 1834 Dundee. Sail; Captain D. Davidson; Davis Strait; DPL.

Eclipse (1) 1852 Peterhead. Sail; Captain J. Gray; East Greenland; Journal kept by J.B. Arbuthnot, a passenger, and titled "Journal of a Voyage to the Greenland Seal and Whale Fishing;" PPL.

Eclipse (2) 1893 Dundee. Steam; Captain Milne; Davis Strait; BCM.

Eclipse (2) 1896 Dundee. Steam; Captain Milne; Davis Strait; BCM.

Eclipse (2) 1897 Dundee. Steam; Captain Milne; Davis Strait; BCM.

Esquimaux 1885 Dundee. Steam; Captain Milne; Newfoundland and Davis Strait; BCM.

Esquimaux 1886 Dundee. Steam; Captain Milne; Newfoundland and Davis Strait; BCM.

Esquimaux 1887 Dundee. Steam; Captain Milne; Newfoundland and Davis Strait; BCM.

Esquimaux 1888 Dundee. Steam; Captain Milne; Newfoundland and Davis Strait; BCM.

Esquimaux 1891 Dundee. Steam; Captain J. Phillips; Newfoundland and Davis Strait; Logbook kept by W. Stenhouse, first mate; BCM.

Esquimaux 1895 Dundee. Steam; Captain Adams; Davis Strait; BCM.

Esquimaux 1899 Dundee. Steam; Captain H. McKay; Davis Strait; Journal kept by A.B. Walker, principal financial backer of the trip, and used as the basis for the book, "Cruise of the Esquimaux," published in 1906; BCM.

Esquimaux 1900 Dundee. Steam; Captain H. McKay; Newfoundland and Davis Strait; Logbook kept by R. Davidson, first mate; BCM.

Fairy 1838 Dundee. Sail; Captain D. Davidson; Davis Strait; DPL.

Hercules 1822 Aberdeen. Sail; Captain T. Fairburn; Published in W. Scoresby Jr., *Journal of a Voyage to the Northern Whale-Fishery; Including Researches and Discoveries on the Eastern Coast of West Greenland, made in 1822, in the Ship Baffin of Liverpool* (Edinburgh, 1823).

Hercules 1831 Aberdeen. Sail; Captain Allan; Davis Strait; Journal kept by medical officer; ULKC.

Jan Mayen 1865 Peterhead. Sail; Captain R. Martin, Jr.; East Greenland; Account written from rough notes by J. Wilson, crewman; PPL.

Jan Mayen 1866 Peterhead. Sail; Captain R. Martin, Jr.; East Greenland; Account written from rough notes by J. Wilson, crewman; PPL.

Leviathan 1822 London. Sail; Captain Shafton; Journal kept by paying passenger, R.P. Gilles, *Tales of a Voyager to the Arctic Ocean*, Vols. 1-3. (London, 1826).

Maud 1891 Dundee. Steam; Captain Milne; Davis Strait; BCM.

Maud 1892 Dundee. Steam; Captain Milne; Davis Strait; Lost in heavy ice off Coutt's Inlet on October 10th, crew rescued; BCM.

North Pole 1837 Leith/Edinburgh. Sail; Captain J. Lyle; East Greenland; DPL.

Polynia 1890 Dundee. Steam; Captain Milne; Davis Strait; BCM.

Princess Charlotte 1853 Dundee. Sail; Captain G. Deuchars; Davis Strait; DPL.

Sir Colin Campbell 1864 Peterhead. Sail; Captain R. Birnie; East Greenland; Account written from rough notes by J. Wilson, Crewman; PPL.

Thomas 1833 Dundee. Sail; Captain Thomas; Davis Strait; Journal (incomplete) kept by medical officer; BCM.

William and Ann 1830 Leith/Edinburgh. Sail; Captain Smith; Davis Strait; SFM.

14

Some Conservation Schemes of the Hudson's Bay Company, 1821–50: An Examination of the Problems of Resource Management in the Fur Trade

Arthur J. Ray

Current interests in the plight of many Indian groups, and the concern with ecologically directed conservation efforts, has stimulated a re-examination of the North American fur trade. As part of this re-examination, the record of the attempts by the Hudson's Bay Company to adopt and effect a conservation policy for beaver and other fur-bearers in the early nineteenth century deserves careful consideration.

In this paper attention will be focused upon the beaver conservation schemes which the Hudson's Bay Company tried to introduce in Western Canada between

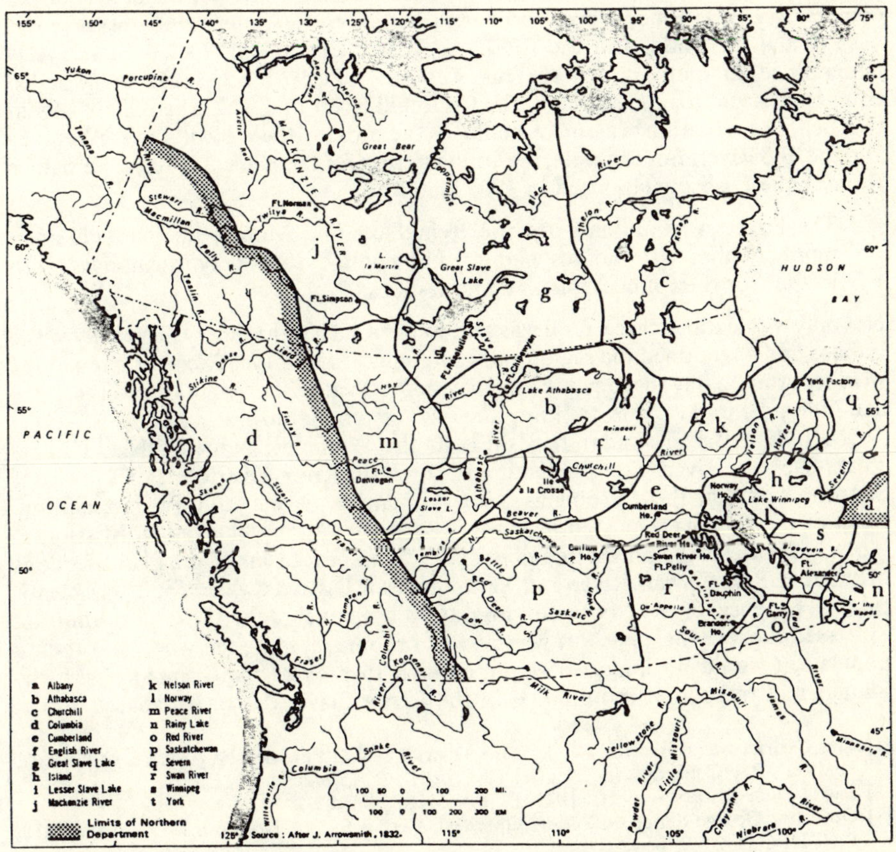

Figure 1. The Hudson's Bay Company's Northern Department.

1821 and 1850 under the direction of Governor George Simpson. For this study the lands which lay within the Company's Northern Department (Fig. 1) have been chosen. The problems which Governor Simpson and the Company faced in their efforts to husband beaver populations in this region were similar to those which they faced in other areas and, or, in the management of other resources.

Trade rivalries and the destruction of fur and game resources 1763–1821

The period between 1763 and 1821 was the time when fur-trading rivalries were sharpest in Western Canada as the Hudson's Bay Company struggled with competitors, most notably the Montreal-based traders operating in groups known as the North West and XY Companies, for the Indians' furs. This competition forced these companies to expand their operations territorially and it encouraged them to increase the numbers of posts which they maintained in the various districts. Therefore, not only were the local fur-bearing animals intensively trapped, but heavy hunting pressure was brought to bear on large game animals also as the Indians attempted to satisfy the mushrooming logistical requirements which this expansion of trading networks generated. Consequently, when the Hudson's Bay and North West Companies merged in 1821 bringing this period of cut-throat competition to an end, whole territories had been laid waste and the resource base of the fur trade, and the food supplies of the Indians had been seriously undermined in many sections. This was particularly the case in the wooded country adjacent to, and south of, the Churchill River in the Northern Department (Fig. 1). The situation was most acute in the area to the east of Lake Winnipeg and the Red River. For example, the annual report for the Hudson's Bay Company's Rainy Lake District submitted in 1822 stated:

> The large game animals are the Rein Deer and Moose, but in such small numbers that the natives cannot kill enough to supply themselves with leather for their moccasins and snow shoes.[1]

Not only was leather scarce, food was said to be in such short supply that many of the Indians had abandoned the region rather than face starvation. Those who remained were reported to have been preoccupied with food-gathering activities and therefore had little time to devote to trapping the few remaining fur-bearing animals.[2]

The Indians faced similar hardships in the Winnipeg, Norway, Island Lake, York, Nelson River, Cumberland and English River Districts and Governor Simpson said that these problems were stimulating an emigration of population. Most of the Indians relocated near the Red River Colony, the Indian Mission at Norway House at the northern end of Lake Winnipeg, or in the grassland sections of the Red River, Swan River and Saskatchewan Districts where food was plentiful.[3] The pace of this outward migration was such that Governor Simpson expressed the fear that unless it was slowed in some manner, the woodland Indian population would no longer be large enough to provide the Company with the hunters and trappers it required.[4] In the Lesser Slave, Peace, Athabasca, Great

[1] This District was called the Lac la Pluie District at that time. Public Archives of Canada, Hudson's Bay Collection (hereinafter PAC HBC) B 105/e/2

[2] Lac la Pluie District Reports, 1822–25, PAC HBC B 105/e/2–5

[3] Governor George Simpson, Letters Outward, Report for the Northern Department, 1835, PAC HBC D 4/102, p. 42

[4] *Ibid.*, 43

Slave and Mackenzie River Districts the depletion of game does not appear to have been a serious problem before the 1830s judging from the letters which the traders in charge of those districts wrote to Governor Simpson.[1]

Conservation strategies before 1841

It is clear that when Governor Simpson assumed control of the Canadian operations of the Hudson's Bay Company in 1821, years of overhunting and trapping had left the trade of many areas on a precarious footing. Yet, he was convinced that effective counter measures could be taken. Reflecting this optimism, in his 1822 report to the Governor and Committee in London he wrote:

> The country is without doubt, in many parts exhausted in valuable furs, yet not to such a low ebb as has been generally supposed and by extending the trade in some parts and nursing others, our prospects are by no means unfavourable.[2]

Since beaver had long been the staple in trade, Simpson focused his efforts on "nursing" them back.

One of the cornerstones of his conservation programme involved the curtailment of trapping operations in overhunted districts while at the same time the Company traders were encouraged to extend the trade into new areas. In the Northern Department this meant that the fur trade was to be vigorously prosecuted in the Mackenzie River District, particularly in the mountainous western and north-western portions of the district, while in most of the territories lying to the south of the Churchill River a variety of approaches were used in an attempt to ease the trapping pressure on the remaining beaver. Recognising the difficulties of obtaining Indian consent to local moratoriums on beaver hunting for reasons which will be discussed subsequently, Simpson began to shift trading posts around within the districts in accordance with the fluctuations of fur returns. For example, in 1824 he closed Fort Dauphin and the Swan and Red Deer River posts in the Swan River District and Brandon House in the Upper Red River District (Fig. 1). To replace these four posts he opened Fort Pelly which was centrally located to serve both districts, including the Qu'Appelle River valley (Fig. 1). Besides cutting the Company's operating costs in these overexploited areas, Simpson believed these actions would, ". . . remove the Indians from exhausted tracts of country which will recruit, to parts which have of late been little hunted".[3] The Governor based this assumption on the belief that the Indians would gravitate towards Fort Pelly, giving the other parts of the district a chance to recover. Similar moves were made elsewhere.[4] Generally, these moves did have a positive effect, and in the Swan and Upper Red River areas for instance, fur-bearers, especially beaver, were reported to be rebounding by 1826. In part this rapid recovery was due to the fact that the closing of Brandon House and Fort Qu'Appelle had the unexpected effect of leading many Assiniboine and Plains Cree to discontinue trapping

[1] Simpson, Correspondence Inward, 1828–30, PAC HBC D 5/3. It should be pointed out, however, that food supplies had always been precarious in the Mackenzie River District where the Indians depended heavily upon hare

[2] Simpson, Letters Outward, 31st July 1822, PAC HBC D 4/102, p. 42

[3] Simpson, Letters Outward, Report to London, Fort Garry, 5th June 1824, PAC HBC D 4/87

[4] *Ibid.*, PAC HBC D 4/87, and Simpson, Report to London, York Factory, 10th August 1824, PAC HBC D 4/87

activities altogether. Instead, it encouraged these two groups to, ". . . follow the Herds of Buffolo wherever they go depending on their Bows and Arrows".[1] Simpson added:

> So many Indians being therefore withdrawn leaves a great portion of that country [Swan River District] to recruit, but the improvement is very slow as of late years it was scarsly[sic] possible to trace a vestige of Beaver on the Assiniboine River in any of its feeders; they now however begin to appear and if left undisturbed for a few years there is no question that important advantages will result.[2]

Besides shifting post locations in response to changing local resource conditions, the traders also attempted to swing the focus of Indian trapping activities from beaver to other fur-bearing species. In doing this, the Hudson's Bay Company traders tried to take advantage of the fact that the populations of all of these animals fluctuated a great deal irrespective of the intensities of hunting pressures. Frequently, the oscillations of population in one species did not parallel those of another, but rather complemented them. Furthermore, there was a considerable variation in the amplitude and duration of these population cycles in the different animal species. Governor Simpson was quick to note this phenomenon and he sought to exploit it by having the Hudson's Bay Company traders encourage the Indians to concentrate their trapping activities on one fur-bearing species as it approached a population peak leaving others, usually beaver, to recoup their losses in the interim.

This practice was particularly well suited to the boundary waters area of Ontario and Minnesota, southern Manitoba and central Saskatchewan. The hydrography of much of this area is characterised by rivers and lakes that have extensive shallow water margins which provide ideal habitats for muskrat and beaver. By the 1820s nearly all of the beaver and other prime fur animals had been trapped-out of this country and the fur trade was heavily dependent on muskrat.[3] Of all of the fur-bearers, this animal exhibited the greatest fluctuations in numbers. Significantly, the oscillations of muskrat populations were closely tied to short-term variations in rainfall and runoff. Living in the shallow waters of lakes and rivers and having no control over local water levels, they were more sensitive to water level changes than were the beaver. High water can limit the growth of aquatic plants which the animal feeds upon, principally rootstocks, cat-tails, bulrushes and arrowhead plants, while low water may cause ponds to freeze to the bottom during winter. Should the latter occur, many muskrat starve to death since, unlike the beaver, they do not store food. Low water was an additional problem since it seems to have favoured the outbreak of disease. For instance, in 1824 Simpson reported that water levels were low in the Cumberland Department (Fig. 1) and, as a consequence, ". . . produced a disease among the Rats that destroyed them by the thousands as heaps were found dead on opening their houses this spring".[4] As a result of this epidemic the muskrat returns of the

[1] Simpson, Letters Outward, Report to London, York Factory, 20th August 1826, PAC HBC D 4/89

[2] *Ibid.*, PAC HBC D 4/89. During this period the Plains Indians became relatively independent of the Company. A. J. Ray, *Indians in the fur trade* (University of Toronto Press, Toronto. 1974 in press)

[3] *Ibid.*

[4] Simpson, Letters Outward, Report to London, York Factory, 10th August 1824, PAC HBC D 4/87

Cumberland Department declined from 150,000 skins in 1823 to 70,000 skins in 1824.[1] On many other occasions low water levels were said to have produced epidemics which killed many muskrat. In all probability, the disease which broke out under these circumstances was Tularemia.[2]

While muskrat populations thus frequently suffered heavy losses, they rebounded quickly with more favourable moisture conditions. The female muskrat may have up to five litters a breeding season and may produce as many as eleven offspring in each litter. In contrast, the female beaver mates in February, bears her young in May and usually nurses only four cubs per litter. The higher fertility rates of the muskrat allowed them to recoup their population losses much more rapidly than the beaver.[3]

Since the variations of muskrat populations were closely tied to moisture conditions, the traders often used fluctuations in local water levels to make predictions about their future muskrat returns and to plan short-term management strategies on a district basis. For instance, as noted previously, it was very dry in eastern Saskatchewan and western Manitoba in 1824 and low water levels along the lower Saskatchewan River led to a precipitous drop in the muskrat trade of the districts adjoining it, particularly the Cumberland Department. In 1825 more humid conditions returned, ending a two-year drought, and spring flooding along the lower Saskatchewan was extensive that year. This flooding led Simpson to predict that the muskrat would soon reappear in considerable numbers in the Cumberland Department. However, because of the severity of the preceeding drought and the intensive trapping pressure which had been exerted on the dwindling muskrat population, he believed that it would not be until 1826 that the recovery would be strongly felt in the returns from the district. Indeed, it was not until 1827 that a marked improvement was reported. By the latter date, higher muskrat returns were being registered throughout eastern Saskatchewan and southern Manitoba.[4]

During this wet phase the Indians were encouraged to turn their attention to muskrat once more in the hope that other fur animals, particularly beaver, would have a chance to recover. For example, when discussing the returns of the Swan River area in 1827 (Fig. 1), Simpson reported:

> This District has improved more during the past year than any other . . . which is to be accounted for by the excellent hunts made in Rats while the returns

[1] *Ibid.*

[2] Tularemia is an epizootic disease which causes hemorrhaging of the heart, lungs and liver of the animal and usually death. Low water levels increase the concentration of Tularemia bacteria and therefore increase the probability that muskrat or beaver will contract the disease. For a discussion of this disease (*Pasteurella tularensis*) see W. L. Jellison *et al.*, Epizootic tularemia in the Beaver, *Castor canadensis*, and the contamination of stream water with *Pasturella tularensis*, *American Journal of Hygiene* **36** (1942) 168–82. For a discussion of its effects on muskrat populations see P. L. Errington, *Muskrat populations* (Iowa State University 1963), and P. L. Errington, *Muskrats and marsh management* (Washington D.C. 1961) 49–53

[3] Errington, *Muskrats and marsh management* 35–7

[4] Simpson, Letters Outward, Report to London, York Factory, 1st September 1825, PAC HBC D 4/88. Besides the low ebb of the muskrat population of the Cumberland District Simpson indicated that many of the Indians had scattered and he believed that this would adversely affect the department's returns for a year or two. In one of his reports for 1827, Simpson wrote that muskrat were increasing in the Cumberland, Swan River and Winnipeg Districts. Simpson, Letters Outward, Report to London, York Factory, 25th July 1827, PAC HBC D 4/90

> of other furs have fallen off materially particularly in the article of Beaver; this reduction . . . is owing to the measures taken by Chief Factor Clarke to protect it, who discovering that the natives could find ample employment in rat hunting since the year 1823/24.[1]

Fortunately for the Hudson's Bay Company, the market for muskrat skins was strong in Eastern Canada and the United States during the 1820s, and in this instance their conservation scheme paid immediate economic dividends.[2]

The principal difficulty of attempting to exploit the oscillations of muskrat populations to give the beaver periodic respites from heavy trapping related to the fact that the amplitudes of the muskrat cycles were never of sufficient duration to allow the beaver to make a substantial recovery. Simpson was aware of this problem. In 1827 he wrote that although the Indians of the Cumberland District had begun to focus their trapping operations on muskrat, he doubted that it would prove to be very beneficial to future beaver returns from that area since the beaver population was so depleted that it would take many years to recuperate.[3] Simpson's doubts proved well founded. 1828 was a dry year once again. This led him to write:

> I much fear that there will be a considerable falling off next year on this branch of our Trade [Muskrat] as the present season has been hitherto unusually dry, and the Marshes and pools which they frequent are from that cause become stagnant which will engender disease and in all probability destroy them. . .[4]

Although Simpson's prediction of poor muskrat returns from eastern Saskatchewan and southern Manitoba in 1829 proved to be wrong, persistent low water levels did result in a substantial decline of this trade throughout the region in 1830.[5]

Besides shifting post locations and attempting to exploit alternative fur resources under favourable conditions, the Hudson's Bay Company employed a number of other strategies to manage the dwindling beaver trade. One of these efforts was directed towards encouraging and cajoling the Indians to stop trapping beaver during the summer season when pelts were of little value. Accordingly, in his 1822 report to London, Governor Simpson wrote, ". . . the Indians have been informed that skins out of season will not be taken off their hands".[6] During the competitive period before 1821 they had been accustomed to bringing in summer beaver to obtain alcohol and other itmes whenever they required them. Thus, the Indians were not willing to readily accept this abrupt departure from well-established trading practices. Because of this Indian reluctance, many of the company's district heads (called factors) apparently made little effort to enforce the new rule in the 1820s and early 1830s. As late as 1836, the London directors

[1] Simpson, Letters Outward, York Factory, 25th July 1827, PAC HBC D 4/90
[2] Simpson, Letters Outward, Report to London, York Factory, 8th September 1823, PAC HBC D 4/86
[3] Simpson, Letters Outward, York Factory, 25th July 1827, PAC HBC D 4/90
[4] Simpson, Letters Outward, Report to London, York Factory, 10th July 1828, PAC HBC D 4/92
[5] Simpson, Letters Outward, Report to London, York Factory, 26th August 1830, PAC HBC D 4/97
[6] Simpson, Letters Outward, Report to London, York Factory, 10th August 1824, PAC HBC D 4/85

complained to Simpson that too many out-of-season skins were still being sent home from the Bay.[1]

In subsequent years, some of the factors did make a more earnest attempt to obtain Indian compliance with the regulation. Some of them—such as George Gladman who was stationed at Norway House in the mid-1840s—resorted to rather dramatic, and questionable, tactics to demonstrate their resolve on this matter to the Indians. In his letter to Governor Simpson dated 21st June 1844, Gladman wrote, ". . . an Indian brought me a Beaver skin the other day/the animal being recently killed/this being against the rule I slapped his face with it. . . ."[2] There can be little doubt that such demeaning treatment would have left a strong impression upon the minds of the local Indians. However, for reasons which will be discussed subsequently, it is unlikely that the Indians discontinued their summer beaver hunts even if they did stop bringing summer pelts to the posts.

In the hopes of curbing the indiscriminate nature of Indian trapping (with respect to ages of beaver) and reducing the effectiveness of their operations, Simpson believed that the use of steel traps and castoreum bait, a trapping technique which came into general usage in the 1790s, should be discontinued.[3] Concerning this subject, in 1822 he wrote:

> The use of Beaver Traps should have been prohibited long ago, they are the scourge of the Country and none will, in the future, be given out except for new Districts exposed to opposition and frontier establishments.[4]

Thus, the further trade of steel traps in the Northern Department was banned except in areas near the United States border and the Red River Colony where the Indians could obtain them from alternative suppliers.

Most of the conservation schemes outlined thus far had little chance of success unless the local Company traders were committed to the Company's long-term goal of placing the fur trade on a sustained-yield basis. Yet there was always the temptation on the part of individual traders to increase their fur intakes on a short-term basis and therefore "look good" in the Company's eyes in terms of profits. Also, the Indians living in one district could trade their furs in another near by if the local trader was too zealous in his efforts to implement conservation measures. In order to deal with these problems, Simpson and the governing council for the Northern Department (which was composed of the chief factors of the various districts) decided upon a more coercive plan of action. In 1826 they introduced a quota system for the beaver trade of the Northern Department (Table 1). Under this scheme, the average annual beaver returns were calculated for each of the fourteen districts of the department for the three-year period between 1823 and 1825. In 1826, each district was allowed to take in a percentage of its annual trade during the base period. In the case of two districts, those of the Lower Red River and Rainy Lake (Fig. 1), beaver quotas were not established because of the close proximity of American traders. In other districts, the fur intake was reduced by from between one-fifth to one-half of what it had been

[1] Simpson, Letters Inward, Hudson's Bay House, London, 9th March 1836, PAC HBC D 5/4

[2] Simpson, Letters Inward, Norway House, 21st June 1844, PAC HBC D 5/11

[3] R. F. Wells, Castoreum and steel traps in Eastern North America, *American Anthropologist* **74** (1973) 479–83

[4] Simpson, Letters Outward, Report to London, York Factory, 31st July 1822, PAC HBC D 4/85

TABLE 1

*Fur quotas for 1826**

Beaver assorted	Athabascan	Saskatchewan Lesser S. Lake Ft. Assiniboine	English River	Cumberland	Swan River and Upper Red River	Lower Red River	Winnipeg	Lac la Pluie	Norway House	Island Lake	Severn	Nelson River	Churchill	York Factory
Outfit 1823	7,726	7,800	1,757	272	663	30	44	826	138	352	660	794	447	504
Outfit 1824	5,479	6,493	1,201	303	644	37	123	735	7	187	368	1,033	317	370
Outfit 1825	6,186	6,896	1,078	397	433	80	76	501	26	152	211	702	342	285
Returns in beaver for 3 years	19,391	21,189	4,036	972	1,740	147	243	2,062	171	691	1,239	2,529	1,106	1,159
Equal to an average per annum of	6,463	7,063	1,345	324	580	49	81	687	57	230	413	843	368	386
Less	$\frac{1}{5}$ 1,292	$\frac{1}{5}$ 1,412	$\frac{1}{2}$ 673	$\frac{1}{2}$ 162	$\frac{1}{4}$ 145	—	$\frac{1}{5}$ 27	—	$\frac{1}{2}$ 29	$\frac{1}{2}$ 115	$\frac{1}{2}$ 207	$\frac{1}{2}$ 422	$\frac{1}{2}$ 184	$\frac{1}{2}$ 193
Not to exceed for outfit or 1826	5,771	5,651	672	162	435	—	54	—	28	115	206	421	184	193

* Source: HBC D 4/89

(Table 1). Governor Simpson indicated that all of these quotas would be rigidly enforced with the exception of that of the Saskatchewan District.

The quota for the Saskatchewan District was not enforced because the beaver returns of that department came largely from the Piegan Indians who obtained their pelts by trapping, trading and raiding south of the international boundary. Few beaver were taken by the Indians living north of the border because the animal had been nearly trapped out in many sections of the Saskatchewan region, especially in the country to the north of the North Saskatchewan River. Also, the Hudson's Bay Company traders were instructed to discourage the Indians living in the latter areas from taking any more beaver. Thus, by not enforcing the quota, Simpson was hoping to underwrite part of the cost of "nursing" the beaver back in the Saskatchewan District by drawing upon the resources of the American territories to the southward.[1]

In brief, during the 1820s the Hudson's Bay Company undertook a series of initiatives which were intended to place the beaver trade on a sustained-yield basis. Yet, in spite of the fact that a variety of approaches were attempted, beaver populations continued to decline.

Opposition to the Hudson's Bay Company's early conservation programme

The failure of the Hudson's Bay Company to achieve any real success in its early conservation programme was due largely to the fact that American and Métis traders as well as the Indians opposed the Company's efforts. The Americans conducted much illegal trade north of the United States border, especially in the northern Ontario, Manibota and Saskatchewan boundary areas. Simpson attempted to deal with this problem by entering into an agreement with the American Fur Company in 1833 whereby the Hudson's Bay Company agreed to pay the former company £300 a year in return for the promise that it would not trade with the Indians living to the north or west of Lake Superior.[2] However, this arrangement was not entirely satisfactory because it did not prevent other independent American traders from dealing with the Indians living in the region. The Americans did not have a long-term investment concern in the resources of the Canadian west and therefore showed little interest in any efforts which were directed towards placing the fur trade on a sustained-yield basis. Indeed, as noted earlier, Simpson regarded the American territories in the same light. Thus, these two opposing groups made every effort to mine the other's territory of its fur resources. Under these circumstances the Indians who lived in close proximity to the international boundary could always find a ready market for any of the furs which the Hudson's Bay Company refused to take.

Besides the Americans, many Métis also were unsympathetic to the Company's long-term objective. The Métis were of Indian-European ancestry and most of them had served as employees of the North West and Hudson's Bay Companies. When these two companies merged in 1821 many Métis were released from service as Governor Simpson attempted to economise the Company's operations. While many of them became provisioners and cartmen, others continued their long association with the fur trade by operating independently. Because of their limited

[1] Simpson, Letters Outward, Report to London, York Factory, 20th August 1826, PAC HBC D 4/85

[2] E. E. Rich, *The fur trade and the northwest to 1857* (Toronto 1967) 255–6

resources, the Métis could not have afforded to encourage the Indians to husband beaver or other furs for long periods of time. Indeed, they stood to profit on a short-term basis from any such efforts initiated by the Company because it in effect meant that the Indians were forced to deal with Métis, American, or any other traders who were present.

As the foregoing discussion has indicated, without a monopoly it was not possible to manage the fur trade on an ecologically sound basis since the primary suppliers of fur pelts, the Indians, did not readily support the Hudson's Bay Company's conservation programme. Granted the programme was launched by the Company for its own benefit and not because of any reverence for nature or long-range concern for the Indians' welfare, nonetheless it clearly would have been in the Indian's best interest to cooperate. An explanation for this seemingly paradoxical reaction on the part of the Indians, who modern writers often portray as conservationalists, is a complex one which requires a consideration of their attitudes towards planning, resource ownership and their economic position at the time.

Nearly all twentieth-century views of conservation maintain that man must consciously plan for the future although there is a range of opinion as to the purpose for this planning. On the one hand, some conservationalists maintain that the objective is to preserve the environment for religious and aesthetic reasons, while others contend that the principal aim is to manage resources on a sustained-yield basis. There is little evidence in the historical record to suggest that planning for either of these purposes was widespread among the Indians of the Northern Department. Therefore, when the Hudson's Bay Company put forward the notion that fur resources should be carefully managed for present and future purposes, it was introducing a concept which was alien to most of the Indians, a concept which many of them accepted only reluctantly over a long period of time.

That such a concept was not well developed appears to have been a function of their way of life, religious beliefs, system of land tenure, and in the case of the woodland Indians, the political system. The nomadic life of the Indian meant that the husbanding of resources for future use was difficult. Being constantly on the move, and having only limited transportation capabilities, excepting the Plains Indians after they acquired the horse, it was difficult to stockpile large quantities of food and furs and subsequently convey them from place to place as the need arose. Instead, they lived for the present and usually experienced a "feast or famine" existence. Many fur traders believed that such an existence resulted from the Indians being basically improvident in nature. However, some of the more experienced and penetrating observers, such as Andrew Graham of the Hudson's Bay Company, understood the economic factors which often motivated such behaviour among the woodland Indians. When writing on this subject in the late eighteenth century Graham observed:

> Several score of deer I have known killed at one time, the natives only taking the tongues, heads, hearts and feet, according as they choose; letting the carcasses go adrift in the river. They have a method of drying meat in the smoke; would they thus preserve the venison they throw away, hunger would less frequently assail them. But frugality and prudence in this respect are not amongst the virtues of these natives. Though to be impartial, it is just to mention that the reason of a conduct, so unaccountable to Englishmen, may proceed from the difficulty that would arise from conveying a stock of

> provisions from place to place in their migratory way of life. We ought not to be rash in our censures.[1]

In the plains area, on the other hand, such wasteful utilisation of resources cannot be accounted for in this manner. The acquisition of the horse by the Plains Indians of Canada during the eighteenth century made it possible for these Indians to move large quantities of food and other supplies over long distances. They often did so in the late summer and autumn when they brought dried meat, robes and hides to the trading posts. Yet these Indians often indulged in the same kinds of excesses which were described by Graham. For instance, during the summer, herds of buffalo were often slaughtered to obtain humps and tongues, regarded as delicacies, for consumption during summer ceremonies like the annual sun-dance ceremony. On such occasions the other parts of the bison were left to rot on the prairies. It may be that the seemingly inexhaustible numbers of bison made conservation practices and the judicious use of the resource appear to be unnecessary to the Plains Indians.

Although the mobile life-style of the woodland Indians coupled with a limited transportation capability made long-range planning difficult, and the abundance of certain basic resources in the plains made it seem to be a perhaps needless excercise, Indian mythologies and religious beliefs also played an important role in thwarting the development of effective conservation strategies when they were needed. For example, according to the explorer David Thompson, the western Cree and Ojibwa believed in the existence of a great spirit which was the master of all life and was always kind to the human race. According to Thompson, these Indians also believed that the great spirit, called "Keeche Keeche Manito", placed the care of all other living things, excepting man, in the care of Manitos or inferior Angels, and each Manito ". . . has a separate command and care, as one has the Bison, another the Deer; and thus the whole animal creation is divided amongst them".[2] Similar beliefs were held by many of the Indians of the northern plains. For example, Alexander Henry the Elder said of the Assiniboine:

> In their religious notions . . . there is general agreement between the Osinipoilles and the Cristinaux [Cree]. They believe in a creator and governor of the world, in a future life, and in the spirits, gods or *manitos*, whom they denominate *wakons*.[3]

Significantly, according to the Indian's view, the destiny of any animal species was not in man's hand, but rather, in that of the appropriate Manito. As Thompson noted:

> On this account the Indians, as much as possible, neither say, nor do anything to offend them, and the religious hunter, at the death of each animal, says, or does, something, as thanks to the Manito of the species for being permitted to kill it. At the death of a Moose Deer, the hunter in a low voice, cries "wut, wut, wut"; cuts a narrow strip of skin off the throat, and hangs it up to the Manito. The bones of the head of a Bear are thrown into the water, and thus of other animals; if this acknowledgment was not made the Manito would drive away the animals from the hunter. . . .[4]

According to this mythology, assurance of future success in hunting and trapping

[1] Glendwyr Williams (Ed.), *Andrew Graham's observations on Hudson's Bay, 1767–1791* (London 1969) 154
[2] Richard Glover (Ed.), *David Thompson's narrative, 1784–1812* (Toronto 1962) 75
[3] Glover (Ed.) "Thompson's narrative" 75–6
[4] *Ibid*

depended upon maintaining the continued goodwill of the appropriate Manitos who permitted the Indians to kill their game. To obtain the blessing of the Manitos, the Indians had to perform the appropriate rituals. Thus, for many of the Indians of the Northern Department, resource management was a ritual activity. Lavish use of resources, such as the slaughtering of herds of buffalo for tongues or humps was permissible, and the careful husbanding of game, as advocated by the Hudson's Bay Company, would have been considered unnecessary, or even futile.

Indeed, the Indians were aware that certain fur-bearing animals were rapidly disappearing as early as the late eighteenth century, but they believed that they were powerless to do anything about it even though it was clear that it would affect them adversely. For example, the Indians knew that beaver were being hunted out of northern Ontario, Manitoba and eastern Saskatchewan by 1800 due to the use of steel traps and castoreum bait. David Thompson discussed this problem with two older Indian men (Ojibwa) living in the Swan River Department. They informed him that according to their legends the beaver were an ancient people who once lived on dry land, but because they had angered the "Great Spirit" were driven into the water to live. They further related:

> . . . the Great Spirit has been, and now is, very angry with them and they are now all to be destroyed. About two winters ago Weesaukejauk [a mythical being] showed to our brethren, the Nepissings and Algonquins the secret of their destruction; that all of them were infatuated with the love of Castorum of their own species, and more fond of it than we are of fire water. We are now killing the Beaver without labor, we are now rich, but [will] soon be poor, for when the Beaver are destroyed we have nothing to depend on to purchase what we want for our families.[1]

Thus, the Indians rationalised the destruction of beaver in terms of their mythology and beliefs. They were simply carrying out the wishes of the Great Spirit by trapping beaver wherever they found them.

Besides traditional attitudes towards planning and mythological beliefs, the Indian concepts of land tenure and resource ownership, and in the case of the woodland Indians, the political system also made local game management impracticable. Management is feasible only when access to resources can be controlled. To control this access land and resource rights need to be clearly defined and an adequate means of enforcement must be present to ensure that they are not violated. However, as research in northern Ontario has shown, intra-tribally, land appears to have been a free good, and consequently, band territories traditionally were not sharply defined. This was also the case in the grassland area.[2] Generally, Indian bands tended to return to the same general areas every year, but the bounds of these areas were only loosely delimited. This form of tenure has been termed the "hunting range system".[3] Significantly, under this system there was no well-developed sense of trespass. Although bands hunted in the same general region in most years, the resources belonged to whoever came first even if they were within an area normally inhabited by another band. Also, if a band was not exploiting a resource in its home area, a neighbouring group could do so.

[1] Glover (Ed.) "Thompson's narrative" 155
[2] Edmonton District Report, 1815, PAC HBC B 60/e/1 and M. J. Herskovits, *Economic anthropology* (New York 1965) 332
[3] E. S. Rogers, The hunting group-hunting territory complex among the Mistassini Indians, *Bulletin of National Museum of Canada*, **195** (Ottawa 1963) 82

This system of tenure meant that the Hudson's Bay Company conservation schemes could not work under competitive conditions. As the district reports for many of the areas of the Northern Department show, whenever a trader managed to convince a local Indian band of the necessity of not trapping beaver for a few years, all too often neighbouring bands would move in and hunt them since unexploited resources could be claimed on a "first come" basis.[1] In fact, the initial effort to implement these conservation measures may have served to increase the amount of hunting which bands normally did in territories occupied by nearby groups. For instance, the Sturgeon and Rainy Lake Bands living in the Rainy Lake District both agreed not to hunt on their respective lands for a few years. However, in 1826, W. Cameron reported that the Sturgeon Lake band was trapping on the lands of the Rainy Lake Indians while they "nursed" their own territory.[2] Similar problems occurred in other districts to the north and west inhabited by Athabascan-speaking groups. For example, in 1841 Colin Campbell reported that the Beaver Indians of the Peace River Valley were unable to preserve beaver in many sections of their country because neighbouring Indians were destroying them at all seasons. Thus, although the Beaver Indians were willing to curtail their hunting activities for a few years to give the beaver a chance to recover, they were forced to trap them in order to prevent other Indians from trapping out all the beaver which remained.[3]

In an effort to deal with the land tenure problem, Governor Simpson began to consider the possibility of assigning territories to particular bands or families in those districts where the Company held a reasonably tight control on the fur trade. However, he realised that this plan would not work in all districts because of varying cultural-ecological conditions. In his 1828 report to London he wrote:

> On the subject of nursing the country . . . the plan suggested in . . . my Dispatch from Moose [Factory] of allotting certain tracts of country to the different Bands can only be carried into full effect in extended Districts such as Albany, where the population is very thin; but in small districts frequented by Rein Deer and where the Fisheries are not numerous the Indians are under the necessity of going sometimes from one extremety thereof, to the other, in search of the means of living. . . . We are endeavouring to confine the natives throughout the country now by families, to separate and distinct hunting grounds . . . and in a few years I hope it will become general, but it is a very difficult matter to change the habits of Indians although they may see the ultimate benefit thereof to themselves and families.[4]

The London directors favoured this plan and thought that it held more prospect of success as a conservation scheme than Simpson's plan of shifting post locations to draw Indian populations away from overhunted areas. The London committee feared that the latter scheme would only lead to the overtrapping of new territories. In support of Simpson's land allotment plan, the directors offered the suggestion that the Indians be paid a bounty for each beaver lodge they preserved in their assigned territories.[5]

[1] Charles A. Bishop, The emergence of hunting territories among the Northern Ojibwa, *Ethnology* **9** (1970) 10–11
[2] Lac la Pluie District Report, 1826–27, PAC HBC B 105/e/7
[3] Simpson, Letters Inward, Fort Chipewyan, 30th December 1841, PAC HBC D 5/6
[4] Simpson, Letters Outward, Report to London, York Factory, 10th July 1828, PAC HBC D 4/92
[5] Simpson, Letters Inward, Hudson's Bay House, London, 16th January 1828, PAC HBC D 5/3

By introducing a new system of tenure Simpson hoped to legitimise restrictions to access to resources. Yet he also realised that the political system of the Woodland Indians lacked any of the coercive institutions which would be needed to enforce the sanctions which might be adopted for that purpose. It was for this reason that he proposed to implement the scheme only in those areas where the Company held a monopoly. Under these conditions, the Company was in a position to force compliance by applying economic pressures. It could thus impose an external authority system upon the Indians.

Although Governor Simpson believed the Company would achieve success in this plan in areas where it held a monopoly, the continuing decline of game in most areas made it increasingly difficult to stabilise Indian populations on the land. Whereas game animals were scarce only in the woodland districts south of the Churchill River and east of Lake Winnipeg in the 1820s, by the late 1830s the problem had begun to plague the Peace River, Athabasca and portions of the Great Slave Lake and Mackenzie River Districts. For example, in 1841, Colin Campbell wrote from Fort Dunvegan in the Peace River District that moose had become too scarce to be depended upon and that bison had been hunted out several years ago.[1] To the northward in the Mackenzie River department, J. Lewis reported similar problems. Writing from Fort Simpson in 1842 he said the region was impoverished of beaver and large game. In subsequent correspondence, Lewis asked Governor Simpson to attach Fort Resolution of the Great Slave District to the Mackenzie River District. His aim was to gain access to the provision trade of the Yellow Knife Indians who hunted the barren ground caribou.[2] Elsewhere to the south and south-eastward, game animal populations continued to decline in woodland areas. Thus, by the 1840s the barren ground caribou area of the northern Churchill and eastern Great Slave Districts and the bison ranges of the grassland portions of the Saskatchewan and Swan River Districts were the only sections of the Northern Department which had surplus game.

Between these two food surplus areas stretched a vast wooded country where food shortages and starvation were becoming commonplace. These hardships served to stimulate and sustain a large-scale migration of Indian population at the very time Simpson was proposing to begin to try to pursuade the Indians to adopt a less mobile life-style. For instance, James Hargrave's correspondence to Simpson from York Factory in 1844 and 1845 indicated that nearly all of the Indians living near the bay had abandoned the area and moved inland to the mission near Norway House and to the Red River colony region where they were certain that they could obtain food in times of need. The only Indians who remained near York Factory were the ones who were too poor to make the move.[3] A similar pattern of movement was underway in the Island Lake, Nelson River and English River Districts.[4]

To the west and northwest, an outward flow of Indian population from the woodlands was also taking place. According to the letters which John Rowand, Chief Factor of the Saskatchewan District, wrote to Simpson in the 1840s, Cree and Chipewyan Indians were leaving the Lesser Slave Lake and upper Churchill

[1] Simpson, Letters Inward, Fort Dunvegan, 21st March 1841, PAC HBC D 5/6
[2] Simpson, Letters Inward, Fort Simpson, 25th November 1842, PAC HBC D 5/7
[3] Simpson, Letters Inward, York Factory, 1st April 1844 and 1st April 1845, PAC HBC D 5/12–13
[4] Simpson, Letters Inward, York Factory, 1st April 1844, PAC HBC D 5/12

River country and moving into his district to avoid starvation during the winter.[1] These movements proved to be the source of strained relations between Rowand and other district factors. For example, R. Mackenzie Simon who was stationed at Ile a La Crosse Lake on the upper Churchill River accused Rowand of actively encouraging these migrations by accepting the furs which the Cree and Chipewyan from Ile a La Crosse brought to him. Since the Company was trying to discourage Indian movements of that sort, and considering that the rules and regulations passed by the council for the Northern Department forbade traders from one district from trading with Indians of another, Mackenzie Simon's charges were serious.

Rowand's response to these charges is instructive because it illustrates the dilemma which many of the traders faced at that time. Rowand admitted that he had in fact traded with Cree and Chipewyan Indians from Mackenzie Simon's district, but he justified his action saying:

> . . . even Crees, freemen and half breeds bred and born about Lesser Slave Lake have left their country, who say they will not go back they say their lands are too poor in large animals they cannot live there. Pray whom must I blame for this great change some one, suppose I say Mr. McKenzie for sending some of his best hunters to hunt furs, and every thing they can find at our Fort Gates [Edmonton] I can give you the names of some who made their last winter hunts at a very short distance from the post of Lesser Slave Lake, a spot left by the Natives of that country to recruit.
>
> Permit me to say that we give no encouragement to Indians of other districts, far from it, I wish their traders would keep them at home, and not to be tormenting us with their Indians. We have enough to do to attend to ours, but I will say to kick or turn a poor Indian out of doors who arrives nearly dead with starvation at the Company's establishment, who expects assistance of a mouth full of provisions, and a few loads of ammunition, to save his life it is more than I am able to do. I could not do it to a dog, much less a human being. . . .[2]

Thus, when confronted with the choice of aiding destitute Indians, or enforcing standing rules and regulations, Rowand opted for the more humane course of actions as did most of the Company's traders. Indeed, since one of the overriding principles of the Company's policy towards the Indians called for such action, it is clear that as long as food was in short supply in the woodlands, the Indians would continue to gravitate towards those areas where there was some prospect of obtaining relief. The implementation of a hunting territory system therefore would be difficult until such time as the outward migration of Indians and the recovery of the animal populations brought about a new ecological balance.

Indeed, John Rowand voiced the opinion that the migrations of Indian populations which were under way might serve to achieve the conservation goals of the Company. He expressed this view in a letter to Governor Simpson on the 4th of January 1843. In this letter he wrote:

> I fear from all what I see and hear the best days of the fur trade is gone bye unless something brighter is hid in store from my view of [the] matter

[1] Simpson, Letters Inward, Edmonton, 4th January 1841, 4th January 1843 and 1st January 1844, PAC HBC D 5/6, 8 and 10

[2] *Ibid.*, 4th December 1844, PAC HBC D 5/12

> & thing. Large animals have become that scarse of late years it is now almost impossible the poor natives can live in the woods, where the most valuable fur bearing animals are to be trapped.
>
> The whole of the strong wood Indians are out in the plains not only those connected with this district but from many other parts of the Country who are now to be found living in the different camps with the plain tribes on the south side of the Saskatchewan River who are afraid of starving to death with their families if they are made to return to their land. . . .
>
> One consolation for those of our friends who are still young, is this, while the trade is deprived of those Indians hunts for the present, by keeping away for a few years their once rich Countries must recruit during their absence, for my part, I know of no better plan to nurse a worn out Country than give it some rest.[1]

Thus, Rowand believed that the abandonment of the forests by large numbers of Woodland Indians would eventually have the same effect that the Company was trying to obtain through its various management schemes.

Eventually a balance between Indian populations and game and fur-bearing animals was achieved. There is little doubt that the population density of the Indians in many forest areas was less than it had been in pre-contact or early fur trade times. Furthermore, the remaining population depended much more heavily on fish and non-migratory small game, especially rabbits. Under these new circumstances, the resource base which the Indians exploited was of a more fixed nature and could be effectively exploited by a relatively sedentary Indian population. It was after these ecological adjustments had taken place that the hunting-territory system advocated by Governor Simpson as a conservation measure began to be adopted by the Indians beginning as early as the late 1820s in portions of the northern Ontario region.[2]

Changes in the conservation programme after 1841

Because of the general failure to achieve any significant success in its beaver conservation programme before 1841, the Governing Council for the Northern Department issued a series of new rules and regulations that were intended to halt the continuing decline of beaver. The quota system which had been introduced in 1826 had failed for the reasons which already have been outlined. Consequently, in 1841, the council decided to impose a more stringent allotment plan. By the 90th resolution in council it was ruled that:

> The Impoverishment of the Country in the article of Beaver is increasing to such an alarming extent that it becomes necessary to take effectual measures for providing an immediate remedy to that and it is [resolved] That the gentlemen in charge of Districts and Posts be strictly enjoined to discourage the hunting of Beaver by every means in their power and that not more than half the number collected Outfit 1839 be traded during the Current and two succeeding Outfits at the undermentioned Districts and Posts. . . .[3]

[1] Simpson, Letters Inward, Edmonton, 4th January 1843, PAC HBC D 5/8
[2] Bishop, *op. cit.*, pp. 10–11
[3] York Factory, Minutes of the Council for the Northern Department, 14th June 1841, PAC HBC B 239/k/2

In order to put "teeth" into this regulation and discourage company men from ignoring it as many had the earlier quota system. the council issued an additional resolution (number 91) which stated:

> ... as a further remedy for the evil, if it be found that gentlemen disregard this instruction, as they have done many others issued from time to time for the same object it is [resolved] That the Governor and Committee be respectfully advised to give notice of retirement from the Service to such Gentlemen as may not give effect to the Spirit and the letter of the resolution now passed for the preservation of Beaver.[1]

In an effort to obtain Indian compliance with this new restriction on beaver trapping, the council decided to offer them a positive inducement to hunt other fur animals (the "small furs"). Accordingly, it was decided that:

> In order to encourage the Indians to greater exertions in hunting other Furs and that they may not suffer any privation in consequence of the proposed restriction it is [Resolved] That all Indians at Posts where the restriction exists and who do not kill Beaver be paid in Goods the value of 10 skins or Made Beaver for every 9 skins in small Furs they trade in the Course of the Year.[2]

Thus, the Indians were offered a premium on the small furs which they trapped.

Since these new regulations virtually banned the trapping of beaver in the Northern Department in the lands adjacent to and north of the Saskatchewan-Nelson River system there was a mixed reaction to them on the part of the Company employees.[3] Donald Ross, who was stationed at Norway House, had advocated the implementation of an effective conservation plan in his letters to Simpson prior to the passage of the resolutions. Not surprisingly therefore, he approved of the new regulations and argued for their continuance in subsequent years.[4] John Lee Lewis who was in charge of Fort Simpson in the Mackenzie River District did not share Ross's view and was an outspoken critic of the new measures. Pressing his case with Simpson, in November of 1842 he wrote:

> ... drawing your attention for the moment to the restriction on killing Beaver within this District, which I hope by the Minutes of next summers Council to see rescinded, and that we be permitted to return to our usual means of making Packs, the McKenzie is still rich in the article of Beaver and our forbidding the Indians killing [beaver] only makes them generally discontented and tends very little to their preservation, and with many of them cannot be prevented in this hard country, a half starved and famished Indian ... be it Beaver, or any other kind of food that come in their way, they must and will kill and if we do not take the skins from them after; they can I do assure you balk little at our refusal, for soon they convert the same into warm and comfortable clothing for themselves and Families, and when on a Winter day facing a cold N.W. blast at the Thermo 50 or 60 below zero, they find that they are the gainers and not us, by the present law and regret not our flimsy Blankets or Capots.[5]

[1] *Ibid.*, PAC HBC B 239/k/2

[2] *Ibid.*, B 239/k/2

[3] The ban did not apply to the Peel's River area of the Mackenzie River District. Simpson, Letters Inward, Peels River, 20th December 1841, PAC HBC D 5/6

[4] Simpson, Letters Inward, Norway House, 10th April 1841, 8th August 1942 and 16th August 1843, PAC HBC D 5/6–8

[5] Simpson, Letters Inward, Fort Simpson, 25th November 1842, PAC HBC D 5/7

Although some of the objections which Lewis raised were legitimate, the governing council went ahead with its initial plan of trying the new scheme for a three-year period. By most accounts, this programme was a success. During the summer of 1842, James Hargrave reported from York Factory that the Indians living in his district were responding favourably to the premium being paid on small furs and he was optimistic that the plan would work.[1] On the 16th of August 1843, Don Ross wrote from Norway House that the trapping restrictions were having all of the desired effects and that beaver cuttings and other evidence of the animal's presence were becoming common along the canoe route between Norway House and York Factory. For these reasons he was in favour of extending the ban another two years beyond the expiry date in the summer of 1844.[2] A similar recovery was under way in the territories lying to the northwest. On the 1st of January 1844, Colin Campbell said that the Indians had informed him that since the beaver hunting restrictions had been put into effect, the animal had made a come back and was more numerous in the Athabasca District than it had been in many years.[3]

During the spring of 1844 the subject of the restrictions occupied the attention of most of the district traders since the trial three-year period was coming to an end. Because of the widespread success of the plan and the mounting Indian pressure to relax the rules, most of the district heads favoured re-introducing a modest quota system. Regarding this topic, R. Mackenzie Simon wrote to Governor Simpson from Ile a la Crosse on 2nd March 1844 informing him:

> Our Chipewyans are now very anxious about killing Beaver. When the Beaver Nursing system was established in -40- I told them it was to continue in force for three years, without killing one Beaver if they could avoid it; but after that period, would be allowed to kill a certain number annually. I would like now that you would be pleased to inform me what Number that should be. . . . We have dealings with 300 Indians in this District, and I think if they could give an average 2 Beaver Skins each, annually, large and Small it would be as much as the District could well afford. . . .[4]

Similar pressures were being felt elsewhere and the same kinds of questions were being asked. Even Donald Ross at Norway House, who had championed a continuation of the ban beyond the original termination date the previous year, had modified his views, and in March of 1844, expressed the belief that a relaxation of regulations would have to be made the next time the council met.[5]

The London directors shared this view and in a letter dated 4th March 1844, they suggested to Simpson that:

> In order to test in a satisfactory manner the improvement reported, it may perhaps be proper to relax for a season, in two or three districts, the prohibition with regard to beaver hunting, and you will determine in council in what districts such a relaxation shall take place, which of course is to be acted upon only in the winter, the issue of traps for all the depots being strictly prohibited.[6]

[1] Simpson, Letters Inward, York Factory, 21st May 1842, PAC HBC D 5/7
[2] Simpson, Letters Inward, Norway House, 16th August 1843, PAC HBC D 5/8
[3] Simpson, Letters Inward, Fort Chipewyan, 1st January 1844, PAC HBC D 5/10
[4] Simpson, Letters Inward, Isle a la Crosse, 2nd March 1844, PAC HBC D 5/10
[5] Simpson, Letters Inward, Norway House, 12th March 1844, PAC HBC D 5/10
[6] Simpson, Letters Inward, Hudson's Bay House, 4th March 1844, PAC HBC D 5/10

Therefore, the Governing Committee in London was willing to permit beaver trapping once again as long as the restrictions against summer hunting and the use of steel traps remained in force.

Having obtained this approval from London, the council for the Northern Department met in June of 1844 and passed a resolution which stated that in order to test the degree of improvement since the restrictions had been in effect:

> . . . the Gentlemen in charge of the several districts of the Northern Department, be permitted to trade all seasoned Beaver, ascertained to have been hunted during the winter and by the use of the Ice Chisel only.[1]

Although the council's resolution permitted beaver trapping in all districts, rather than in a selected two or three as the London directors suggested, the latter did not oppose it. They were willing to wait a year and assess the results.[2]

The decline of the beaver trade

Ironically, at the very time the Company's conservation programme was beginning to bear fruit, the market for beaver pelts in Europe started to decline as the silk hat replaced the felt hat. Having long been accustomed to viewing beaver pelts as a trade staple, the London directors of the Company were slow to adjust to change. Rather, they initially regarded the shift away from beaver felt hats as temporary. Thus, on 1st April 1843, they wrote to Simpson informing him:

> From an extrodinary break of fashion, the article [beaver pelts] . . . has of late fallen much into disuse in hat making, silk hats being principally worn at present, the consequence is that its value has greatly decreased in the market. . . . This depression is but temporary as no doubt exists that Beaver hats will soon again come into general use, when of course amendment may be expected in the price[3].

The London directors continued to hope that the beaver hat would regain its lost popularity. In fact, they were optimistic that their conservation efforts would bear fruit in such large returns of beaver pelts that it would lower the market price of beaver felt hats to a level that would make them cheaper than silk hats. However, this never happened and the beaver hat failed to regain its lost popularity. The Company directors then attempted to market beaver in other forms, principally as a fur similar to otter, mink and other pelts, but these efforts were disappointing. Therefore, shortly after the Hudson's Bay Company began to achieve success in its effort to conserve beaver and conduct this trade on a sustained yield basis, beaver pelts ceased to be a staple trade item.

Conclusion

After 1821 the Hudson's Bay Company initiated what must have been one of the earliest attempts in North America to put a primary resource industry on a sustained-yield basis. In endeavouring to achieve this goal the company faced a number of serious problems. In some areas, particularly near the international

[1] York Factory, Minutes of the Council for the Northern Department, 10th June 1844, PAC HBC B 239/k/2
[2] Simpson, Letters Inward, Hudson's Bay House, 4th March 1844, PAC HBC D 5/10
[3] Simpson, Letters Inward, Hudson's Bay House, 1st April 1843, PAC HBC D 5/8

border, it was confronted with opposition from American and Métis traders who did not have a long-range investment concern in the region, and consequently, were unsympathetic to the Company's aims. Under these competitive conditions, which characterised the trade in most areas to the south of the border, management was not possible and therefore the Hudson's Bay Company did not make a concerted effort to husband beaver in border regions. Rather, its efforts were concentrated in those lands where its monopoly was relatively secure.

Yet even in areas where it held a firm hand on the trade, the Company was confronted with considerable obstacles since the fur trade was an industry in which two very disparate cultures took part. The Indians and the Europeans had fundamentally different views regarding future planning and the nature of, as well as the need for, game management. Many Indians believed that game would always be sufficient provided that the appropriate rituals were followed. In contrast, the European traders maintained that game supplies could be assured only if man consciously husbanded them. Besides holding different views toward the need and nature of game management, the Indian and European concepts of land tenure differed. On an intratribal basis, land was a free resource and bands were free to hunt on each other's lands whenever the need arose, or whenever a given band was not exploiting a resource that was valued by a neighbouring group. Hence, the notion of trespass held by Europeans was not held by the Indians in the Northern Department in precontact and early historic times. Thus, there were no traditional rules or customs among the Indians which a band could appeal to in order to justify limiting the access of their neighbours to resources found on their own lands. Furthermore, the egalitarian band organisation of the Woodland Indians meant that there were no coercive Indian institutions that could have forced individual or group compliance to such rules had they existed. Under these circumstances, management was nearly impossible even in areas where the Hudson's Bay Company held a monopoly. It was in an effort to deal with these problems that the Company attempted to alter Indian concepts of land tenure. Also, by using its economic power under monopoly conditions, the Company imposed the authority system and regulations which were needed to control hunting activities. Thus, by pursuing these policies for reasons of self-interest and paternalism, the Company played a major role in changing the Indian's way of life in some fundamental ways.[1]

[1] A shorter version of this paper was presented at the 1973 Algonquian Studies Conference in Green Bay, Wisconsin, 6th and 7th April 1973. The author would like to thank the Hudson's Bay Company for granting him permission to consult and quote from the Company's records. I would like to thank the York University Cartography Office, especially Robert Ryan, for drafting the map

15

Forest Management and Exploitation in Colonial Java, 1677–1897

Peter Boomgaard

In 1619 the Dutch East India Company (Vereenigde Oostindische Compagnie or VOC) conquered Jakatra on the northern shore of western Java and renamed it Batavia. The Dutch retained a presence in this area for over three centuries, until 1949, when the Netherlands Indies ceased to exist and became Indonesia. From the 330-year period of Dutch presence, this article singles out the 220 years between 1677, when the Dutch acquired sovereign power over most of western Java, and 1897, when the new forest regulations were published.

This article explores whether the Dutch introduced what we now would call a system of sustained and stable yields, and if so, when and under what systems of management and exploitation of the forests. It focuses on the eighteenth century, a period that has never received the attention due to it, having been treated almost uniformly as the dark ages of Dutch colonial forestry.[1]

At the end of the colonial period, forestland made up about 24 percent of the total 132,000 square kilometer surface area of Java.[2] The literature on the Java forests distinguishes two main forest types: teak forests (ca. 27 percent) and junglewood forests (ca. 73 percent). This way of categorizing is in itself revealing, for the attitude of the colonial state in Java (or, for that matter in British India, where the same distinction was made) reflects an overriding emphasis on economic gain: valuable teak versus "worthless" junglewood. Although this attitude changed around the end of the nineteenth century, by then the labels could no longer be removed.

The VOC at the Mercy of the Javanese Rulers: 1600–77

There were many reasons for the Dutch to establish their headquarters on Java. The most-often cited are Java's strategic position in the southeast Asian trade routes and its large quantities of rice, but another reason must surely have been the availability of timber.

Even before the Dutch had founded Batavia (1619), they had started the construction of a wharf on the island

2 Peter Boomgaard

Table 1 ▪ Teak contingents, central Java, 1733–1809

Year	No. of logs	Year	No. of logs
1733	10,500	1776	11,300
1758	20,000	1777	10,600
1761	22,800	1780	9,800
1764	16,900	1801	7,800
1765	13,300	1809	8,800

of Onrust, opposite what then still was called Jakatra.[3] The wharf accommodated a large shipyard, which was a major consumer of timber for shipbuilding. In 1675 this shipyard employed two hundred men; in 1700, almost six hundred; and in 1755, six hundred fifty. After 1685 it had two wind-powered sawmills. In Batavia itself there were private shipyards, most of them run by Chinese shipwrights, and at least one private water-powered sawmill (dating from 1659). Beginning in 1627 there was also a wood market or timber yard, to which both the VOC and private entrepreneurs had access.

Timber was needed not only for ships, but also for building in the city, for fuel, and for furniture. Batavia was expanding rapidly during the seventeenth century, and almost all buildings were wooden constructions. Wood was, of course, also used as fuel, either in its natural state or as charcoal. The demand for charcoal was stimulated by the presence of powder-mills in the city, the first of which was established in 1659. In 1787 there were seven of these mills. Finally, wood was needed for furniture; up to 1650 much teak was employed for furniture making.

Because the VOC had conquered the area sword in hand, it felt free to exploit the region's resources as it saw fit. The company was reasonably successful in obtaining as much wood for fuel as it needed from the area lying immediately around Batavia, the so-called Environs of Batavia, but good timber, particularly teak, was harder to obtain. The Environs produced sufficient teak for the construction of houses at least until 1644, but teak for house building was being shipped from Jepara (central Java) to Batavia as early as 1622. In 1636 a flotilla of indigenous ships was sent to the mouth of the river Citarum to the east of Batavia, in order to obtain teak from Krawang, where the company had no territorial rights.[4] By 1650, although the Environs was still supplying some timber, the VOC had to rely almost entirely on the willingness of the rulers of Java's north coast for its supply of teak.

The VOC always tried to do business directly with the producers of timber (such as the Chinese woodcutters in Krawang) or at least with the timber merchants (such as the Chinese in Jepara), but it needed a permit from a local ruler (called "regent" by the Dutch) to get the merchandise on board its ships. This often proved to be problematic, either because the local rulers wanted a larger share in the proceeds, or because the sultan of Mataram, overlord of these rulers, forbade the sale of timber to the VOC.[5]

To ease these problems, the VOC shifted part of its ship-building activities to the teak-production areas. Production centers included Demak, Semarang, Torbaya, Jepara, Rembang, and Lasem, all on the north coast of central Java. In these areas the company hired local labor, often with the help of the regent, and sometimes under the guidance of Dutch shipwrights.

In 1652 a major stumbling block for the VOC was eliminated when the ruler of Mataram promised the company that it could buy all the timber it needed. Nevertheless the company's relationships with the sultan and the local regents remained a problematic issue, often affecting the supply of teak and the construction of ships.[6]

The Acquisition of the Teak Areas: 1677–1745

Between 1677 and 1685 the VOC was able to secure a more regular supply of timber and ships. In 1677 the ruler of Mataram ceded the Priangan (western Java, to the south and east of Batavia) to the company, including the teak-producing districts of Krawang, Ciasem, Pamanukan, and Sumedang. In the same year he granted the VOC the right to establish its own wharfs on the north coast of central Java and to recruit as much labor as it needed. In 1681 the VOC became the overlord of the Cirebon sultanate (western Java, to the east of Priangan), thereby acquiring another teak-producing area.

The steps undertaken by the VOC in the ceded Western Java territories in order to secure the timber supply foreshadowed later actions in the main teak areas of Central Java. The company first acquired preferential rights to all the teak produced in these areas. Thereafter, it established a monopoly, and finally it imposed on all regents a fixed annual delivery, the so-called timber contingent, at a price set by the company. At this stage, the company established forest ranges under European overseers working with indigenous corvée laborers. The overseers of the forest ranges worked to make sure that the required amount of teak timber was cut in the dimensions preferred by the company.[7] Although these European overseers were often soldiers, probably recruited more for their military capabilities than for silvicultural knowledge, in at least one case (Cirebon, 1684) the company sent a number of German woodcutters as supervisors to a forest range. It was not the last time that German woodsmen played a role in Dutch colonial forest exploitation in the Indies.

Although the ruler of Mataram in 1677 had granted the VOC the right to establish shipyards and to employ as many shipwrights as it saw fit, the timber-supply problem had not yet been solved. In 1678 the company made contact with a high Javanese nobleman, Tumenggung Lamongan, appointed by the *sunan* (as the ruler of Mataram was called now) as head of the *kalang* (a "tribe" of woodsmen, timber merchants, and carpenters). This official promised the company that he would order his people to fell timber for them; conditions of sale, however, had to be agreed upon by the kalang themselves. Up to then the company had been doing business with the harbormaster of Demak, who bought timber from the kalang in order to sell it to the company. By cutting out the middleman the VOC would get its timber cheaper, eliminating at the same time the potentially dangerous hold of a high Javanese official over the com-

4 **Peter Boomgaard**

pany. Soon, however, Tumenggung Lamongan became a greater danger: he apparently cheated both the kalang and the company. For the next few years, while the VOC tried to employ as many kalang as it could, the sunan and the officials in his service tried to keep them from working for the company.[8]

Glossary

blandong: compulsory forest service
boschganger: forest overseer
kalang: a "tribe" of woodmen, timber merchants, and carpenters.
Sunan: ruler of Mataram.

In 1680 the legal basis for a solution was reached: a contract between the sunan and the VOC specified that all the sunan's subjects were permitted to sell timber to the company. Somewhere between 1680 and 1685 a number of decisive steps were taken: the sunan ceded thirty-six kalang villages near Rembang to the company with the explicit aim of providing the VOC with timber. The VOC sent twenty-five Europeans to Rembang as supervisors over the wood-cutting kalang. At last the company had succeeded in establishing its authority over the timber producers, though only in one teak district.[9]

During the years to come the VOC failed to acquire territorial rights on the north coast of central Java—the thirty-six kalang villages should be regarded as a "prebend" and not as ceded territory, but the company was nevertheless able to get a firmer grip on the supply of timber. A 1705 contract between the sunan and the company ordered all regents to help the VOC with kalang and timber. A 1709 contract gave a number of regents explicit orders to supply the company with as much timber as it needed, at the going price, and put the kalang under the direct authority of the company.

Finally, in a 1733 contract the sunan agreed to an annual delivery of 10,500 logs.[10] The 1733 contract specified which areas would supply how many logs, and the text of the contract makes it clear that the institution of fixed numbers of logs per area had already been in existence for some time. As no mention of this had been made in the contracts of 1705 or 1709, it must be supposed that these set contingents had been established gradually between 1709 and 1733.[11]

In 1743, after a civil war in which the company was heavily involved, the VOC became the territorial ruler of the most important teak-producing areas, namely the forests of Surabaya, Rembang, and Jepara. At the same time the company was granted a monopoly (or rather monopsony) to all teak sold in Java. Two years later the company obtained rights to the other teak areas on Java's north coast: Tegal, Pekalongan, and Semarang. The VOC had won the struggle for control of Java's teak forests.

Forest Exploitation under the VOC (1745–1808)

Production and Producers

Table 1 presents the figures for the teak contingents of central Java.[12] These figures seem to suggest rapid growth between 1733 and 1761, and decline between 1761 and 1801. In fact, however, the official contingents do not reliably reflect real production.

In 1764 the governor-general wrote to the governor of Java's north coast, the man in charge of the main teak-producing areas, that actual production in most years did not surpass 8,000 logs, far below the official contingent, and that the company did not have sufficient ships to transport timber in the amount of the official quota to Batavia.[13] The reductions granted successively after 1761 are therefore attempts to close the gap between the official quota and the actual production. If the governor-general's figures are correct, the average actual production for the eighteenth century was very modest.[14]

Teak was harvested not only from the north coast, but from western Java as well. Information on the contingents from these districts is less abundant, but can be estimated as follows: Krawang, 1,000 logs (1695); Ciasem and Pamenukan, 1,000 logs (1803); Cirebon, 5,000 logs (1808).[15]

This suggests a contingent for western Java of 7,000 logs, or about 5,000 cubic meters and a total annual production of Java as a whole of 15,000 logs, or 11,000 cubic meters. Compared to an average annual production of 200,000 cubic meters or more during the first decades of the twentieth century, when the forests of Java were being managed by a professional forest department, such amounts are negligible. Thus, if these estimates reflect Java's real production, the VOC's reputation as the great destroyer of Java's teak forests seems questionable, to say the least.

At least after 1764, however, when a number of bottlenecks were removed, actual production was certainly much higher. Prior to 1764, there were at least two reasons for the regents' inability to fulfill their teak-delivery obligations: deterioration of forests and lack of adequate labor. The first problem was solved by moving to forests further afield, even into regions where the VOC had no territorial rights. The second problem was solved by employing more axmen and water buffalo for dragging timber from the range to the depot.

In 1747, two years after the acquisition of the entire northern coast, the company had drawn up a list of the number of corvée laborers, by regency, to be sent to work in the forests. The company optimistically expected the regents to send 6,800 people,[16] but this was by far too heavy a burden for the central Java regencies. Table 2 shows the actual number of corvée laborers and buffalo working in the Javanese forests.[17] As table 2 shows, the number of people and buffaloes increased considerably between 1776 and 1809, though the numbers required by the company in 1747 had not been reached even in 1809.

Table 2 ■ Laborers and buffalo working in the forests, 1776–1809

Year	People	Yokes of buffalo
1776	1,600	1,000
1787	2,200	1,370
1800	2,500	1,650
1809	5,050	2,500

The VOC also continued to work on its transportation problems. The company had experimented with using rafts instead of ships for transport as early as 1764, but these experiments had failed, except for the short stretches between Batavia and Krawang, and Indramayu and Cirebon. Between 1764 and 1787, however, the Dutch and/or the Javanese must have mastered the art of timber-raft construction for long distances or, alternatively, the art of navigating such rafts. A publication dated 1787 mentions an annual raft from Rembang to Batavia, and the Memorandum of Transfer by the governor of Java's north coast, dated 1791, mentions the annual fleet of timber rafts, seven altogether.

The rafts were made of bamboo or other light species of wood and were towed by indigenous ships. Even on the high seas they must have been relatively stable, because in 1804 seventy-nine cows were sent with the rafts of Demak, Karawelang, and Rembang to Batavia. However, the rafts were not always safe from attacks by pirates: in 1792 three rafts lying in the roadstead of Cirebon were snatched away.[18] Thereafter the north-coast rafts were accompanied for security by a convoy of armed boats.

After solving its labor and transportation problems, the VOC was able to send more timber to Batavia. The company was always interested in obtaining surplus, over and above the contingents, for which they were willing to pay more. Since such surplus quantities were not registered, it is impossible to estimate how much teak above the official quota was delivered to the company.

The VOC was not the only producer of Javanese timber. The Residents, the highest local representatives of the company on the north coast, lined their pockets with the proceeds of illegal sales of teak and even built their own ships. In Java this was widely known, but seldom appeared in the official reports. Then, too, there were an enormous number of private shipyards (next to the VOC shipyards) lining the coast from Jepara to Tuban. By 1776 a number of forests in Kudus, Lasem, and Tuban had been farmed out, predominantly to Chinese entrepreneurs. Around 1800 sixty-seven villages in Rembang, Lasem, and Tuban had been farmed out to private timber entrepreneurs, including the Resident of Rembang.[19] In the forests still belonging to the native rulers—particularly in Blora—the Chinese had a free hand. Because the harvests of private entrepreneurs were not included in the official contingent figures, it is clear that my production estimate of 15,000 logs per annum, based on the contingent figures, is much too low. Thomas Stamford Raffles, the British lieutenant-governor of Java (in office from 1811 to 1815) and author of *The History of Java*, might have been right in estimating the annual production for private consumption of central Java alone at 50,000 to 60,000 logs.[20] Adding in the production of western Java, a conservative estimate of the annual production of Java as a whole during the late eighteenth century might be 75,000 logs, or 55,000 cubic meters. This is still far less than the production of the early twentieth century, but given the wasteful wood-cutting practices and the primitive methods of forest management of the eighteenth century, such a production level may well have been very destructive.

Forest Management

It is not entirely clear how specific the VOC's timber requirements were, but there were apparently fixed minimum dimensions for the logs to be delivered, namely 18 to 20 Rhineland feet (one Rhineland foot equals approximately 15 inches) by 9 to 10 inches. In some cases they required special dimensions from specific teak-producing regencies, such as Tinkam planks from Pemalang.

The daily practice of the forest people, at least up to 1777, seems to have been as follows: the regent concentrated his people in a certain forest, where they established a logging camp under the direction of their own headmen and the Dutch *boschganger* (forest overseer). There they cut all trees from which logs of 18 to 20 feet and over could be made. Clearcutting was not practiced: trees that did not meet the required minimum dimensions were left standing. In at least one regency (Jepara), even trees exceeding the desired dimension were excluded from the cut: a 1776 survey of these forests showed an abundance of young trees (under 18 feet) and large trees (over 25 feet), but nothing in between.[21]

Most logs were squared locally, but logs for the Jepara water sawmill may not have been trimmed. Until after 1800 no saws or European axes were used in the logging camps: all work was done with Javanese axes, adzes, and machetes.

When the timber had been cut it was dragged by water buffalo over a temporary or permanent timber slip either directly to a coastal timber depot, or to a depot on a river from which the logs could be floated to the coastal depot. The distances involved could be huge, as could the number of water buffalo: from some inland areas such as Blora overland haulage of logs to the Rembang timberyard could take twenty days, and around 1800 Van Ysseldijk personally witnessed the arrival, at the same timberyard, of a particularly large log dragged by eighty yoke of buffalo, of which ten had died in the process. Only after 1800 was a timber wagon used.[22]

In 1777 a different system of felling was attempted: regents and European overseers now spread the forest people and their buffalo out from concentrated logging camps to all the forests of a regency, where they were instructed to fell only the oldest and largest trees. This system of selective felling was reinforced with financial incentives: instead of a fixed payment per log irrespective of dimensions, payments were now made according to size, ranging from sixteen stivers for a log of 18 feet by 9 inches to eight rijksdaalders (or twenty-four times as much) for a log of 85 feet by 24 inches.[23] Although data on wages and prices are scarce, the few data that we have suggest that a normal wage-laborer could earn about five stivers per day. From this, he would have to buy a daily subsistence of rice (one *kati* or about 1.3 pounds), which would cost one or two stivers.[24]

This new felling system was probably more detrimental to the forests than the old logging-camp system. Felling timber destroys more trees accidentally when only the largest specimens, probably distributed more or less evenly, are cut down. Also, the new system required clearing more temporary timber slips, which again led to the unnecessary destruction of large tracts of forest. On the other hand, logging-camp exploitation can wipe out entire forests if no attention is paid to rejuvenation, but selective felling leaves a chance of survival, as long as it is not carried out too carelessly or too often.

It is not clear to what extent the

Cutting a teak tree in Java, circa 1920. Photo courtesy of the Royal Tropical Institute.

system of selective felling was introduced. At least in Jepara, in 1797 the forest people were concentrated in one large forest, where they had been working during the last five years, which sounds remarkably like a logging-camp system.[25]

By the end of the century the Dutch had apparently become aware of the destructive character of selective felling. A 1797 draft instruction for an inspector of forests introduced the idea of clearcutting: the inspector was instructed to divide every forest into parcels and to clear one lot entirely per year, so that the first parcel to be clearcut would have mature timber when the last parcel had been cut.[26] This is the first formulation of the principle of sustained yield during the period of Dutch colonial forestry in Java. When, however, in 1801 two inspectors were appointed, this clause was not written into their instructions, and the idea of clearcutting was not mentioned again until 1808, when the official instructions for the boschganger contained a clause prescribing clearcutting, albeit not a very clear one.[27]

In managing the forests, the VOC were faced with a number of indigenous forest customs. One such custom was the annual burning of the teak forests.[28] Such an activity was entirely anathema to trained European foresters, but as the VOC forest overseers were generally soldiers and sailors—laymen in forest affairs—they did not bat an eyelid when confronted with these practices. The motives of the Javanese (or, perhaps, kalang) were apparently to keep wild animals (especially tigers) at bay, to keep down weeds, to remove refuse from felling timber and dry leaves with a minimum expenditure of time and energy, to fertilize the forests cheaply, to make the forests more accessible, and to replace old *alang alang* (Imperata cylindrica) with young shoots, which were excellent for grazing. Finally, teak seed supposedly sprouted more easily if it was heated before being sown or planted. In 1801 the "burning" of teak seed was even officially recommended.[29]

But Dutch officials gradually qualified their tolerance for the burning of teak forests. In 1787 they made a distinction between burning dry leaves with "small fire," which they applauded, and the wholesale burning of refuse after felling, which they deemed harmful to young teak trees. An 1801 report instructed the inspectors of forests to encourage the burning of teak forests, provided it did not harm the young trees, but an 1809 report again emphasized the detrimental effects of burning. It may therefore come as a surprise that the Forest Regulations of 1829 were still in favor of burning. Only much later, in 1857, was it finally forbidden.[30] Here again the similarities with experiences and debates in British India are striking.[31]

Under the logging-camp system that existed at least through 1787, tracts that had been cleared were sown or planted with teak seeds. During the 1776 forest survey, carried out in the 256 separately named forests of central Java's north coast and lasting six months, almost all districts reported reforestation efforts. Unfortunately, more detailed information on this reforestation is lacking, but apparently in most cases the seeds were sown and not planted, and no weeding, thinning, or interplanting took place.

The Publication of 30 October 1787, which deals with the forests of Rembang, Lasem, and Tuban, contains a specific clause stipulating that empty spaces in the forests be used not for cane, cotton, or other crops, but for teak plantations. A detailed report of a survey of the forests of Jepara and Kudus, dated 1797, mentions well-cared-for teak plantations, as does a report from the Resident of Pekalongan, who in 1795 had started a teak plantation in a "wilderness" where no teak had grown before.[32] The Memorandum of Transfer of 1801 seems to regard the fact that the Residents of Pekalongan and Jepara had laid out teak plantations (in the case of Jepara, two plantations of 10,000 trees each) as a novelty, or at least something worthy of explicit mention. So either the planting of teak, reported in 1776, had fallen into disuse, perhaps owing to the introduction of selective felling, or the first real plantations are the ones mentioned around 1795.

The 1795 plantations still existed in 1807, when the Resident of Rembang started planting 40,000 trees. During an inspection of the forests in 1818 the Pekalongan and Rembang plantations are mentioned again, but nothing is said about Jepara. On the other hand, this inspection mentions plantations in Tegal, dating from the years 1804 to 1807.[33] Although perhaps not all plantations laid out just before or after 1800 were equally successful, it seems that at least a serious start had been made.[34]

The most efficient and successful method of forest conservancy was, however, not laying out teak plantations, but closing the forests to felling. The first instance of such a measure is the fifteen-year closing of the forests of Sumedang (Priangan, western Java) in 1722, because they had been cut out. The fifteen-year period was afterward extended to thirty-three years.[35] The first examples of forests in central Java being closed date from 1752, when all felling was forbidden in the forests of Pemalang and Demak. The Pemalang forest remained closed until felling was resumed in 1788, apparently too early, because in 1790 the forests were closed again, this time for a much shorter period. In 1801, shortly after felling in the Pemalang had been resumed, the governor of the north coast asked permission for another temporary closure, which was apparently still effective in 1818.[36]

Another interesting example is the forest of Pidi in Batang. In 1776 it contained young teak trees only (five to seven inches in diameter), no doubt because all trees of larger dimension had been cut, and it was closed, with all the other Batang forests, in 1780, for a period of twenty-five years. This term was apparently extended, because it was still closed in 1818. By then it had recovered so well that it was singled out for the establishment of a forest range to be exploited under the new system of clearcutting.[37]

Of course, the closure system did not always work as well as in the case of the Pidi Forest, but the examples cited here seem to suggest that the VOC tried to avoid the entire degradation of its teak forests by closing them, that it regularly evaluated the condition of its forests, and that terms were extended if a forest had not recovered sufficiently. Nevertheless, it is also clear that between 1776, when the first survey covering the entire north coast of central Java was conducted, and 1818, the year of the first survey after the return of the Dutch administration, the total area of teak forests had been reduced, and the quality of the remaining forests left much to be desired.

Overall, the archival evidence shows that during the eighteenth century the VOC did, indeed, allow its forests to deteriorate, partly because it was unable to control all forest exploitation, and partly because its methods of management in the forests where it could control exploitation were primitive and inefficient. However, the archival evidence does not support the often-repeated accusation that the VOC did nothing to stop the process of deterioration. In fact, just before the turn of the century, the VOC had formulated, if only on paper, the principle of sustained yield.

The First Forest Departments: 1808–26

The VOC ceased to exist formally in 1800, and the Dutch republic took over its possessions. Historiographical orthodoxy has it that nothing much changed until the arrival of Governor-General and Marshall Herman Willem Daendels (1808–11), who had been sent to the Indies with the explicit purpose of putting the administration of these possessions on a new footing. But as far as the forests were concerned, Daendels's reforms built on those of his immediate predecessors. The forest surveys held in 1797 had revealed the extent of damage to the forests and had indicated that the current methods of forest management could function satisfactorily only if strictly adhered to and closely supervised. This could be brought about only by an expansion and reorganization of forestry personnel.

A central forest department did not exist at that time. The system was generally as follows: A European Resident was the superior of a number of native regents, who were responsible for the delivery of the timber contingent of their regency. The regents sent the required number of forest people[38] and buffalo to the logging camps or to the locations where selective felling was taking place. The only European official in the field, supervising the activities of the forest people in a number of regencies, was the boschganger, who was answerable to the Resident. The system was slightly different in Rembang, where the regent did not figure in the line of command. In Rembang the forest people worked under the authority of four native forest officials, whom the Dutch called *boschhoofden*. These officials and their people answered not to the regent of Rembang, but directly to the Resident, who delegated actual supervision, as elsewhere, to a boschganger. This was the so-called *blandong* system.[39]

Between 1797 and 1808 plans were drawn up for an improvement of the existing system. These plans involved three major components: (1) centralization of the forest administration; (2) expansion of the number of European forest officials; (3) generalization of the blandong system.

The first blueprint for this reorganization dates from 1797. For the time being it remained on the drawing board. In 1801 the first modest step toward implementation was taken with the appointment of two inspectors of forests, who were directly answerable to the governor of Java's north coast. In 1803 the creation of a forest service (*Collegie van Houtvesters*) was announced, with the governor of the north coast as its

12 Peter Boomgaard

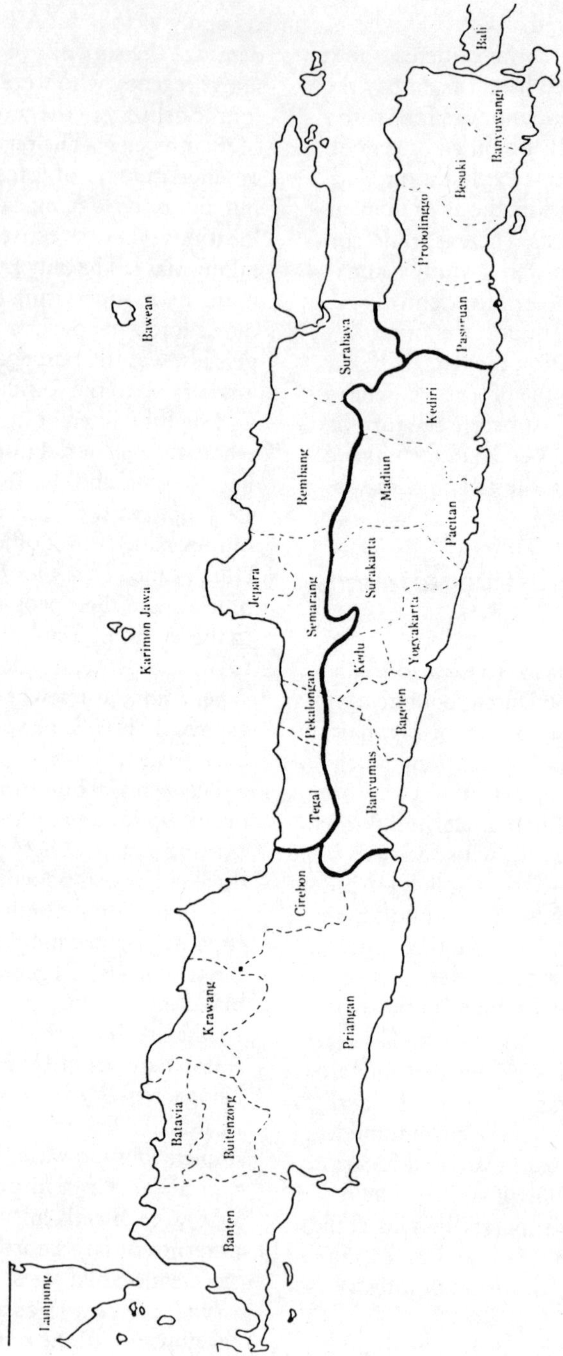

Map of Java. Reprinted from Changing Economy in Indonesia: A Selection of Statistical Source Material from the Early Nineteenth Century to 1940, *edited by Peter Boomgaard (Amsterdam: Royal Tropical Institute, 1991).*

president. At the same time all forests were declared to be Crown lands. Instructions and regulations were to be drawn up by the governor, but this was never done.[40]

Daendels, arriving in 1808, created an inspector-general of forests and a Board of Forests. The board consisted of five high officials who were answerable to the inspector-general and the governor-general himself. Daendels also appointed twelve forest overseers who were to work under the direction of the inspector and the board. The overseers gave orders to the heads of the blandong.

Under Daendels, the blandong system was generalized over all the teak-producing districts, and it was even introduced in eastern Java, where no compulsory forest service had existed before. Instead of a per-log payment, which was often kept by the regents and did not reach the forest people, the forest people were now given rent-free rice fields, which they could work during the three or four months when work in the forests came to a standstill (November to February). In addition, forest workers were now sent home for eight to fourteen days after eight to fourteen days of work in the forests. The number of forest people expanded considerably between 1808 and 1810, reaching over 7,000 people in 1810—not 100,000 families, as reported by Thomas Raffles.[41]

Daendels's instructions gave clear-cutting official status, mentioned the laying out of and care for teak plantations (at least 100,000 trees per annum), and also mentioned forestry practices including girdling, the use of the saw, and the burning of refuse. Either under Daendels or a few years before his arrival all leases of villages and forests that had been rented out to private entrepreneurs were cancelled. Daendels forbade all private forest exploitation, including the sale and transport of any timber that had not been bought from the forest board.[42]

Daendels was, in fact, the creator of the system of forest management and exploitation by the government. Although some have argued that such a system was exactly what the forests needed at that time, it had the side-effect of bringing the entire ship-building industry to a standstill. The blandong were producing four times as many logs in 1809 as they had done under the company (40,000), but the government had fixed timber prices too high.[43]

If Daendels was the great creator of the government's management system, Thomas Stamford Raffles, the British lieutenant-governor of Java, was the great destroyer. In 1813 he abolished the forest board, which he replaced by a superintendent of forests, whose functions were restricted to Rembang. He drastically lowered timber prices and introduced a license system, not unlike the one that did so much damage to the forests of British India and Burma, under which private entrepreneurs were allowed to fell as much timber as they needed. This, of course, was a boon to the ship-building industry, and it boosted the timber export, but it was also destructive to the forests.[44] Finally, Raffles acquired, by treaty, a number of districts with large teak forests, up to then in the hands of the native rulers.

When the Dutch administration returned in 1816, the new rulers, the commissioners-general, reinstated the system of government exploitation and management begun by Daendels.[45] The acting superintendent of forests and the inspector of forests were now commissioned to carry out a survey of all teak forests and to try to find wage laborers for those activities that had been carried out as compulsory blandong services.

In 1819 the commissioners-general had collected sufficient information to formulate a new forest policy, which consisted of the following basic elements:

- reestablishment of a forest board, consisting of one director of forests and two inspectors; their personnel to consist of seventeen European overseers or rangers, twenty-one native foresters and ninety-six guards, apart from a number of clerks and retainers;
- reestablishment of local authority in the Residents, a departure from the principles that had guided forest reform policy since 1797, which had been directed at a total separation of local authority and forest administration;
- compulsory blandong services as the rule; wage labor, the exception (only in Surabaya and Pasuruan, eastern Java, as an experiment);
- reinstitution of clearcutting and the laying out of teak plantations.[46]

This new policy, in some aspects better than Daendels's, proved even more short-lived than his. This time the promising new initiatives were undermined not by the invasion of a foreign fleet, but by the arrival of a new commissioner-general carrying explicit instructions for the implementation of drastic budget cuts.

The Age of Destruction: 1826–65

In 1826 the new commissioner-general abolished the forest board, and the responsibility for all forest affairs reverted to the Residents. Given that four years later, in 1830, the so-called cultivation system was introduced by the central bureaucracy in Batavia, the Residents had things on their minds other than forest conservation.

The cultivation system was a successful attempt to encourage cash-crop production for the European market, but it required all the time and energy of the Residents to make it successful. Under this system a large number of sugar factories and establishments for processing coffee, indigo, and tobacco were established, the construction of which required enormous amounts of timber. In addition, sugar factories were voracious consumers of wood for fuel, which was also true of the other industrial establishments, though to a lesser extent.

In the meantime much timber was needed for the construction of new fortifications, the improvement of harbors and other infrastructural facilities, and the construction of carts and ships in order to satisfy the increased demand for transportation. The increased demand for transportation also prompted an ambitious program of road construction, often to the detriment of the forests.

The new crops required arable land, and this meant large-scale land clearing. Coffee, particularly, was supposed to thrive on newly cleared forest soils. At the same time, average annual population growth was higher than ever before, and although the total amount of arable

Table 3 ■ *Blandongs* and teak production, 1837–65

Period	*Blandong* laborers (avg. annual no.)	Teak production (avg. annual no. of logs)
1837–40	16,300	100,000
1857–61	n.a.	160,000
1863–65	n.a.	175,000

land was increasing somewhat more slowly than the population, clearing for subsistence agriculture was still considerable.[47]

As this short description indicates, the forests were the victims of a massive onslaught, which was not less destructive for being unintentional. At the same time conservation measures were virtually nonexistent. The system of clearcutting was fairly soon abolished, and the old VOC system of selective felling reinstated.

Because of the almost total absence of a forest service and an equally absent concern for the well-being of the forests, qualitative information on this period is scarce. Available figures on forest activities during this period come from the Directorate of Cultivation, which was an avid producer of figures, albeit not always reliable ones (see table 3).[48]

Archival figures indicate that Daendels employed over 7,000 blandong in 1810, and that his workforce had produced 40,000 logs in 1809. I estimated that production during the late eighteenth century was 75,000 logs. As table 3 indicates, annual production during the period between 1837 and 1840 was 100,000 logs. It seems therefore safe to conclude that the number of forest people and production increased considerably under the new system, the more so if we take into account that the data for 1837 through 1840 reflect the officially registered production only.[49]

But one aspect of forest conservation was not being neglected: the laying out of teak plantations. Between 1837 and 1842 an average annual number of 375,000 trees was planted; and between 1856 and 1865 quantities of one to two million per annum were reported.[50] Although in many instances the sites selected for these plantations were far from favorable and the upkeep of the young trees left much to be desired, a survey carried out around 1880 found almost three hundred square kilometers of teak plantations, laid out before 1865, still in existence.[51]

In 1849, which may be considered a turning point during a period of almost-unremitting gloom, the government appointed three fully qualified German foresters to serve in the Javanese forests. As both England and the Netherlands had neglected their own forests, it is not surprising that scientifically based forest management in British India and Java was introduced by Germans.[52] One of the arguments used by the minister of colonies to convince King William to send the German foresters to Java was that Java needed to institute a system of sustained-yield forest production. This apparently was the first time since 1797 that the notion of sustained yield had been mentioned.

Other positive developments during this period were the gradual reintroduction, starting in 1854, of clearcutting and renewed attempts, dating from 1856, to replace compulsory labor by wage labor. The number of trained foresters gradually expanded, and in 1858 the appointment of an inspector of forests heralded the rebirth of a forest service.

But there were also negative influences on the forests, notably the start of railway construction during the 1860s. Large tracts of good teak were felled to give way to rails, and trees were also felled for sleepers, which had to be replaced every four to five years. On top of it all, transport of illegally cut timber was made much easier by the very presence of the railways.

During this period too, another development began that would influence the mode of forest exploitation for a long time to come. In 1840 one contract had been concluded with a private entrepreneur for the exploitation of thirty-six square kilometers of teak forest, and during the latter half of the 1850s this policy was continued and expanded. A system of government management and private exploitation was gradually taking shape.[53]

In 1860 a commission was appointed to conduct an accurate survey of the forests. The findings of this survey, completed in 1871, represent the situation that existed under a regime of maximal exploitation and minimal conservancy.

The survey found the total surface area of teak forests on Java (exclusive of Madura, Banyuwangi, Yogyakarta, and the Mangkunagara lands in Surakarta) to be 6,400 square kilometers. In 1940 the teak forests of the whole island (including Madura) occupied 8,300 square kilometers, which implies a considerable improvement in a period of seventy years, even taking into consideration the areas omitted in the 1871 survey.[54] The reader might well ask how the 1871 figure compares to the surface area of earlier periods. There is a probably-not-very-reliable figure for 1840: 8,500 square kilometers. This figure is exclusive of a number of areas included in the 1871 figure, which should be reduced for purposes of comparison to 6,000. This implies a loss of teak lands of about 30 percent in thirty years.[55]

The only other available estimate dates from 1776 and is unfortunately expressed in a rather problematic unit of measurement: distances are given in hours.[56] In the eighteenth century an hour as a measure of distance was about 5 to 5.5 kilometers, but even if we accept this range, we must recognize that it is highly unlikely that the forests were rectangles. Allowing for a reduction due to surface areas with ellipsoid forms, we might set a figure of 4.5 kilometers for an hour. Thereby, we would arrive at an estimate of 3,650 square kilometers for the forests surveyed in 1776.[57] The figure for 1871 for the same areas was 1,550 square kilometers, or a loss of 55 to 60 percent in ninety-five years. If we apply these ratios to all teak forests of Java, we find a loss of 40 percent between 1776 and 1840 (sixty-five years), and about 30 percent between 1840 and 1870 (thirty years).

State Management, Private Exploitation: 1865–97

In 1865 the government instituted a set of new regulations, published as Ordinance of 10.9.1865 (*Indisch Staatsblad* 96), which abolished the system of government exploitation. Private enterprise would take government's place. The new regulations brought a formal end to the blandong services, at least in principle: wage laborers were hereafter to carry out all forest work. The 1865 regulations are usually seen as a watershed between the old system of forest exploitation, intimately linked with the cultivation system, and a new, "liberal" phase of forest management.

There were also new regulations (*Indisch Staatsblad* 97) for the administration and management of the forests. These regulations divided the forests into three groups: (1) teak forests under a regular system of conservancy; (2) teak forests not under a system of conservancy; (3) junglewood forests. This division, which gave effective protection to the first category only, soon proved impractical. A new Ordinance of 14.4.1874 (*Indisch Staatsblad* 110) brought all teak forests under a regular system of conservancy.

The junglewood forests had been mentioned in the regulations of both 1865 and 1874, and to some extent their protection was provided for, but the government did not create official junglewood forest reserves until 1890. Under the 1890 regulations, all junglewood forests above 5,000 feet in western Java and above 4,000 feet in central and eastern Java were declared reserved forests.[58]

The 1870s and 1880s witnessed the introduction of the so-called forest arable farming method of teak cultivation, under which contracts were made with the local population to cultivate food crops between the rows of young trees in forest plantations. The forest arable farming method was a satisfactory arrangement for both contract parties.[59]

Another important measure was the establishment in 1869 of a fixed structure for the forest service (*Indisch Staatsblad* 11). Re-introduced after a forty-three-year absence, the new forest service was headed by two inspectors and thirteen conservators who held sway over twenty-three rangers, forty-four foresters, and one hundred ten forest guards. In 1870 the forest service got its own division within the Civil Service Department. In 1874 the teak forests were divided into thirteen forest districts, each under the direction of a conservator.[60]

In the meantime private enterprise gradually took over the exploitation of the forests, as the 1865 regulations meant them to do. Between 1865 and 1875 the number of concessions increased from seven to twenty-five, and the percentage of wood being cut by private enterprise increased from 43 to 70 percent. The international depression during the late 1870s and early 1880s caused a drop from twenty-five to eighteen concessions (1886) and from 70 to 40 percent of wood cut by private enterprise (1885). When the international economy picked up again, the number of concessions increased rapidly to an all-time high of seventy-eight (1896) and the percentage of wood cut by private enterprise to 95 percent (1894).[61]

This, however, was the beginning of the end of private enterprise in the Javanese forests. When the forest service, and afterward the government, became aware that the aims of private exploitation and state management were incompatible, they took steps to stem the tide of private exploitation. In 1897 the teak forests were divided into two categories: forest districts, organized provisionally; and forest ranges, with a definite organization. Forest ranges were to be exploited according to the principles of sustained and stable yields. The first forest range was created directly in 1897. In the forest ranges the principle of state exploitation was accepted.[62] The final triumph for the forest service came in 1914, when the provisionally organized forest districts were also thrown open to state exploitation. Private enterprise, so important at the turn of the century, was slowly but surely crowded out.

Conclusion

Between 1677 and 1745 the VOC gradually tightened its grip on teak production on Java. Between 1745 and 1808 the company was officially the only institution managing the teak forests in the areas under Dutch sovereignty (in the areas still under native rule the VOC was not involved in management), but the company had to share its exploitation with private enterprise.

Although the VOC had neither the personnel nor the expertise to manage its forests successfully, it did try to preserve them as well as it could by laying out plantations and by closing forests that had been cut out. On paper, the VOC had accepted the principle of sustained yield as early as 1797. But the production levels of the nineteenth century continued to be extremely high, far outpacing production in the late eighteenth century.

Between 1808 and 1826 there were two short-lived attempts to create a proper forest service, which was to be in charge not only of forest management but also of forest exploitation. But the understaffed services were not able to produce sufficient timber for the government and the private sector. Between 1813 and 1819 only a skeleton service remained, and low timber prices temporarily made shipyard owners and other timber consumers happy. No centrally organized forest service, with a fixed structure, was in existence between 1826 and 1869, although a first inspector was appointed in 1858, heralding the eventual reemergence of such a service.

During the period of hardly any forest management, the forests were exploited both by government and private enterprise. Teak areas decreased rapidly. The rate of decrease of teak areas was higher between 1840 and 1870 than it had been between 1776 and 1840, and production of teak was also much higher.

Between 1865 and 1897 state management and private exploitation went hand in hand, but it became increasingly clear that state management, aimed at sustained and stable wood production, was incompatible with private exploitation. When, therefore, the decisive step toward implementing sustained-yield forestry was finally taken in 1897, the government also decided that under the new system, forest exploitation would be carried out by the forest service. This was the beginning of the end of private enterprise in Java's forests. The result was production that was higher than ever before, without the damaging effects suffered during earlier periods. ▲

Notes

1. For a very negative but undocumented judgment on the Dutch East India Company's forest management, see Jan Willem Hugo Cordes, *De djati-bosschen op Java: Hunne natuur, verspreiding, geschiedenis en exploitatie* (Batavia, Netherlands Indies: Ogilvie, 1881), p. 189; Ch. S. Lugt, *De bosschen van*

Nederlandsch-Indië, hunne benutting en verzorging (Baarn, the Netherlands: Hollandia-drukkerij, 1915), pp. 9–10.

2. C. J. J. van Hall, *Insulinde: Werk en welvaart* (Naarden, the Netherlands: Uitgeverij "In den Toren," n.d. [ca. 1940]), p. 117.

3. Most details of this section are taken from F. de Haan, *Oud Batavia* (Bandoeng, Netherlands Indies: Nix, 1935). The principal city, present-day Jakarta, was known as Jakatra before 1619, then Batavia, and finally Jakarta.

4. De Haan, *Oud Batavia*, p. 477; F. de Haan, *Priangan: De Preanger-regentschappen onder het Nederlandsch bestuur tot 1811* (Batavia, Netherlands Indies: G. Kolff, 1912), 3:5.

5. E. H. B. Brascamp, "Uit het Koloniaal Archief XXXIV," *Tijdschrift Bataviaasch Genootschap* 60 (1921): 351, 356; de Haan, *Oud Batavia*, p. 495.

6. Brascamp, "Uit het Koloniaal Archief XXIX," *Tijdschrift Bataviaasch Genootschap* 59 (1919/20): 452; Brascamp, "Uit het Koloniaal Archief XXXII," *Tijdschrift Bataviaasch Genootschap* 60 (1921): 138–39; Brascamp, "Uit het Koloniaal Archief XXXIV," p. 351; W. Zwart, "Uit de boschgeschiedenis van Java en Madoera, III," *Tectona* 31 (1938): 80; Luc Nagtegaal, "Rijden op een Hollandse tijger: De noordkust van Java en de VOC 1680–1743" (Ph.D. diss., State University of Utrecht, 1988), pp. 139–42.

7. Brascamp, "Uit het Koloniaal Archief XXXVI," *Tijdschrift Bataviaasch Genootschap* 61 (1922): 178; Brascamp, "Uit het Koloniaal Archief XXXVIII," *Tectona* 15 (1922): 958–59; de Haan, *Priangan*, 3:376–78, and 4:30, 40.

8. Friedrich Seltmann, *Die Kalang: Eine Volksgruppe auf Java und ihre Stamm-Mythe* (Stuttgart, Germany: Franz Steiner Verlag, 1987), pp. 16–20. On the kalang, see also C. Guillot, "Les *Kalang* de Java, rouliers et prêteurs d'argent," in *Marchands et hommes d'affaires asiatiques dans l'Océan Indien et la Mer de Chine 13E-20E siècles*, ed. D. Lombard and J. Aubin (Paris, France: Editions EHESS, 1988), pp. 267–78.

9. Brascamp, "Uit het Koloniaal Archief XVII," *Tectona* 13 (1920): 1087; Seltmann, *Die Kalang*, pp. 21, 413. According to Nagtegaal, "Hollandse tijger," pp. 140–41, the woodcutters' villages had already been ceded in 1677, but it seems plausible that the actual cession did not take place until after the war (1680).

10. Most sources on timber to be delivered to the company refer to squared logs (*balken*); there are only occasional references to specific kinds of timber.

11. Brascamp, "Uit het Koloniaal Archief XVI," *Tectona* 13 (1920): 668–69; Brascamp, "Uit het Koloniaal Archief XXXV," *Tijdschrift Bataviaasch Genootschap* 61 (1922): 140; Brascamp, "Het contract met de Soesoehoenan," *Tectona* 16 (1923): 639–40.

12. Brascamp, "Koloniaal Archief XXXV," p. 136; Brascamp, "Uit het Koloniaal Archief IX," *Tectona* 12 (1919): 383; Jan Karel Jakob de Jonge, *De opkomst van het Nederlandsch gezag in Oost-Indië* ('s-Gravenhage, the Netherlands: M. Nijhoff, 1878), 10:330–35; Brascamp, "Uit het Koloniaal Archief VI," *Tectona* 10 (1917): 1045–55; Resolution Semarang, 14 April 1777, Collection van Alphen/Engelhard, 1900, 166, 2, Algemeen Rijks Archief ("General State Archives," hereafter cited as ARA), The Hague, the Netherlands; Memorandum of Transfer, Governor of Java's Northeast Coast, 1780, Coll. van Alphen/Engelhard, 1916, 78, ARA; Memorandum of Transfer, Governor of Java's Northeast Coast, 1801, Coll. van Alphen/Engelhard, 1900, 191, ARA; Report of von Winckelman on the Compulsory Forest Services, 23 March 1809, Coll. van Alphen/Engelhard, 1900, 262, ARA.

13. J. A. van der Chijs, *Nederlandsch-Indisch Plakaatboek*, 16 vols. ('s-Gravenhage, the Netherlands: M. Nijhoff, 1885–1900), 7 (1890): Publication 11 May 1764; Brascamp, "Koloniaal Archief VI," pp. 1048–53.

14. This calculation is based on the assumption that squared logs measured, on average, 7.35 meters by 30 centimeters (see A. E. J. Bruinsma, "Het boschwezen op Java, voorheen en thans," *Tectona* 8 [1915]: 746).

15. De Haan, *Priangan*, 4:37 and 40. Report of von Winckelman, 23 March 1809. I could find no data on Sumedang, but as Krawang probably had stopped producing teak before the end of the eighteenth century, my estimate of the total contingent for western Java may be fairly accurate.

16. Van der Chijs, *Plakaatboek*, 5 (1888): Publication 30 November 1747.

17. Report of a General Survey of the Teak Forests of Central Java, 19 December 1776, Coll. van Alphen/Engelhard, 1900, 166, 1, ARA. For Rembang the number of people is not

20 Peter Boomgaard

given, and I have used a figure for 1797 (576). For 1787 we only have the number of buffalo, but I have assumed that the ratio between people and buffaloes did not change; Van der Chijs, *Plakaatboek*, 10 (1892): Publication 30 October 1787; Report of von Winckelman, 23 March 1809.

18. De Haan, *Priangan*, 4:32; "Cheribon in den goeden ouden tijd," *Tijdschrift voor Nederlandsch-Indië*, n.s. 8, no. 2 (1879): 192; Memorandum of Transfer, 1791, Collection Nederburgh, 384, ARA; Letter from Governor of North Coast to Governors-General, 30 June 1804, Coll. van Alphen/Engelhard, 1900, 210, ARA.

19. Report on General Survey of Teak Forests of Central Java, 1776, Coll. van Alphen/Engelhard, 1900, 166, 1, ARA; Report of von Winckelman, 23 March 1809.

20. Thomas Stamford Raffles, *The History of Java* (London: John Murray, 1830), 1:203.

21. Report of a Survey of the North Coast Forests of Central Java, 19 December 1776, Coll. van Alphen/Engelhard, 1900, 166, 1, ARA.

22. Letter from Governor of North Coast to Governors-General, 27 May 1803, Coll. van Alphen/Engelhard, 1900, 204, ARA; Brascamp, "Koloniaal Archief XXVIII," p. 1109; Herman Willem Daendels, *Staat der Nederlandsche Oostindische bezittingen, onder het bestuur van den GG—in de jaren 1808–1811* ('s-Gravenhage, the Netherlands:, 1814), p. 51; Cordes, *De djati-bosschen*, p. 165.

23. Resolution Semarang, 14 April 1777.

24. Van der Chijs, *Plakaatboek*, 10 (1892): Publication 15 September – 31 December 1778 (the rixdollars mentioned in the source should be read as stivers); Publication 9 March 1779; Publication 10 July 1781.

25. Report of a Survey of the Forests of Jepara and Kudus, 3 November 1797, Collection Nederburgh, 379, 1, ARA.

26. Several Draft Instructions for the Forest Commission, 1797, Coll. van Alphen/Engelhard, 1900, 179, ARA.

27. Van der Chijs, *Plakaatboek*, 13 (1895): Publication 21 August 1801; 15 (1896): Publication 21 August 1808.

28. Report of a Survey of the North Coast Forests of Central Java, 19 December 1776.

29. Cortes, *De djati-bosschen*, pp. 58–61; Van de Chijs, *Plakaatboek*, 10: Publication 30 October 1787; 13 (1895): Publication 21 August 1801; Letter from Governor of North Coast to Governors-General, 1803, Coll. van Alphen/Engelhard, 1900, 204, ARA.

30. Van der Chijs, *Plakaatboek*, 10 (1892): Publication 30 October 1787; 13 (1895): Publication 21 August 1801; Report of von Winckelman, 23 March 1809; Decree 14 December 1829, 21 (*Indisch Staatsblad* [*IS*] 125); Decree 4 January 1857 (*IS* 1).

31. C. A. R. Bhadran et al, *First Centenary of Forest Administration in Madras State 1856–1956: Souvenir* (Madras, India: Forest Service, 1959), p. 23.

32. Report of a Survey of the Forests of Jepara and Kudus, 3 November 1797, Collection Nederburgh, 379, 1, ARA; Memorandum of the Former Resident of Pekalongan, 1798, Collection Reinward, 19, ARA. The resident was mistaken about his "wilderness"; this was the so-called Weleri Forest, where at least in 1776 teak was growing, although even then large parts had been cut. So much for the replanting practices of 1776!

33. General Report on Java's North Coast, 1807, Coll. van Alphen/Engelhard, 1900, 244, ARA; Report on Teak Forests to Commissioners-General, 1 November 1818, Collection Baud, 40, ARA.

34. This seems to be a conclusive refutation of the assertion that the Connolly plantations of 1842 were the "earliest teak plantations in the world" (Bhadran, *Centenary*, p. v).

35. Van der Chijs, *Plakaatboek*, 4 (1887): Publication 16 June 1722; De Haan, *Priangan*, 4:34.

36. Memoranda of Transfer of Java's North Coast, 1791 and 1801; Report on Teak Forests to Commissioners-General, 1 November 1818.

37. Report of a General Survey of the Teak Forests of Central Java, 19 December 1776; Report on Teak Forests to Commissioners-General, 1 November 1818.

38. The term *kalang*, used around 1700 for woodcutters, disappears as such after 1750. When the kalang are mentioned after 1750, it is as carpenters, carters, or cartwrights. After 1750 the people working in the forests were indicated with the term *boschvolkeren* (forest people), or in the case of Rembang with the Javanese term *bland(h)ong*. After Daendels's reforms, the term blandong was used for all compulsory labor in the forests.

39. Or rather "bland*h*ong," but as all the literature refers to "blandong," I will stick to this spelling. Almost all literature on the eighteenth-century forests calls all the forest people "blandong," but it is clear that in the sources only the Rembang forest people, operating under this special system, are called "blandong," at least up to the first years of the nineteenth century.

40. Several Draft Instructions for the Forest Administration, 1797, Coll. van Alphen/Engelhard, 1900, 179, ARA; Van der Chijs, *Plakaatboek*, 13 (1895): Publication, 21 August 1801; 13: Publication 8 September 1803.

41. Raffles, *History*, 1:204.

42. Van der Chijs, *Plakaatboek*, 15 (1896): Publication 21 August 1808; 15: Publication 29 August 1808; Coll. van Alphen/Engelhard, 1900, 262, ARA; Report on the Blandong, 1809, Coll. van Alphen/Engelhard, 1900, 262, ARA; Daendels, *Staat der Nederlandsche Oostindische bezittingen*, pp. 50–51.

43. G. H. van Soest, "Het boschwezen op Java," *Tijdschrift voor Nederlandsch-Indië*, 3d ser. 3, no. 2 (1869): 165.

44. Henry David Levyssohn Norman, *De Britsche heerschappij over Java en onderhoorigheden, 1811–1816* ('s-Gravenhage, the Netherlands: Belinfante, 1857), pp. 161–64; Cordes, *De djati-bosschen*, pp. 200–201; Stebbing, *Forests of India*, 1:206.

45. *IS* 1817, no. 48.

46. *IS* 1819, nos. 17 and 18; Cornelius Theodorus Elout, *Bijdragen tot de kennis van het koloniaal beheer, getrokken uit de nagelaten papieren van wijlen* ('s-Gravenhage, the Netherlands: M. Nijhoff, 1851), pp. 57–60.

47. G. H. van Soest, "Boschwezen op Java," pp. 456–59; Cordes, *De djati- bosschen*, pp. 205–10; Peter Boomgaard, *Children of the Colonial State; Population Growth and Economic Development in Java, 1795–1880* (Amsterdam, the Netherlands: CASA/Free University Press, 1989).

48. All data are taken from the unpublished, annual General Cultivation Reports (Cultuur Verslag, 1836–50, in ARA, Archief Ministerie van Koloniën [AMK]) and the published, annual Colonial Reports (since 1849, *Koloniaal Verslag*, The Hague, the Netherlands: Ministry of Colonies).

49. Van der Chijs, *Plakaatboek*, 15: Publication 21 August 1808; 15: Publication 29 August 1808; Report on the Blandong, 1809, Coll. van Alphen/Engelhard, 1900, 262, ARA; Daendels, *Staat der Nederlandsche Oostindische bezittingen*, pp. 50–51. Van Soest, "Het boschwezen op Java," 165.

50. General Cultivation Reports and Colonial Reports.

51. Peter Boomgaard, "Colonial Forest Policy in Java in Transition: 1865–1914" (Paper prepared for the conference on "The Socioeconomic Foundations of the Late Colonial State in Indonesia, 1880–1930," Wassenaar, the Netherlands, June 1989), p. 11.

52. Bhadran, *Centenary*, p. 7; Stebbing, *Forests of India*, 1:207; Cordes, *De djati-bosschen*, p. 214.

53. Cordes, *De djati-bosschen*, pp. 146, 165, 214, 220, 229–30, and 246.

54. Data for 1871 from Cordes, *De djati-bosschen*, p. 154 (with some corrections); for 1940 from C. J. J. van Hall, *Insulinde: Werk en welvaart*, p. 117. In both cases the areas measured include "empty," "devastated," and (other) non-teak parcels within the official boundaries of the teak forests.

55. Report of a General Survey of the Teak Forests of Central Java, 19 December 1776.

56. Cultivation Report, 1840, ARA, AMK, Verbaal 17 September 1842, no. 15.

57. The figure for the forests of Jepara had to be corrected because the original figure was entirely out of proportion, probably due to a scribal error. The correction is based on the average ratio 1776:1870 excluding Jepara.

58. Boomgaard, "Colonial Forest Policy," pp. 9–13.

59. The Dutch term is *bosveldbouw methode*; later on it was usually called contract row-cultivation (*contractrijen-cultuur*), and nowadays *tumpang sari*; in the international literature it is called *taungya*, a Burmese word. See E. J. Wind, "Het boswezen in het tijdvak 1850–1950," in *Een eeuw natuurwetenschap in Indonesië, 1850–1950* (Jakarta, Indonesia: 1950), p. 141; Nancy L. Peluso, "Rich Forests, Poor People, and Development: Forest Access Control and Resistance in Java" (Ph.D. diss., Cornell University, 1988), p. 68.

60. Boomgaard, "Colonial Forest Policy," pp. 9–13.

61. H. J. Kerbert, "Geschiedkundig overzicht van de hout-vervreemdingspolitiek op Java," *Tectona* 12 (1919): 629.

62. Kerbert, "Geschiedkundig overzicht," pp. 634–35; J. Nirschl, *Die Forstwirtschaft in Niederländisch-Indien* (Leipzig, Germany: K. F. Koehler, 1920), pp. 37–38; A. E. J. Bruinsma, "Het boschwezen in Nederlandsch-Indië," *Indische Gids* 16 (1894): 870–78.

16

Conserving Eden: The (European) East India Companies and their Environmental Policies on St. Helena, Mauritius and in Western India, 1600–1854

Richard Grove

The history of tropical forest change over the last millennium is difficult to chart with any confident degree of accuracy. Indeed, systematic attempts even for the last hundred years have been made only recently. In general, more is known at present about the history of tropical forests in Asia and Southeast Asia than forests in Africa or South America.[1] This lack of knowledge is partly due to the fact that the causal factors behind the erosion of tropical forests area are particularly difficult to disentangle. However, important connections can be made between European expansion, the penetration of capitalist economic forces, and the transformation of tropical environments.[2] Above all, the spread of market relations in the tropics has served to encourage the rapid clearance of forests for agriculture. The history of global deforestation has probably been closely associated at many of its fastest stages with the dynamics of the forces of industrialisation and the expansion of a European-centered world-system. Nevertheless, there have also been phases of

I have benefited greatly in writing this paper from suggestions by William Beinart, Vinita Damodaran, Mariodi Gregorio, and Quentin Cronk. The research for it was generously supported by the Royal Society; Clare Hall, Cambridge; the British Academy, and the estate of J. C. Smuts.

[1] See J. F. Richards, J. R. Hagen, and E. S. Haynes, "Changing Land-Use in Bihar, Punjab and Haryana," *Modern Asian Studies, 19* (1985), 699–752. In some respects a reliance on official sources produced between 1870 and 1970 has led to a neglect of the critical but little understood period of forest clearance between 1780 and 1850, colonial perceptions of which led to the developments described in this paper.

[2] For some initial attempts in this direction, see I. Wallerstein, *The Modern World System* (Academic Press, New York, 1974). There remain, of course, some major problems to be encountered in equating any expansion of the "European world system" with processes of ecological change. It is now well established that the activities of indigenous peoples in Australia and East Africa, for example, caused widespread ecological change long before the advent of the European. The same was true in many Pacific islands; see the chapter on Easter Island in Clive Ponting's *A Green History of the World* (London: Sinclair Stevenson, 1991).

rapid change in zones as yet unaffected by the dynamics of capitalist transformation as they are normally understood. The conceptual problem that arises has less to do with the difficulties of explaining the dynamics of rapid ecological change than with the problems entailed in using a monolithic and Eurocentric world system theory to describe patterns of proto-capitalist and indigenous consumption. Thus, at various times the relatively rapid clearance of large areas of forest in much of the Pacific area, in tropical Africa, and in Northern India has taken place outside the context of the expansion of any world system. Significantly, some phases of non-European clearance have led to the development of important and ambitious state initiatives in forestry and environmental control, although very few have ever been properly investigated.[3]

Since the Second World War, tropical deforestation has taken place so rapidly that the prospect of the disappearance of the remaining core areas of continental forest has led to a widespread contemporary apprehension about its consequences. This is frequently expressed in terms of anxieties about global climatic change and the likely disappearance of a high proportion of existing species of fauna and flora. Such concerns are, in fact, not new and, in their essentials, constitute a re-statement of anxieties expressed for more than two centuries.[4] Furthermore, interventionist responses to rapid forest destruction have a long and coherent intellectual history. Many emerged from the specific circumstances of colonial expansion and associated ecological change, although there have been some important exceptions. In particular, western anxieties about the possible connections between deforestation and climatic change emerged quite specifically in the colonial rather than the metropolitan European context. This essay intends to identify, briefly, some of the main milestones in the history of early western conservationist responses to the ecological effects of the colonial transformation of the landscape and to the destruction of tropical forests in particular.

Some historians have suggested that the environmentally destructive effects of colonialism were not only economic but had their roots in ideologically imperialist attitudes towards the environment.[5] This does not seem an extraordinary thesis to advance, particularly because the evidence seems to indicate that the penetration of western economic forces (not all necessarily synonymous with colonialism) did indeed promote a rapid ecological transforma-

[3] An outstanding exception is found in Conrad Totman's survey of forest conservation in Japan in the seventeenth century. See *The Green Archipelago* (Berkeley: University of California Press, 1989).

[4] See R. H. Grove, "Conservation and Colonial Expansion: The Development of Environmental Attitudes and Conservation Policies on St. Helena, Mauritius, and in Western India" (Ph.D. thesis, Cambridge University, United Kingdom); and "The Origins of Environmentalism," *Nature*, May 23, 1990.

[5] This argument is put forward in D. Worster, *Nature's Economy; A History of Western Ecological Ideas* (Cambridge, 1985).

tion in some instances,[6] which was especially true in the early nineteenth century in India and Southern Africa.[7]

On closer inspection, however, the hypothesis of a purely destructive environmental imperialism constituting a complete break with the pre-colonial past does not stand up well at all. Indeed this notion apparently arose out of a misunderstanding about the contradictory, heterogeneous, and ambivalent nature of the colonial state's workings. Above all, a number of scholars have been generally unaware of the extent to which colonial governments were peculiarly open to the pressures of contemporary environmental lobby during the first half of the nineteenth century, a time of great uncertainty about the role and long-term security of the colonial state. Thus, although it undoubtedly promoted widespread ecological destruction, colonial enterprise also helped to create a context conducive to rigorous analytical thinking about the processes of ecological change and to the formation of a conservation ideology.[8]

Ironically, too, the colonial state in its pioneering conservationist role provided a forum for controls on the unhindered operations of capital for short-term gain which, it might be argued, constituted a fundamental contradiction to what is normally supposed to have made up the common currency of imperial expansion and profit maximisation.[9] Moreover, the absolutist nature of colonial rule made it possible to introduce interventionist forms of land management that would have been very difficult to impose in Europe, even though many of them provoked active indigenous resistance. Colonial expansion also promoted the rapid diffusion of new scientific ideas and environmental concepts among colonies and between metropole and colony over a large area of the world. Such ideas sometimes acquired a potent momentum of their own independent of the apparatus of colonial rule while still exerting influence on its actors.

Conservation in its more modern sense has often been perceived as a phenomenon with antecedents in late nineteenth- or early twentieth-century North America.[10] This interpretation regards Henry D. Thoreau, John Muir,

6 For a useful survey, see R. P. Tucker and J. R. Richards, eds., *Global Deforestation and the World Economy* (Durham, N.C.: Duke University Press, 1983).

7 See R. H. Grove, "Early Themes in African Conservation; The Cape in the Nineteenth Century," in D. Anderson and R. H. Grove, *Conservation in Africa; People, Policies and Practice* (Cambridge, 1987), 22 47.

8 For discussions of the later development of conservation ideologies, see R. Grove, "Scottish Missionaries, Evangelical Discourses and the Origins of Conservation Thinking in Southern Africa, 1820–1900," *Journal of Southern African Studies*, 15:2 (January 1989), and W. Beinart, "Soil Erosion, Conservationism and Ideas about Development: A Southern African Exploration, 1900–1960," *Journal of Southern African Studies*, 11:1 (1984).

9 For a useful analogous discussion of colonial state hostility to capital interests, see D. Washbrook, "Law, State and Agrarian Society in Colonial India," *Modern Asian Studies*, 15 (1981), 648–721.

10 E.g., Worster, *Nature's economy;* S. Hays, *Conservation and the Gospel of Efficiency; The Progressive Conservation Movement, 1880–1920* (Cambridge, Mass., 1959); D. Lowenthal, *George Perkins Marsh, Versatile Vermonter* (New York, 1958).

and George Perkins Marsh as the leading originators of modern environmentalism. In contrast, scholars dismiss colonial conservation and forestry regulation, especially in India, as mere disguises legitimised by a subordinate colonial science for resource exploitation and land seizure by the state.[11] Although some elements of truth in such doctrinaire accounts are normally coupled with doubtful notions of pre-colonial Golden Ages of ecological balance, they have all tended to overlook the remarkably innovative nature of early colonial conservationism, the characteristics of which have been very much neglected until recently. They also tend to underplay the highly developed trend towards exclusionist forms of state forest control developed in pre-colonial states in South Asia, as well as the well-established history of rapid, and often state-sponsored, deforestation in the centuries before the onset of East India Company rule in the sub-continent.[12]

As a result the notion of North American primacy in conservation thinking has remained unquestioned. In fact, a good deal of evidence indicates that complex notions of state intervention in natural resource protection, many of them strongly connected with new and highly anthropomorphic valuations of the environment, emerged and were extensively promulgated in the colonial context more than a century before George Perkins Marsh published his famous *Man and Nature* in 1864.[13] What is more, Marsh apparently drew much of his inspiration from a detailed knowledge of the history of important colonial experiments in environmental intervention. Indeed, the very speed of degradation induced by colonization helped to evoke such experimentation.[14]

Colonial conservationism probably did emerge not so much because colonial annexation and capitalisation brought about extensive ecological changes as because the conditions inherent in early Dutch, French and British colonial rule promoted the rapid rise of a distinctive group of professional naturalists and scientists to disproportionate influence. Many of these early experts constructed an effective critique of the environmental effects of colonial rule, often as a surrogate for more direct but less politically palatable social com-

[11] See especially R. Guha, *The Unquiet Woods; Ecological Change and Peasant Resistance in the Himalaya* (Delhi: Oxford Universtiy Press, 1989), and M. Gadgil, "Towards an Ecological History of India," *Economic and Political Weekly*, vol. XX (1985), 1909–18. Similar interpretations appear in V. Shiva, "Afforestation in India; Problems and Strategies," *Ambio*, vol. 14:6 (1985), and M. Gadgil, *Deforestation; Problems and Prospects* (New Delhi: Society for Promotion of Woodlands Development, 1989).

[12] For some details of early phases of pre-colonial deforestation, see Makhan Lal, "Iron Tools, Forest Clearance and Urbanisation in the Gangetic Plains," *Man and Environment*, 10 (1984):83–90. Widespread deforestation in the Ganges valley during the fourteenth century led to water-table declines and extensive soil deterioration.

[13] G. P. Marsh, *Man and Nature* (New York, 1864).

[14] Marsh was well aware of the deforestation history of St. Helena and early attempts to control the process. He was largely unaware, however, of the history of Indian conservation with which he became acquainted only through correspondence with Dr. Hugh Cleghorn during the late 1860s. See Cleghorn/Marsh correspondence (Marsh Papers, Archives Department of Vermont, Burlington, Vt., 1865–82).

mentaries on colonialism itself. Pre-eminent among these professional naturalists, especially in India before the 1851 rebellion, were the medical surgeons of the East India Company, whose relatively influential status permitted a very rapid diffusion of their environmental diagnoses and interventionist prescriptions. Their outlook was strongly influenced by Hippocratic and Physiocratic ways of thinking. Both of these medically based philosophies dealt implicitly with the relations between the external environment and the health of man.[15] Their impact was, in environmental terms, more keenly felt in the tropics than in contemporary Europe, where vegetational changes in more temperate conditions were generally much slower and less easily observed in their effects. Although the Hippocratic interpretation of environmental processes was innate to the training of all western physicians, the Physiocratic environmental critique (itself derived from medical antecedents), was much more narrowly confined, at first, to the response of French colonial science to tropical conditions in the late eighteenth century. By the 1760s in Mauritius (and a little later in India and the Cape Colony), these two environmental discourses had been translated into a set of highly interventionist conservation policies, preceding the emergence of conservationism in either Europe or the United States.[16]

Thus, as early as 1840, one can clearly distinguish the emergence of fully developed environmental concerns and conservation policies strongly reinforced by what were considered to be scientific interpretations of environmental interactions. These interpretations were based largely on the writings of French and German Romantic scientists and, later, on the work of Alexander von Humboldt.[17] Such concerns had not developed in isolation. From the late eighteenth century, particularly in Mauritius (the Isle de France), environmentalist theories developed in close association with political or even revolutionary radicalism, anti-slavery, and overt anti-colonial sentiment, as well as with a naturalistic Orientalism. Such interconnections have been a characteristic feature of the early history of conservationism. Along with Alexander von Humboldt, other thinkers also serve to exemplify the apparent dualism of environmental and social reform in the critically formative years between 1770 and 1870. Pierre Poivre, Jean Jacques Rousseau, Ernst Dieffenbach, P. E. Strzelecki, Hugh Cleghorn, Edward Balfour, John Croumbie Brown, and Henry David Thoreau all stand out as exemplars of this phenomenon.

15 For an extended introduction to the evolution of Hippocratic environmental psychology see Clarence Glacken, *Traces on the Rhodian Shore* (Berkeley: University of California Press, 1969).

16 See R. H. Grove, "Conservation and Colonial Expansion," (PhD thesis, University of Cambridge, 1988, Ch. 5). It has to be said that the more utopian and transcendental elements in conservationism during the period were partly suppressed, for fairly obvious reasons. Nevertheless a careful inspection of the writings of Hugh Cleghorn in particular indicates similar underlying preoccupations, reinforced by the Humboldtian antecedents of much of early conservation thinking in colonial India.

17 For a specific characterisation of this genre of science, see A. Cunningham and N. Jardine, eds., *Romanticism and the Sciences* (Cambridge: Cambridge University Press, 1990).

Only in its later stages of development did conservationism, as a highly bureaucratised justification for state control of land use, start to enforce its own inflexible and alienating logic. Moreover, conservationism then evoked widespread popular resistance in the context of the development of the more narrowly defined interests of late nineteenth-century imperialism.[18] By then the original reformist ideologues of colonial conservation had almost all long since faded in influence. Their ideological successors were, instead, to be found at work far more in the metropolitan context, where state conservationism developed much more slowly in response to the pressures of urban growth and industrialisation.[19]

EARLY EUROPEAN COLONIAL EXPANSION AND THE ORIGINS OF WESTERN CONSERVATIONISM

The essential precepts of modern conservation actually arose from the conditions of colonial rule and colonial notions of the state's role. Indeed conservationist ideas developed almost in tandem with the growth of the material demands of nascent capitalism, whether in its European form or in indigenous forms in, for instance, pre-colonial India, China, or Japan. The growth of market relations and urbanisation had, of course, long been associated with deforestation in Europe and the Middle East, particularly in the more southern margins of these regions.[20] Indeed, it was in the course of the deforestation of classical Greece that Theophrastus of Erasia, an early ecologist, as well as the keeper of Aristotle's botanic garden and his archivist, first put forward his precocious theories of desiccation. These linked deforestation and declines in rainfall with the destruction of trees around the perimeters of the Greek city-states.[21] In temperate northern Europe, anxieties about the impact of deforestation developed much later than this and for reasons more straightforwardly connected with timber depletion. During early medieval phases of European expansion eastward into the forests of Germany and central Europe, some pioneer attempts were made to limit agricultural clearance under the German kings. Sometimes these involved very clear-sighted notions of sustainable timber use and, in very rare instances, an awareness of the value of forest

[18] See Richard Grove, "Colonial Conservation, Ecological Hegemony and Popular Resistance; Towards a Global Synthesis," in J. Mackenzie ed., *Imperialism and the Natural World* (Manchester: Manchester University Press, 1990).

[19] For details of the growth on early metropolitan environmental lobby groups, see J. Sheail, *Nature in Trust* (London: Dent, 1976). The most important of these was the Commons Preservation Society founded by Octavia Hill, Robert Hunter, John Stuart Mill, T. H. Huxley, and George Shaw-Lefevre. See Lord Eversley (G. Shaw-Lefevre), *Forests, Commons and Footpaths* (London, 1912).

[20] See J. Perlin, *A Forest Journey; The Role of Wood in the Development of Civilization* (New York: Norton, 1989) 35–145.

[21] J. D. Hughes, "Theophrastus as Ecologist," *Environmental Review*, 4 (1985), 291–307. See also remarks on Theophrastus in Glacken, *Traces on the Rhodian Shore* (Berkeley, 1967); and H. Rubner, "Greek Thought and Forest Science," *Environmental Review*, 4 (1985), 277–96.

protection in preventing soil erosion.[22] More systematic insights into the mechanisms of environmental degradation had to await the circumstances of the much more extensive deforestation and soil erosion episodes associated with the rise of the great European maritime empires. In particular, the mercantile and colonial expansion of Venice gave rise to a phase of very rapid and destructive deforestation in Northern Italy, Dalmatia, and Crete.[23] The military and strategic demands of the empire implied a need for a continuous supply of wood for ship and barge building, most of which was carried out at the *Arsenale* in Venice itself.[24] One of the first signs of the progressive deforestation of the Venetian hinterland, the movement of the barge-building industry inland from Venice, was well under way in the second quarter of the fifteenth century. During the 1450s, the high rate of deforestation apparently became recognised as a danger to the Venice lagoon because it increased the amount of silt brought down by the rivers. To prevent the lagoons from filling up, the Council of Ten ordered the re-planting of all cut over woods at the edges of streams of flowing water.[25] A definite policy designed to conserve and increase the supply of oak from state woodlands was formulated between 1470 and 1492. Although innovative, these measures failed almost entirely to conserve or replenish the timber supply or to curb the ruinous destruction of the oak woods, which continued unabated until the sixteenth century. This failure contributed in no small measure to the decline of Venice and its trading and military displacement by maritime powers with easier access to relatively undepleted forests.[26] However, while the Venetian experience of contending with soil erosion was certainly important, far more holistic or comprehensive speculations about the processes of environmental degradation were taking shape on the new European island colonies. These speculations eventually focussed on the connections between deforestation and climatic change.

From the twelfth century onwards the Spanish and Portuguese colonisation of the Canary islands, Madeira, and, after 1492, the West Indies started to lead to a developing awareness of the capacity of the colonist to decimate both indigenous populations and forest cover. The high rate of deforestation in Madeira and the Canary islands was directly related to the remarkable growth in the emergent European urban market for sugar products and the opportunities for profits presented by the development of sugar plantations. Faced with the consequences of deforestation and the loss of hydraulic storage

22 See E. Mummenhoff, *Altnurnberg* (Bamberg, 1890), 55–57.

23 Ellen C. Semple, *The Geography of the Ancient Mediterranean* (London, 1932).

24 For a wider discussion of these matters, see F. C. Lane, *Venetian Ships, Shipbuilders and the Renaissance* (Baltimore: University of Baltimore, 1934).

25 See Venice Archives (Arsenale); basta 8, fl./9/10

26 The Venetian records show that local peasant opposition to state conservation, featuring even active sabotage and incendiarism, contributed to this policy. The phenomenon of popular resistance to colonial forest policy was repeated over and over again in the context of much later colonial conservation policy, especially in Africa and India. See Richard Grove, "Colonial Conservation, Ecological Hegemony and Popular Resistance," 15–57.

capacity, the colonists and planters had to resort to a wholesale development of artificial irrigation systems constructed by Genoese engineers but based on the designs of Moorish irrigation systems in Spain.[27] The new level of deforestation also gave rise to a more generalised theoretical response. In particular, Theophrastus' classical theories were circulated once more with the Renaissance, while the publication of his works in Latin in 1483 assured their wider dissemination.[28] Christopher Columbus, for example, is reported by his son, Ferdinand, to have warned constantly of the dangers of forest clearance on West Indian islands, such as a decline in rainfall.[29] Such notions may have arisen from the explorer's own observations of the acute water shortages in the Canary islands after the forests were cut over. These observations are likely to have been coupled with a knowledge of Theophrastus' theories.

Thus on the earliest European island colonies the destructive effects of capital-intensive economic activity first became fully apparent and elicited an environmental critique. Significantly too, at about this time the classical idea of the island as an Eden or social refuge was revivified, much as Theophrastus' ideas had received a new lease of life at the time of Columbus. Indeed the conviction that an Eden might be found across the ocean may have sustained Columbus in his transatlantic exploit.[30] Long before this, in *Purgatorio,* Dante had already transferred classical notions of Paradise or the Hesperides to the Atlantic in the form of an "earthly paradise" or "a island in the Southern Ocean."[31]

During the fifteenth century too, the developing Renaissance conception of the botanic garden as a symbolic location for the recreation of Paradise took firm root, displacing the simpler apothecarial connotations of the early medieval herbal gardens.[32] In this way the Italians (and shortly afterwards the Dutch, French and English) adapted and elaborated upon Arabic, Persian, and Mughal garden models, thus structurally incorporating notions of *Paradaeza* (meaning, literally, in Farsi/Ancient Iranian, an enclosed garden) associated with the gardens of, for example, Mesopotamia and Moorish Spain.

All these changes formed part of an important Renaissance shift in attitudes away from identifying earthly Nature with the fallen state of man and towards a new conception of nature as an intended and beneficent part of divine

[27] R. Bryans, *Madeira, Pearl of Atlantic* (London, 1954), 30; and A. M. Watson, *Agricultural Irrigation in the Early Islamic World* (Cambridge: Cambridge University Press, 1983).

[28] Theophrastus' *Historia Plantarum* was translated into Latin in 1483, according to I. H. Burkill (*Chapters in the History of Indian Botany* [Government of India Press, Delhi, 1965]).

[29] K. Thompson, "Forests and Climatic Change in America; Some Early Views," *Climatic Change, 3* (1983), 47–64.

[30] For a detailed discussion of this, see Ernst Bloch, "Geographical Utopias," ch. 2 of *The Principle of Hope* (Oxford: Basil Blackwell, 1986).

[31] See A. Keymer, "Plant Imagery in Dante," (M.Phil. dissert., Faculty of Modern Languages, University of Cambridge, 1982).

[32] J. Prestt, *The Garden of Eden; The Botanic Garden and the Recreation of Paradise* (New Haven, 1981).

purpose. By about 1550, therefore, many of the major intellectual and metaphorical themes had emerged which were later to form the infrastructure for a new kind of environmentalist response to evidence of the Europeans' destructive capabilities in tropical climates. Although the conditions of European protocapitalism (with its nascent urban market for luxury products from the tropics) stimulated the onset of tropical plantation agriculture and the tropical timber trade, a new valuation of nature (symbolised at first in the re-creation of Paradise in the botanic garden and closely associated with the geographical search for the Garden of Eden) provided a philosophical and quasi-theological basis for an interventionist response to environmental destruction. As a part of this response, it is noteworthy that two European precedents for land management, new irrigation technology and the antecedents of the botanic garden, played a significant part. More important, the paradise metaphor, particularly in its stereotyped association with newly colonised and previously uninhabited desert islands, acquired a transitional symbolic role in the more complex use of the image of island as a device for representing religious or social utopias. Although Thomas More had pioneered the use of this utopic form as the basis for social critique, it received a new lease of life with the spread of a distinctively Calvinistic willingness to perceive and create earthly paradises, a development reflected in the emergence of a whole genre of desert island literature in English and French. Bishop Francis Godwin's *The man in the moone* is a case in point. The book conceived of the island of St. Helena as a paradise and pointed to the limitations of its resources in sustaining visiting Europeans. By fictionally representing his hero, Diego Gonzales, as a man who moves physically away from island (a convenient metaphor for the world), circles the world in a flying machine, and then views the world from the moon, Godwin underlined the new physical and interpretative power of man relative to the world and the cosmos. Implicitly, then, Godwin portrays man as having acquired both a new capacity for destructiveness and a capacity, in the terms of the seventeenth-century scientific revolution, to understand and control his actions. Appropriately, both these new capacities began to be displayed on St. Helena in the late seventeenth century. However, the development of a utopic, yet empirically based, literary genre became of great consequence in conceptualising the growing human impact, especially among the adherents of such utopian writers as Defoe and his devotee, Jean-Jacques Rousseau.

The rapid expansion of European plantation agriculture in the West Indies did promote some attempts at conservation during the seventeenth century. Often, however, the response to environmental degradation involved a progressive westward displacement from one Caribbean island to another as these environments were successively exhausted by plantation cropping. It soon became clear that the very destructive and visually dramatic impact of plantation agriculture on the soils of a tropical island required either outmigration and relocation of the whole plantation economy or the development of a systematic conservationist response. Proximity to other islands and to South America meant that the former course was often adopted in the West Indies. The Dutch, for example, in a number of instances, shifted their plantation activities from island to island and then, eventually, to Brazil.[33] As David Watts has recently shown, conservationist forms of soil erosion control did develop in the West Indies during the seventeenth century.[34] However, they were devoted more to ensuring the continuation of plantation activity than the survival of any forest cover. Those colonial officials, such as the Governor of Barbados in the 1670s, who expressed more comprehensive concerns and plans for conservation, were generally ignored by their masters at the metropole.[35]

The emergence in the seventeenth century of a new kind of capitalist structure, the joint stock trading company, as a new, flexible, and more capital-intensive element in European expansion helped to promote plantation and trading activities at a much greater distance from Europe than before. This provided, too, the conditions for further far-reaching environmental changes. The new companies, often reinforced by the growing ambitions of European states, began to transform some of the often uninhabited oceanic islands, earlier treated simply as stopping places for the ships of many nations, into the locations of more permanent kinds of settlement for military and plantation development, as well as timber extraction. By 1670 St. Helena, Mauritius, and Formosa had all been colonised by the Dutch or the English. It was on these islands, located along the main trading routes to India and China, that sophisticated forms of state conservation first emerged. In some respects the Cape of Good Hope can also be treated physiographically as an island, with a parallel and early growth in the state's awareness of environmental constraints.[36]

The tropical island colonies were generally small and the consequences of deforestation, especially in terms of streamflow decline, could be quickly observed and thus acted upon. In this way insular environments stimulated both innovative interventionist responses and theorisation in environmental terms. The history of St. Helena illustrates this particularly well.[37] Annexed by the English in 1659, the island was extremely isolated. This extreme

[33] J. Edel, "The Brazilian Sugar Cycle of the Seventeenth Century and the Rise of West Indian Competition," *Caribbean Studies, 9* (1969), 24–44.

[34] D. Watts, *Environmental Change, Slavery and Agricultural Development in the Caribbean since 1492* (Cambridge, 1985).

[35] This stood in stark contrast to the official response developing outside the European domain, particularly in Japan. Here, as Conrad Totman has recently revealed, rapid population increases, urban growth, and fuelwood demand had led to a widespread ecological crisis by the mid-seventeenth century. The Tokugawa Shogunate responded by initiating an elaborate forest conservation and reafforestation programme designed both to safeguard timber reserves and prevent soil erosion. These measures do, however, not appear to have become known to European contemporaries. See C. Totman, *The Green Archipelago*.

[36] See Grove, "Early Themes in African Conservation," 21–39.

[37] For an overall historical treatment, see P. Gosse, *St. Helena, 1502–1938* (London, 1938).

isolation, combined with its central importance as a secure watering and provisioning point for the East India Company ships, placed the island's resources under heavy pressure. By 1700 St. Helena had become extensively deforested, and soil erosion and deep gullying were widespread. The history of this progressive degradation is recorded in minute detail in Company diaries and in correspondence between the governor and the EIC Court of Directors in London,[38] which also elaborately record the story of the innovative early attempts made to impede further deterioration.[39]

In less remote locations the island would have been quickly abandoned. Indeed, in 1716, a newly arrived Governor of St. Helena canvassed the possibility of resettling the inhabitants on Mauritius, which the Dutch had recently vacated.[40] The continuing importance of St. Helena to the Company as a replenishment point meant that this could not realistically be allowed to happen. Between 1670 and 1770, therefore, almost every successive governor on the island attempted to control the ecological changes which were taking place. The most ambitious efforts were those made by Governor Roberts in 1709, who set up forest reserves and afforestation programmes, introduced an irrigation system, and attempted to exterminate the goat population, which he had identified as peculiarly destructive.[41] Furthermore, he and his successor attempted to transplant specimens of the increasingly rare St. Helena redwood in the fear that it was about to become extinct.[42] The efforts of the Governors, amongst the earliest made in recognition of the likely extinction of an endemic species, were successful. Later Governors, particularly John Byfield in the 1730s, imitated Roberts' conservationist methods and philosophy.[43] In so doing they found themselves constantly at odds, as Roberts had been, with the relatively unschooled opinion of the East India Company's Court of Directors. Indeed, throughout the eighteenth century the Company seemed quite incapable of understanding the environmental consequences of introducing plantation schemes into the island and was especially unwilling to support the Governors in their efforts to regulate the environment of the island through protecting the forests and corralling animals. Each succeeding Governor quickly found himself siding with the settler freeholders against the Company on a wide variety of land-use issues, not the least being the recognition of the need for communal land-use practices, to which the islanders had consistently

38 For a detailed compilation of the St. Helena records, see H. R. Janisch, ed., *Extracts from the St. Helena Records* (Jamestown, St. Helena: Government Printer, 1908).

39 For a detailed account, see Grove, "Conservation and Colonial Expansion," ch. 3; see also Q.C.B. Cronk, "The Historical and Evolutionary Development of the Plant Life of St. Helena," (PhD thesis, no. 13567, Cambridge University, 1985, 53–88.)

40 "Letter," St. Helena Council to Court of Directors, dated 19 Feb., 1715, in Janisch, *St. Helena Records*, p. 113.

41 Council to Court of Directors, dated November 1708, in Janisch, *St. Helena Records*, p. 85.

42 *Ibid.*, 88.

43 St. Helena Council Diary, dated Oct. 23, 1745, in Janisch, *St. Helena Records*, p. 183.

adhered despite all the efforts of the Court of Directors.[44] Despite the conservation policies pursued by Roberts and his successors, the sheer scale of the erosive processes at work on St. Helena and the lack of any relevant environmental insights derived from the outside world meant that, when Sir Joseph Banks visited the island in 1771, he was shocked at what he saw and contrasted it unfavorably with the situation at the Cape Colony. There, he asserted, the Dutch colonists had made a "Paradise of a desert," while the Company had allowed the "Paradise" of St. Helena to "become a desert."[45]

The central difficulty for the Governor-conservationists of St. Helena was that they had to deal with a Company hierarchy that, before the 1790s, possessed no insights into the impact of European capital-intensive and slave-employing agricultural methods on a tropical landscape or into the destructive hydrological consequences of tropical deforestation. In direct contrast to the French state at the same period, the EIC had not at this stage yet developed any serious commitment to the commercial or practical cultivation of scientific expertise.[46] A further problem related to the divergent traditions of the state's involvement in forestry in England and France. In England the state's role in forest regulation, particularly during and after the Commonwealth period, became severely restricted. As a result, that country was forced increasingly to rely upon timber supplies from the Baltic region and North America.[47] In France, by contrast, the state itself acquired a more direct role in forest control under Colbert's direction.[48] Thus in the period from 1660 to 1760, as a consequence of these differences, two quite separate models of tropical forest management developed on St. Helena and Mauritius respectively. A distinctive forest conservation system was also developed at the Cape but was cut short by the onset of British rule.

In France, the provisions of the 1669 Forest Ordinance ensured that forest management was treated in conjunction with water management and administered by Inspecteurs des Eaux et Forets.[49] Thus, from the outset, French approaches to the environment of Mauritius involved a well-developed awareness of the possibility of ecological regulation and control. The Dutch had already exploited intensively the ebony for the market in luxury island

[44] *Ibid.*, 105–6.

[45] Sir Joseph Banks, diary entry, dt May 1771 in J. C. Beaglehole, ed., *The "Endeavour Journal" of Sir Joseph Banks, 1768–1771*, vol. 6 (Sydney: Sydney University Press, 1968).

[46] Probably the East India Company's earliest formal scientific appointment was that of Dr. Johann Koenig as "Company Naturalist" at Madras in 1778. Koenig had earlier held a post as "Naturalist" to the Nawab of Arcot. The much earlier instance of patronage by the Company represented by the sponsorship of Edmund Halley on his St. Helena expedition in 1676 should perhaps be discounted for this purpose.

[47] J. Thirsk, ed., *Agricultural History of England and Wales*, vol. 5, 375–6 of (Cambridge: Cambridge University Press, 1985).

[48] John Croumbie Brown, *The French Forest Ordinance of 1669* (Edinburgh, 1876).

[49] Jean-Claude Wacquet, *Les Grandes maitres des eaux et forêts de France de 1689 à la Revolution* (Paris: Librarie Droz, 1978).

woods during the earlier period of Dutch rule.[50] By contrast, the Dutch management of the Cape environment, at least on paper, was highly regulated and cautious, mirroring the careful organisation needed to husband and manage the limited resources of the Netherlands.[51] Dutch Mauritius, on the other hand, had been treated merely as a source of high-value timber and was abandoned once the forests easily reached near the coast were depleted. Indeed, the exclusively depletionary management of Mauritius seems to have been the exception to the general rule of Dutch caution. Deliberate and systematic working of the teak forests in Java, on an apparently sustainable basis, had already been commenced by the Dutch East India Company (VOC) in 1677.[52] Such a policy clearly depended on a considerable degree of territorial control of a kind not yet achieved in Asia by the French or British during this period.

When the French annexed Mauritius in 1721, the commercial rivalry between Britain and France had begun to sharpen. This led directly to a intensive contest for control over sources of marine timber for the navies needed for the growing struggle for global trade and influence, particularly in the Indian Ocean.[53] It was in this setting that the financial collapse of the Compagnie des Indes after 1763 elicited the direct intervention of the French Crown in the government of Mauritius and a more direct attempt at intervention in the control of the strategically important forests.[54] The fact that the French possessed no equivalent to the vast forests of North America, which were now accessible to the British, further stimulated a cautious attitude towards timber resources, particularly on Mauritius (the Isle de France) and Reunion (Bourbon).

An analogous process had been taking place in India itself. From as early as AD 800 a gradual pattern of state forest control had been emerging.[55] Rein-

50 J. R. Brouard, *The Woods and Forests of Mauritius* (St. Louis, Mauritius: Government Printer, 1963), 1–12.

51 T. R. Sim, *The Forests and Forest Flora of the Colony of the Cape of Good Hope* (Edinburgh, 1907), 76–80.

52 For an overall literature survey, see P. Boomgard, "Forests and Forestry on Java, 1677–1941," in J. Dargavel, ed., *Changing Tropical Forests, Historical Perspectives on Today's Challenges*, (Canberra: Australian National University, Centre for Resource and Environmental Studies Special Publication, 1989). Far more detailed accounts are contained in numerous articles by E.H.B. Brascamp, in particular "Hourlevanties onder de O. I. Compagnie, De Aatc van 21 Juin desor de soeshanan aan de O. I. Compagnie verleend tot het Kappen van hout-weiken in de bosschen van blora [uit het koloniaal Archief No. XLX11]," *Tijdschrift voor Indische taal, land en Volkenkunde van het Koninklijke Bataviaasch Genootschap van Kusten en Wettenschappen*, 52 (1932), 108–12.

53 See R. C. Albion, *Forests and Sea-Power; The Timber Problem of the Royal Navy, 1652–1862* (Harvard, 1926).

54 The dynamics of French naval timber demand are dealt with in P. W. Bamford, *Forests and French Sea-Power, 1660–1789* (Toronto: Toronto University Press, 1956), 88–102.

55 Some idea of the early stages of this process is contained in Diprakanjan Das, *Economic History of the Deccan* (Delhi: Munshiram, 1976), 105–115.

forced by the advent of Mughal rule, this shift away from communal forest use towards state intervention was actually accelerated by the ascendancy of the successor states to the Mughals, particularly on the West Coast of India, where timber harvesting was most intensive. Thus, as early as the 1740s, the rulers of the Maratha empire had found it expedient to acquire control over large tracts of coastal forest and to set up plantations, both for shipbuilding and revenue.[56] In Cochin and Travancore similar monopoly controls over forest were initiated.[57] However, the most extensive pre-colonial programmes for afforestation and forest protection were carried on, between 1770 and 1840, by the Amirs of Sind.[58] Such pre-colonial policies of state forest control in India are historically important, not least because the British used them extensively to legitimate new attempts at forest reservation in the Bombay Presidency during the 1840s.[59]

Thus, by about 1770, throughout many regions of South Asia, the Mascarenes, and the East Indies, a widespread emergence of state forest controls had become clearly apparent. This phenomenon developed as relatively integrated forces of economic penetration began to respond to new levels of trade and demand, particularly in the search for timber to satisfy the increasingly complex demands of both western and indigenous navies and growing urban centres.[60] In Mauritius the demands of the colonial state took a very distinctive form. A continuing timber shortage in France and the growing size of the French navy meant that Labourdonnais, Governor of Mauritius after 1731, found it worthwhile to construct and repair ships locally, utilising Mauritius teak. A fleet of these ships was then used, successfully, to attack Madras in 1746.[61]

[56] *Report of the Bombay Forest Commission, 1887*, vols. 1 and 2 (Bombay: Government Press, 1887). The taxation features of Maratha forest policy are an unexplored field in research terms; however, for a cursory survey, see H. B. Vashishta, *Land Revenue and Public Finance in Maratha Administration*, (Delhi, 1975, 138–46).

[57] Until the end of the eighteenth century, the forests of Cochin were under the control of the feudal chiefs of Nadivazlis, who owed allegiance to the Rajah of Cochin. In 1813 a forest department was set up under a Mellei Melvicharappan [Mountain Superintendent]; see H. Viswanath, ed., *Working Plan for Chakakuan* (Trivandrum: Forest Department, 1958), pp. 12–13 [Ref. Trav. 0. 3, Indian Institute, University of Oxford]). For details of the Travancore timber monopolies and early forest conservancy, see F. Bourdillon, ed., *Report on Travancore Forests* (Trivandrum: Government Printer, 1886), 15–16.

[58] A short account of these is given by H. T. Lambrick in *Sir Charles Napier and Sind* (Oxford: Clarendon, 1952), 22–24, 192–3.

[59] E.g., see Alexander Gibson, "Description of the System Adapted for the Forest Conservancy of the Bombay Presidency," in A. Gibson, ed., *A Handbook for the Forests of the Bombay Presidency*, (Byculla: Government Printer, 1863). Gibson states, for example, that "Teak and "Junglewood" have been carefully preserved ever since this tract of country came under the rule of the Angria, and even, I think, dates from the taking of Sevendroog by Admiral Watkins."

[60] A detailed account of the impact of new naval demand is found in R. A. Wadia, *The Bombay Dockyard and the Wadia Master Builders* (Bombay, 1957).

[61] Brouard, *Woods and Forests of Mauritius*, 12.

However, naval demand was not the only destabilising factor on Mauritius. Under Labourdonnais the ecological distortions that had developed in an island environment subjected to the demands of a plantation economy gave rise, as they had on St. Helena, to anxieties about the capacity of this climatically unreliable island even to produce sufficient food for its settler population and slave-labour force.[62] These fears led Labourdonnais to develop a new botanic garden with the specific objective of breeding drought-resistant crops to tide the population over the leaner years. This type of institutional concept was later imitated and elaborated by Colonel Robert Kyd as justification for founding a botanic garden at Calcutta in 1786, an innovation which later became vital in providing the infrastructure for the scientific monitoring of environmental changes in India.[63]

The most significant intellectual departure in French environmental policy on Mauritius, the installation of a forest reservation system, was explicitly based on the fear of the impact of deforestation on rainfall and soil erosion. This was a direct consequence of the assumption of direct political control by the Crown and the rise to political influence in government, especially in the Navy ministry, of a group of men strongly attracted by Physiocracy, a political and economic philosophy which had originated in the writings of economist Richard Cantillon and physician Francois Quesnay. Significantly, it attached particular importance to the working of the "laws of nature" in economics.[64]

The political ascendancy of the Physiocrats facilitated the appointment of Pierre Poivre as Commissaire-Intendant of Mauritius in 1766. Poivre had evolved a conservationist approach to the environment essentially as matter of moral economy in a manner characteristic of the Physiocrats. A proper understanding of the basis of the environmental moral economy which he advocated necessitates an appreciation of the antecedents of the man himself.[65] Poivre was at one time a Jesuit priest and had first visited Mauritius in the 1740s. He then used the island as the base for a series of expeditions to collect spice tree specimens from the Moluccas.[66] A primary part of this exercise involved attempts to transplant imported species of spices in Mauritian soil. This had

62 R. Toussaint, ed., *Dictionary of Mauritian Biography*, 154. See also; Grove, "Conservation and Colonial Expansion," 88–102.

63 R. Kyd to Company Board, Fort William, Madras, dated 15 April 1786, quoted in "Pro ceedings of the Supreme Council relative to the establishment of a botanic garden at Mackwa Tannah," Calcutta, in India Office Library, Home Miscellaneous, No. 799, pp. 1–207.

64 H. Higgs, *The Physiocrats, Six Lectures on the French Economistes of the Eighteenth Century* (London, 1897).

65 For biographical details of Poivre, see L. Malleret, *Pierre Poivre* (Paris: L'Ecole Francais D'Extreme Orient, 1974).

66 These journeys are detailed in Madeleine Ly-Tio-Fane, *Mauritius and the Spice Trade: The Odyssey of Pierre Poivre* (Port Louis: Mauritius Archives Publication Fund, 1958); and "Pierre Poivre et l'expansion français dans l'Indo-Pacifiques," *Bulletin de l'Ecole Français d'Extreme Orient* (Paris, 1967) [Extrait du Tome L111 Fasc. n. 2].

led Poivre directly to an interest in soil conditions, soil humidity, water table levels, and the desirability of maintaining an extensive protective tree cover so that agriculture on the island might prosper. He had become especially appreciative of the attempts made by the Dutch at the Cape in soil erosion prevention and in planting shelter belts.[67]

Poivre became very critical of what he saw as the profligate and destructive felling of trees on Mauritius, maintaining that the lack of control contrasted badly with Dutch practices at the Cape. Poivre had, furthermore, made himself familiar, during a period of enforced detention in Canton, with Chinese horticultural methods.[68] Later, too, he became acquainted at Pondicherry and in Malabar with indigenous Indian techniques for irrigation and planting trees and referred extensively to this knowledge in his writings.[69] Poivre's return to France coincided with the rise in prominence of the *agronome* and a period of intense activity in Enlightenment science. After his arrival, he added to his unrivalled knowledge of tropical land-use practices in a wide-ranging survey of travel and scientific literature on the Americas in which he probably first encountered dessication theories.[70] Significantly, many of his writings during this period contained biting critiques of the Europeans' treatment of indigenous peoples in the Americas, and his extensive lectures on tropical land use and agriculture aimed particularly scathing criticism at colonial land-use methods.[71] These criticisms soon came to the attention of the Physiocratic political circle surrounding the Duc de Choiseul, who was himself closely associated with Physiocratic thinking and attracted by a philosophy which set great store by notions of the moral superiority of agricultural versus financial prosperity and by the virtues of scientific agriculture versus short-term profit-making.[72] The Duc de Choiseul had, however, found it politically impossible to assert his reforming Physiocratic zeal in any substantial way in France itself, so he increasingly turned to the colonies in the hope of exercising his reforming ideals, much as Lord Dalhousie did in India in the next century.

The encounter with Mauritius gave both de Choiseul and Poivre the opportunity to experiment with their ideas. Choiseul could exercise the role of the

[67] Pierre Poivre, *Travels of a Philosopher; or Observations on the Manners of the Various Nations in Africa and Asia* (Dublin, 1770).

[68] On the value of empirical observations of the Orient, see P. Poivre, "Utilite d'un voyage dans l'Orient," in Poivre, *Citoyen du Monde; ou lettres d'un philosophe à ses amis dans l'Orient* (Amsterdam, 1763), 172.

[69] L. Malleret, ed., *Un manuscrit de Pierre Poivre; les memoires d'un voyageur*, (Paris: Ecole Francais d'Extreme Orient, 1968), 113–4.

[70] P. Poivre, *D'Amerique et des Americains* (Berlin, 1771).

[71] Poivre's seminal metropolitan interlude is covered in Y. Laissus, "Note sur les manuscrits de Pierre Poivre (1719–1786) conserves a la bibliotheque centrale du Museum National d'Histoire Naturelle," in *Proceedings of the Royal Society of Arts and Sciences of Mauritius*, vol. 4, Pt 2, (1973), p. 37 and p. 37, n. 2.

[72] H. Higgs, *The Physiocrats*, 78.

state in reform by putting Physiocracy into practice, and Poivre could carry out a programme of transplantation and conservation. "If the forests do not regenerate in the island," Choiseul wrote in his instructions to Poivre,

> rain will be less frequent, the over-exposed soil will be burnt by the sun. . . timber shall be your particular attention. . . . Complaints have been made that this very important matter has not received all the care and attention which it deserves. Messrs Dumas and Poivre should hasten to control it with a good policy; they will exercise the existing regulations on the subject, study the exact conditions of the forests, exploit and utilise them in the most economical way possible and only allow people to cut them if they ensure their conservation.[73]

On his arrival in Mauritius in 1767 Poivre immediately made a series of speeches and proclamations to the French colonists, many of which specifically emphasised the high priority attached to preserving the island forests.[74] This, he asserted, would assure the colony a reliable level of rainfall. A morality expressed in land use, Poivre implied, would complement the religious morality of the individual. These environmental homilies were later published as revolutionary tracts in Paris in 1797, probably less for their revolutionary environmentalism than for their class-based critique of colonial absentee landlordism.[75] The desire to make a clean sweep of colonial conventions did not stop with plans for environmental reform. Poivre also took initial steps in a campaign to abolish slavery on the island, a cause which he was unable to carry through. Indeed, he experienced much opposition both in this respect and in his conservation programmes, not least from a succession of island Governors who felt threatened by his attempts to stamp out timber and property speculation by colonial officials for private gain. At the end of 1768 the departure of Governor Dumas, disgraced for his corrupt involvement in land sales, allowed Poivre to carry on his conservation programmes unhindered. The first Forest Conservation Ordinance, or *Reglement Economique*, passed in November 1769,[76] proved to be a model of its kind. During the next century, its essential elements were incorporated in future statutes in, for example, St. Vincent (West Indies), the Cape Colony, Natal, and India.[77]

73 R. Brouard, *Woods and Forests of Mauritius*, 10.

74 *Discours prononces par M. Poivre, Commissaire du Roi, l'un a l'assemblee generale des habitans de l'Isle de France lors de son arrivee dans la colonie, l'autre a la premiere assemblee publique du Conseil Superieur nouvellement etablie dans l'isle* (Imprimerie Royale, Port Louis, 1768, 50 pp). [This publication is very rare, but a copy is at the Auguste Brunet Library, Toulon.]

75 See J. Salles, ed., *Oeuvres completes de Pierre Poivre* (Paris, 1797); A copy of this book is held at the University Library, Cambridge, United Kingdom.

76 The separate articles are compiled in (Anon.) *Reglement economique sur le défrichement des terres et la conservation des bois de l'Isle de France* (Imprimerie Royale, Port Louis, 1769); [Auguste Brunet Library, Toulon.] The rules were signed by Governor Desroches as well as Poivre himself.

77 Details of this transferrence are found, for example, in Government of Natal, *Report of the Committee Enquiring into the Extent and Condition of the Forest Lands of the Colony* (Pietermaritzburg: Government Printer, 1880) [held at Natal Provincial Archives, Pietermaritzburg, Natal, South Africa]. Evidence of the diffusion of knowledge of Mauritius forest protection

In his pioneering conservation efforts, Poivre enlisted the aid of two men in particular, Philibert Commerson and J. H. Bernardin de St. Pierre. Poivre contrived to secure an appointment on the island as a government scientist for Commerson, who had arrived in Mauritius in November 1768 as the naturalist on the Bougainville expedition. Officially Commerson was intended to carry out a pharmacological and timber resource survey of the Mauritius forests.[78] These surveys provided the justification for Poivre's far-reaching conservation legislation of 1769, who was also backed in his efforts against the anti-conservation interests on the island and in Paris by the Chief Engineer of the colony, Bernardin de Saint Pierre. Later better known as one of the first French romantic novelists and orientalists, Saint Pierre was a pioneer figure in campaigns against slavery, a practice which he had first witnessed on Mauritius.[79] It was Saint Pierre who first pointed out to Poivre the extent to which deforestation on the island was contributing to the rapid siltation of the main rivers and harbours of the island. This observation led Poivre to develop the twin concepts of the mountain reserve and the river reserve, the intellectual antecedents of ideas which were widely applied in India after 1846. These steps were only partially successful in inhibiting the rate of deforestation on the island which, towards the end of the eighteenth century and increasingly during the nineteenth century, came under exceptional pressure as a result of the activities of sugar plantation owners. However, had Poivre not introduced his conservation system, there is little doubt that deforestation would have proceeded at a much more rapid pace. Even today the boundaries of some of the reserves established by 1800 can still be seen, particularly along the west coast of Mauritius, adjacent to Black River and Tamarin.

Most of the conservationist initiatives taken by Poivre were essentially locally derived. The critical innovation contained in the new conservation measures, while based in a broad sense on Colbert's 1669 Forest Ordinance, consisted of the articulation of new insights into the relationship between forest cover, stream flow, soil erosion, and rainfall. Poivre's most effective argument to justify the sterilisation of forest capital as reserved land involved a conviction that deforestation would cause a decline in rainfall. However, in evolving a methodology for conservation in 1769, Poivre's precedents were only partly European. His tree-planting policies, for example, were based entirely on Indian precedents, and he utilised species imported from Malabar and Bengal, especially *bois noir*.[80]

---

regulations into India as early as 1844 is provided in W. H. Sleeman, *Rambles and Recollections of an Indian Official* (London: J. Hatchard, 1844), 450.

[78] A certificate releasing Commerson to the service of Poivre's administration was delivered to him by Bougainville on 15 November 1768. For details, see Series designated 'Col-E.', Commerson's File, F1., Archives Nationales, Paris.

[79] See J. H. Bernardin de Saint Pierre, *Voyage a l'Isle de France* (Paris, 1773). I have referred to the edition published in London in 1800 as *A Voyage to the Isle of France*.

[80] This was probably *Dalbergia latifolia*.

Although for political purposes Poivre put forward economic and pharamacological reasons for preserving forests and contriving legislation, other motivations which were almost certainly more significant to the Intendant himself were also involved. These fell roughly into two kinds: the scientific and botanical and the unashamedly utopian. The presence of one of the largest and most elaborate botanical gardens in the world, sometimes known as a "wonder of the East," and the presence of Commerson, one of the most experienced and well-travelled French botanists of the period, meant that the rich and endemic flora of the island was soon recognised as scientifically important in world terms, highly valued aesthetically and worthy of preservation.[81] This alone constituted a significant impetus for the embryonic conservation lobby. Further encouragement to protect the island in its natural state was provided by the individual predilections of St. Pierre and Commerson. Both were avid field botanists steeped in the writings of Rousseau (another contemporary botanist, who is better known in other roles).[82] Both were also social rebels who valued the natural state of the island as integral to their conception of the colony as a practical and mental refuge from the political and, for them, moral turmoil of metropolitan France. As a member of Bougainville's expedition, Commerson had the opportunity to observe the inhabitants of Tahiti (which he called *utopie*) and believed their society idyllic in social and environmental terms. His writings indicate that by 1768 he was already very critical of metropolitan France and its prevailing social values. He had little desire to return to France and, indeed, never did so before he died on Mauritius in 1773. Bernardin de Saint Pierre, trained in botany by Poivre himself, returned to France to write *Paul et Virginie* and to become the main confidant of Rousseau, whom he had long admired and corresponded with. His manifestly ecological views were set out in three major texts, *Journey to the Isle de France, Studies in Nature* (in which *Paul et Virginie* formed the fourth volume) and *Harmonies of Nature*. Poivre, Commerson, and Saint Pierre all arrived in Mauritius at a time when the French preoccupation with the "Robinsonades" (imitations of Defoe's *Robinson Crusoe*) was developing into a fully developed cult or mythology of the South Sea tropical island. The desire of all three men to preserve an unspoilt utopian island environment necessitated some form of environmentalism, simply to ensure the continuity of the Rousseauian return to nature to which they were dedicated.[83] None of these motivations would have been transmuted into the conservation policies pursued by Poivre had it not been for the opportunity which the Physiocrats

81 This valuing was of a very personal kind on Commerson's part: "My plants," he wrote, "my beloved plants have consoled me for everything; I found in them, Nepenthes, curare, dulce" (S. Pasfield-Oliver, *The Life of Philibert Commerson* [London, 1909], 202).

82 For details of Rousseau's botanical enthusiasms, see J. J. Rousseau, *Letters on the Elements of Botany Addressed to a Lady*, T. Martyn, trans. (London, 1782).

83 See Grove, "Conservation and Colonial Expansion," 154–82.

had provided to Poivre in political terms. This permitted him free rein to pursue his environmental programme and his innovatory state patronage of scientific expertise.

The emergence of conservationism on Mauritius under Poivre might be considered of merely antiquarian interest, but it was not so. The Isle de France lay intellectually and geographically, as did Tahiti at the same period, at the very core of the outward-looking intellectual and Romantic projects of late-eighteenth-century France. One of the earliest major literary figures in the French Romantic reaction, Bernardin de Saint Pierre above all ensured that Mauritius became a central and symbolic focus of contemporary aesthetic and Orientalist-leaning perception.[84] As a product of this intellectual climate, Poivre's conservationism itself left an intellectual legacy of environmental attitudes both in Mauritius itself (which remained in the forefront of conservation ideas until the 1870s) and throughout a much wider colonial context.[85] After Poivre left the island for France in 1772, his legislation for forest conservation was strengthened in 1777 and 1795, then much further elaborated in 1804, at a time when the colony had acquired effective political autonomy under Governor Decaen.[86] Political isolation and naval blockade then encouraged the passing of additional pioneering environmental legislation in the fields of water pollution, fishery protection, and safeguarding watersheds. As was the case much later in India, the state botanic garden set up by Poivre at Pamplemousses provided an essential part of the intellectual and technical infrastructure needed for these innovations. Furthermore the notion of the colonial scientific *Académie,* first conceptualized in 1770 by Philibert Commerson, ensured a continuity of scientific tradition developed further in the form of the Royal Society of Arts and Sciences of Mauritius.[87]

Both in Mauritius and later in India and Southern Africa the existence of such local scientific institutions proved a critical stimulus to the formulation of colonial responses to environmental change and provided a platform for conservation propagandists, such as Louis Bouton in Mauritius and Edward Balfour in Madras.[88] In contrast to the situation on St. Helena until the 1780s,

[84] See B. de Saint Pierre, *Studies of Nature,* 5 vols. (London, 1796); *Harmonies of Nature,* 3 vols (London, 1815); *Paul et Virginie* (London, 1841); *La Chaumière Indienne* (Calcutta, 1866).

[85] For specific details of the links between Poivre's conservation policy on Mauritius and the environmental philosophy of Edward Balfour, the original proponent of forest protection in the Madras presidency, see IOL, V/27/568/107 [Correspondence files on trees and the incidence of rainfall].

[86] The main articles were Arrete of the 13th Messidor, Art. 111 (July 1795); and Arrete of the 14th Vendemiaire, Art X111 (October 1804). It was this last law which the British incorporated on their annexation of the island in 1810.

[87] M. Ly-Tio-Fane, "Notice historique" (Port Louis, Mauritius: Royal Society of Arts and Sciences of Mauritius, 150th Anniversary. Commemorative Publication, 1986).

[88] Bouton's conservation ideas were put forward in such publications as: "Sur le decroissement des forets a l'Ile Maurice," *Le Cernien,* April 12–14, 1838; "Note sur le decroissement des forets," in *Rapports de la societé d'histoire naturelle de l'Ile Maurice* (1846), 10. For Balfour's

the environmental analysis of the consequences of deforestation on Mauritius was based not only on local observation but on insights derived from a whole range of intellectual concepts developed in regions as diverse as France, the Cape Colony, China, and Malabar. However, Poivre had synthesized a coherent body of environmental thought, stimulated largely by the specific conditions of colonial rule in the tropics. However, until the 1820s no comparable intellectual development of an environmental consciousness had yet emerged in metropolitan Europe.

The first application of Poivre's ideas outside Mauritius itself took place, perhaps not surprisingly, on St. Helena. Surveys carried out by Alexander Beatson, an army officer and subordinate of Arthur Wellesley in Mysore, on Mauritius in 1794 indicated that by then the scientific rationale behind forest reservation on Mauritius was becoming more widely known.[89] Thus, in that same year the EIC somewhat belatedly instructed the St. Helena government to protect the forests on the island as a way of ensuring that the rainfall would not decline.[90] The establishment of botanic gardens at Calcutta and St. Helena in 1788, which imitated the French establishment on Mauritius, may have heightened EIC sensitivity to the potential damage that dessication might cause. Certainly the famines in Bengal between 1770 and 1784 had been a major anxiety to the Company, particularly to Captain Robert Kyd, who persuaded the EIC to fund a Calcutta Botanic Garden specifically to breed and test drought-resistant crops.[91] In 1808 Alexander Beatson was appointed Governor of St. Helena, where his programmes to plant trees and prevent soil erosion became well known through his own publications, particularly the *Tracts on St. Helena* (1816).[92] Knowledge of this activity and the improvements made by Beatson's programmes to the hydrological state of the island became important when the foundations were being laid for forest protection in India during the late 1840s.[93]

views, see E. G. Balfour, "Notes on the Influence Exercised by Trees in Inducing Rain and Preserving Moisture," *Madras Journal of Science and Literature*, 25 (1849), 402–48. Balfour had been stimulated in his advocacy of forest protection by his observations of the efficiency of the Mauritius conservation ordinances and by his readings of the works of Joseph Priestley on the water-holding dynamics of the atmosphere (see IOL V27/560/107, pp. 1–11).

89 A. Beatson, "Account of Mauritius," (unpublished ms, British Museum, BM13868, London, 1784).

90 Letter, EIC Court of Directors to Governor, St. Helena, dated Jan 23, 1794, in H. R. Janisch, *Extracts from the St. Helena Records and Chronicles of Cape Commanders* (Jamestown, St. Helena, 1908).

91 IOL, Home Misc., no. 799, letters on pp. 1–201.

92 A. Beatson, *Tracts Relative to the Island of St. Helena, Written during a Residency of Five Years* (London, 1816).

93 For details of J. D. Hooker's visit to Ascension island, see E. Duffey, "The Terrestrial Ecology of Ascension Island," *Journal of Applied Ecology*, 1 (1967), 219–36. A brief discussion of earlier afforestation programmes under the East India Company and the impact of tree-planting on rainfall in St. Helena is found in the *St. Helena Almanac*, vol. 1. (1848), 24.

THE INDIAN MEDICAL SERVICE AND THE EMERGENCE OF CONSERVATIONISM IN INDIA

The employment of medical surgeons as superintendents of botanic gardens in India had already expanded the professional role of the surgeon outside the medical service and laid the foundations for a technical infrastructure of scientific expertise. This infrastructure relied not only on the diverse medical and botanical skills of the surgeons but on their high level of university education and the considerable social status conferred by the possession of medical knowledge in a disease-ridden tropical environment. In Mauritius, in a similar fashion, the ascendancy of the scientific lobby and its success in putting its preferred policies into operation was based primarily on the ability to demonstrate convincingly that uncontrolled deforestation might threaten the security of the state itself. The high status afforded to science by the French Physiocratic interest in government facilitated this development.

On a much larger scale, but much more slowly, the views and influence of India's medical elite eventually acquires a similarly high status. Only gradually did this scientific lobby come to concern itself with the effects of forest clearance. However, when the medical surgeons did finally come to define deforestation as a matter requiring serious policy initiatives and an expansion in the role of the state, their opinions carried considerable weight. This was partly due to the long history of the medical service's consultancy to the state on all matters relating to health, water supply, scientific botany, and forest resources. The tours of Dr. A. P. Hove in the Bombay presidency (1787–88) were an early exemplar of the performance of this consultancy role.[94] Moreover, as early as 1778 the Company had begun to employ eminent surgeon-botanists, such as J. G. Koenig, Patrick Russell, and William Roxburgh in official capacities as naturalists, with the active encouragement, after 1784, of Sir Joseph Banks, the president of the Royal Society.[95] The subsequent systematic development of links between the EIC medical service and the scientific and botanical establishment in Britain served to reinforce the authority of the surgeons as environmental commentators.

By 1830 the vast majority of the 800 Company surgeons in India were Scottish.[96] Many had been trained in botany by Professors John Hope in Edinburgh and William Hooker in Glasgow, both keen advocates of rigorous field observation, holistic approaches to nature, and tree-planting pro-

[94] A. P. Hove, "Tours for Scientific and Economical Research Made in Guzerat, Kattiawar and the Conkans in 1787–1788," *Bombay Selections* (Bombay: Government Printer, 1855), vol. XVI, 50–185.

[95] For biographical information on the surgeons employed by the EIC, see D. G. Crawford, *The Roll of the Indian Medical Service* (Calcutta, 1930).

[96] H. H. Spry, *Modern India* (London, 1837), 55–8.

grammes.[97] Wherever attempts were made to establish conservancies or teak plantations in Malabar, in Bengal, or in Burma between 1805 and 1822 (in connection with war-time timber shortage), the Medical Service quickly became involved. As early as the 1820s some of the surgeons lobbied heavily against the deforestation taking place during that decade and argued in favour of plantation programmes. Nathaniel Wallich, the director of the Calcutta Botanic Garden, was pre-eminent among these early campaigners. "Unless speedy provision is made," he stated to a parliamentary committee in 1831, "for the renewal of timber, we shall, within a very short time, find a most painful falling off. I should say it is quite time that means should be resorted to, to preserve the forests that are remaining, and new plantations should be made."[98] In spite of these early efforts, a direct connection between decline in forest area and apparent regional increases in desiccation in India was not put forward as an argument for controlling deforestation until the end of the 1830s.

This delay can be partly attributed to a lack of evidence about rates of deforestation in India as a whole. Even so, throughout the 1820s and 1830s anxieties about deforestation were already frequently expressed. Bishop Heber, for instance, warned in 1824 that excessive deforestation, of the kind he had observed in the Siwalik foothills, might lead to a more general aridity.[99] A second reason for the delay may have been the paucity of articles in sufficiently authoritative technical publications on the links between deforestation and climate, despite the EIC's warning about desiccation to the St. Helena government as early as 1794.[100] Moreover, at this stage Mauritius's experience was still not widely known. In both colonies the essential limitations of resources on an island undergoing visually apparent deforestation and soil erosion stimulated action as well as theorisation. In contrast, no comparably coherent or comprehensive evidence of the human environmental impact was available at a sub-continental scale in India until the mid-1830s. Consequently, a more convincing body of evidence collected on an all-India basis was still required, as well as a theoretical basis for adjudging the likely consequences of deforestation. When a widely accepted analysis of the pos-

97 See I. H. Burkill, *Chapters in the History of Indian Botany* (Calcutta, 1965).

98 N. Wallich, Evidence to Select Committee 1831–1832, British Parliamentary Papers, Colonies, East India, (1831), vol. 10, Col. 735.

99 R. Heber, *Narrative of a Journey through the Upper Provinces of India from Calcutta to Bombay, 1824–5* (London, 1828), 274.

100 In that year the Court of Directors had warned the island Council that "we are of the opinion that encouraging the growth of wood is of utmost consequence to this Island not only from the advantages to be derived from it as a fuel, but because *it is well known that trees have an attractive power on the clouds,* especially when they pass over hills so high as those on your island and we are inclined to believe that the misfortunes the island has been subject to from drought may in some measure have been averted had the growth of wood been properly attended to," (my emphasis).

sible dangers of deforestation did finally emerge in the 1830s, it was based on the coincidence of several scientific and political factors. The most potent was the linkage of the concept that deforestation would cause serious streamflow and rainfall changes with a political awareness of the occurrence of famine in the late 1830s throughout India. But there were other contributory developments. By the mid-1830s, the medical service became increasingly concerned about the rise in the frequency of episodes of serious disease taking place in the rapidly expanding cities, especially in Calcutta, the political capital, in part because of an increasing medical interest in epidemiology and the connections between water supply and disease transmission, an interest articulated rather earlier in India than Britain.

Anxieties about the state's responsibility in this field were increasingly raised in moral and practical terms. In 1836 Surgeon Ranald Martin produced a pioneering report (which considerably preceded the comparable efforts of Edwin Chadwick in Britain) on the need for public health measures and the universal provision of clean water in Calcutta.[101] He further proposed the production of a whole series of medico-topographical reports on India by the medical service.[102]

In this way the ambit of urban public health concern became much more widespread to embrace the whole rural topography of India, extending concerns about the provision of water in cities to include villages in regions known to be subject to drought. In this way Surgeon Donald Butter, in his *Topography of Awadh*, drew attention to the connections between the growing aridity of India and that which had taken place historically elsewhere.[103] Significantly, Butter relied heavily for his insights on Joseph Boussingault's translation of Alexander von Humboldt's work.[104] These translations had recently appeared in scientific journals read widely by Scottish surgeons in India. After 1839 Humboldt's arguments linking deforestation, increasing aridity, and temperature change on a global scale were widely quoted by the medical service in India in its increasingly determined efforts to elicit government controls on deforestation and as part of a wider campaign for public health reforms. Furthermore, the publication of the medico-topographical

[101] R. Martin, *The Sanitary Condition of Calcutta* (Calcutta, 1836).

[102] See discussion of Martin's campaigns in D. G. Crawford, *A History of the Indian Medical Service*, vol. 2, ch. 2. Further medico-topographical reports were compiled for Dacca (1840), Kumaon (1840), Jessore (1837), Assam (1837), Sora (1839).

[103] D. Butter, *Outlines of the Topography and Statistics of the Southern Districts of Oudh and the Cantonment of Sultanpur, Oudh* (Benares, 1839).

[104] See especially J. B. Boussingault, "Memoir Concerning the Effect which the Clearing of Land Has in Diminishing the Quantity of Water in the Streams of a District," *Edinburgh New Philosophical Journal*, 24 (1838), 88–106; and passages in Humboldt's *Personal Narrative of Travels to the Equinoctial Regions of the New Continent, 1799–1804*, H. M. Williams, trans. vol. 4: 134–9 (London, 1838).

reports sponsored by Martin and the Calcutta authorities coincided with the appearance of another set of lay reports which were, by 1839, causing anxiety to the Presidency governments, particularly to the Bombay government.[105]

These reports, which were received from Collectors throughout Southern India, indicated very rapid rates of deforestation since the early 1820s. Initially the Presidency governments, heeding the advice of the Revenue services, which stood to gain from agricultural clearance more than forest conservation, had ignored such warnings, particularly those of the early 1830s which had originated principally with the Rajah of Nilumbur in Malabar. The famines of 1838–39 finally put paid to this attitude, and government officials began to worry about the social instability which might result from famines exacerbated by deforestation. This fear was only distantly related to much less pervasive concerns about possible timber shortages.[106] During the 1840s, such concerns were considerably sharpened by a spreading awareness of the severity of Ireland's famine. Faced by this situation, the governments of the Bombay and Madras Presidencies began quite deliberately to seek relevant technical expertise, giving Company surgeons the opportunity to put themselves forward as expert technical arbiters, a role that some surgeons readily accepted.

In the Bombay Presidency, Dr. Alexander Gibson had already lobbied the government extensively on the subject of deforestation. In Madras, Assistant-Surgeon Edward Balfour had also started, by 1839, to try to persuade the authorities, on his own initiative, of the need for conservation.[107] However, Gibson's thinking initially carried the most weight, principally because he had already acquired a high reputation while a Vaccinator in Khandeish during the 1830s.[108] Furthermore, he had even received gold medals from the Company in recognition of his achievements in pharmacological and epidemiological research. In 1838 Gibson was appointed Superintendent of the Botanic Garden at Dapoorie, near Poona, a position carrying an equivalent weight to the post of Superintendent of the Calcutta Botanic Garden. This appointment meant that he became the obvious source of expertise for the Bombay government in 1839 on forest matters. During a series of subsequent government-sponsored tours through the Presidency, Gibson grew increasingly disturbed

105 E. P. Stebbing, *The Forests of India*, vol. 1 (Edinburgh, 1922), 72–81.

106 See, for example, J. E. Thomas, "Notes on Ryotwar or Permanent Money Rents in S. India, and on the Duty of Government in Periods of Famine," *Madras Journal of Literature and Science* (1838), 53–78, 200–21; and Grove, "Conservation and Colonial Expansion," 258–329. The remainder of this paper is based heavily on the latter pages, which should be referred to for more detailed primary sources.

107 E. G. Balfour, "Notes on the Influence Exercised by Trees in Inducing Rain and Preserving Moisture," *Madras Journal of Literature and Science*, 25 (1849), 402–8. See also IOL V/27/560/107 for details of the correspondence between Balfour and Gibson.

108 See (Anon.), Obituary of Alexander Gibson in the *Proceedings of the Linnaean Society* (1867), p. 33. See also entry on Gibson in D. G. Crawford, *The Roll of the Indian Medical Service* (Calcutta, 1930).

by the full extent of deforestation, particularly on the slopes of the Western Ghats. In March 1841, for example, he wrote to Joseph Hooker that

> the Deccan is more bare than Guzerat and the clefts of the Ghat mountains are the only situation where trees are to be found in any quantity and even they are disappearing fast under the increased demand for land for spade husbandry . . . they are too steep for the plough. It is a matter of regret for the naturalist [sic], perhaps also for the economist, that the woods are in such rapid progress of destruction.[109]

Together with two other surgeons, Hugh Cleghorn (stationed at Shimoga in Mysore) and Edward Balfour (stationed at Madras), Gibson was able to systematically propagandise a forest conservation programme with the government. He backed it up consistently with the threat of famine and constantly inferred that the responsibility for future famines would be laid directly at the door of government, were it not to implement the conservationist prescriptions put forward by the medical service. The political dynamics of the transition in the relations between scientists and colonial government which took place in India between about 1838 and 1847 (the latter being the date of the formal establishment of the Bombay Forest Department under the Conservatorship of Gibson) were complex. In practical terms and underlying the scenario of cataclysmic climatic change which the surgeons put forward, a fundamental debate opened up. In essence, this debate concerned the proper role of the state in environmental control and the extent of the legal right of government to assert control over forest tracts and the cutting of teak forest. The fact that state controls might pose a potentially decisive barrier to the activities of venture capital in the timber market, for which the state itself was the largest customer, was only admitted at a later date.

In 1824 the state had definitively abnegated pre-colonial state forest controls on the basis of a private rights and laissez-faire orthodoxy favoured by Thomas Munro as Governor of Bombay.[110] After 1837 the Company authorities changed course and sought instead to legitimate an increased degree of government intervention in forest control, arguing repeatedly that this new policy was supported by the powers supposedly exercised by pre-colonial Indian states over the forests of the Presidency. Alexander Gibson seemingly was largely responsible for the form of this initiative. He certainly foresaw the opposition to government controls from the new generation of largely indigenous forest landowners and from the European interests benefiting from the lack of regulation of the timber trade between 1824 and 1838. The whole

[109] Letter, Alexander Gibson to J. D. Hooker, dated 1 March 1841, Letter no. 21, India letters, Kew Archives, Richmond, Surrey. This correspondence records the first direct involvement of Hooker in the tropical deforestation issue, a role subsequently reinforced by his visits to Ascension Island and St. Helena in 1843, and India in 1847–50. Hooker's simultaneous involvement in early conservation in the Cape Colony is outlined in R. Grove, "Early Themes in African Conservation," 22–34.

[110] "Timber Monopoly in Malabar and Canara"; entry dated 26 November 1822, in *Minutes of Sir Thomas Munro*, (Bombay: Government Printer, 1881), 178–87.

question of the status of forests as a common property resource also became a matter of discussion and legal debate. Consideration of this problem was stimulated by the developing conflict between the new environmental anxieties of the Company and the forest claims of the private landowners. The Bombay government, now increasingly reliant on the advice of the medical service and on that of Gibson in particular, moved steadily towards a solution involving the formal alienation of forests to Company reserves and the progressive exclusion of private capital and shifting cultivation from the whole forest sector. This course of action soon brought Bombay into direct conflict with the wishes of the Supreme Government in Calcutta, which still favoured the protection of the rights of private landowners at all costs. Both sides tended to ignore the critically important needs of peasant users of the forest. They became a political football between government and commercial timber interests, especially in the neighbourhood of Bombay and later formed a core of popular resistance to British rule.[111] Although Gibson himself was well aware of the danger of alienating customary rights, his advocacy of peasant needs (specifically involving the setting aside of village forests) did not outlast his own personal influence with government or the imposition of the notoriously unjust Forest Act of 1878. Indeed the new Conservator's own fears about the likely social consequences of restricting customary access to forests were vividly fulfilled in 1851 when he narrowly escaped death at the hands of a crowd of angry forest residents in Thana protesting the imposition of new forest regulations.[112]

Between 1840 and 1846 the authorities in Calcutta continued to be most reluctant about funding the establishment of a formalised forest conservancy. They foresaw, quite correctly from hindsight, that such an establishment would imply a severe reduction in large short-term revenue income in favour of high expenditure on long-term and unquantifiable conservationist benefits. Gibson set himself the task of breaking this bureaucratic procrastination at the highest level of government by demonstrating that the penalty for further neglect of the forests could not be afforded politically or economically. Initially he attempted to do this by emphasising the difficulties that deforestation was causing in supplying firewood for residents in urban areas. At one stage Gibson reported that in the city of Mangalore, "firewood, formerly so abundant, is now one of the chief items of expense to the poorer classes of people and is a deprivation severely felt by them."[113] He had previously argued, in 1843, that "a careful examination of forests would disclose other productions of value . . . from a scientific point of view."[114] Although the Bombay gov-

[111] Richard Tucker, "Forest Management and Imperial Politics, Thana District, Bombay," *Indian Economic and Social History Review*, 16 (1979), 27.

[112] This incident was recounted by Gibson in *the Bombay Forest Report* (1856) (Bombay: Government Printer), 13, par. 78.

[113] Stebbing, *Forests of India*, 120. [114] *Ibid.*, 111.

ernment was very ready to accept these ideas and was especially concerned about the firewood problem, none of the arguments were found convincing in Calcutta, where the Supreme Government continued to refuse to consider the cost of running a formal Conservancy.[115] Significantly, arguments for conservation based either on local needs or long-term biological potential still proved to be politically ineffective in India itself in 1846.

At this juncture Gibson decided, as a last resort, to use the "climatic threat" as an argument, not with the apparently hopelessly intractable Calcutta authorities but with the EIC Court of Directors itself. He was encouraged in this by a series of letters from Indian correspondents, including one in February 1846 from Mahabaleshwar, which stated the "common belief that with the removal of wood, the small streams . . . had dried up."[116] Gibson was inclined to take the environmental knowledge of the indigenous population of the Western Ghats far more seriously than most non-medical officials. Thus the Mahabaleshwar letter of February 1846 made up his mind. By March 6, 1846, only a few days after receiving the letter, Gibson finished compiling a manuscript report on the destruction of forests in the Konkan in which he specifically mentioned the climatic and economic consequences of deforestation. "A change of climate," he wrote, "is by no means limited to the mere district in which clearing has taken place, but its influence extends far inland." Deforestation in the Ghats, in the headwards of rivers watering the fertile western coastal lowlands of the Presidency, would, Gibson felt, be clearly affected by such changes. Significantly, the Company attached a great importance to these regions at the time because of their extensive cotton-growing projects. The potentially formidable indirect disadvantages of deforestation were therefore unmistakable. Gibson sent the finished report, dated March 6, 1846, directly to the Court of Directors in London in the form of a letter, thus bypassing the Bombay and Calcutta authorities altogether. He then used material contained within the report as the basis for a short but prominent paper published in the *Transactions of the Bombay Medical and Physical Society*, a journal with a considerable professional readership among officials in the medical service and in civil positions throughout India.[117]

The political significance of these unilateral actions cannot have been lost among officials in government circles in Calcutta. Gibson had appealed over the heads of both the Bombay authorities, more particularly the Government of India, directly to the Court of Directors on the one hand and the Indian scientific community on the other. This was an indicator, if one were needed, of the political weight which Gibson now felt that his professional scientific status conferred on him. Even so, the action was a gamble. However, it

[115] *Ibid.*, 118.

[116] MS Letter, written to Gibson from Mahabaleshwar (IOL V/27/560/107).

[117] A. Gibson, "Report on Deforestation in South Conkan," *Transactions of the Bombay Medical and Physical Society*, 37 (1846), 11–22.

quickly paid off by forcing the decisive intervention of the Court, which compelled the Supreme Government in Calcutta to take serious heed of the advice of this new scientific service. Moreover, Gibson had backed the gamble by documenting his findings in detail among his own medical peers, in the specifically scientific forum of a medical journal. The Company was compelled before long to capitulate to the power of the medical lobby. Thus on December 17, 1846, the Government of India authorised "the employment of an establishment for the management of the forests under the Bombay Presidency, at a monthly charge of 295 Rupees." This enabled the Bombay Government, on March 22, 1847, to appoint Gibson as Conservator of Forests, in addition to his duties as Superintendent of the Dapoorie Botanic Garden, and authorised him to "entertain the establishment which had been sanctioned by the Government of Bombay," for which he had long campaigned. However, the conversion of the East India Company authorities to the utility of the conservationist case actually went much further. A few months later, in July 1847, the Court of Directors issued a remarkable Despatch indicating an unprecedented official awareness of the global threat apparently posed by artificially induced climatic change. "The subject," the Despatch argued,

> is one having a strong practical bearing on the welfare of mankind and we are anxious to obtain extensive and accurate information in regard to it. We desire, therefore, that you will furnish us with any you may possess, and that you will institute inquiries in such quarters as may be likely to lead to the acquisition of particular facts bearing upon the question. It has been suggested that the circumstances of the district of Azimghur afford some illustration of the subject and we shall be glad to receive a correct report of any facts relating either to that district or others which may be calculated to throw light upon the subject of our enquiry.[118]

By 1847, then, the supremacy of the medical service in its hold over government environmental policy had become well-established and was further institutionalised in Gibson's Superintendency of the Bombay Forest Department. This department was formed with the prime aim, not of securing a steady supply of timber (though this was still a pressing concern), but of inhibiting the whole range of environmental and social consequences which deforestation might cause.[119] The fear of these consequences, especially the fear of widespread climatic change, forced the colonial state to comply with a conservationist prescription. Although these environmental considerations were soon supplemented by other political and commercial claims, especially after the end of Company rule in 1857, the Hippocratic and Humboldtian

[118] Despatch no. 21, Court of Directors, EIC, to Government of India, dated 7 July, 1847, quoted in Balfour papers, V/27/560/107. Azimghur was a district mentioned specifically by Donald Butter in his *Topography of Awadh*. The Court had clearly referred to the work.

[119] For a fuller outline of the philosophy involved, see A. Gibson, "A Description of the System Adopted for the Forest Conservancy of the Bombay Presidency," in A. Gibson, ed., *A Handbook to the Forests of the Bombay Presidency* (Byculla: Government Printer, 1863), 53–114.

basis to the conservationism of the surgeons remained a decisive factor in the evolution of colonial forest policy in India and elsewhere. If one seeks a particular historical turning point in the policies of the Company in this respect, the period from 1847 to 1850 is the most critical.

The decisive turning point was probably reached when the Court of Directors, as we have seen, followed up Gibson's report on the Konkan by circularising a Despatch intended to try to establish direct evidence of deforestation and rainfall change. The process culminated in January 1850 with a remarkable statement by Lord Dalhousie, made to the Agri-Horticultural Society of India at Lahore. "During the last season," he wrote,

> I have traversed the plains of the Punjab. . . . There is one characteristic of the wide tract which could not fail to strike the least observant traveller. . . . I refer to the almost total absence of forest trees and of even fruit trees and of bushes leaving the whole territory unadorned by the foliage which is its natural cover, nor stocked with timber requisite for a thousand purposes in the everyday life of the people who dwell in it. . . This is a manifest and will shortly be felt to be an increasing evil unless some measures are taken to provide at present a remedy for the future . . . the Government should provide some means to that end and should bring them into operation without delay. . . . I feel strongly the urgent duty of endeavouring to give this country the clothing of forest trees from my knowledge of the well ascertained and beneficial effect which trees produce on the health and fertility of the tracts in which they are found. No power has been more clearly established than this salubrious and fertilising effect of foliage in an Indian climate. It has been the subject of much enquiry and demonstrated in every report submitted from different parts of India, many of which have passed through my hands, and one of which I forwarded to the local government in Punjab some time ago. . . . None of us can live to see the complete result of that which we now propose to commence. Few of us will gather the fruit where now we plant. But if we succeed in framing this design and advance it in some degree towards completion, we may at least enjoy the satisfaction of the feeling that we shall leave behind us an heritage for which posterity will be grateful.[120]

This statement, an especially revealing one for Dalhousie to make, highlighted the revolution in official perceptions that had occurred as a consequence of the medical service's determined lobbying. Indeed the Lahore memorandum represented, effectively, an unprecedented manifesto for forest protection and sustainable development. The notion of a monolithic state forest sector was clearly one that appealed to Dalhousie's undoubted interventionist and utilitarian ambitions. However, neither this nor the desire for a sustainable timber supply can account for the fundamental change in Company forest policy continued under Dalhousie. Indeed, the roots of the Governor-General's own interest in the forest question lie in his own very personal relationship with the increasingly powerful scientific, especially botanical, lobby.

[120] "Minute by the most noble the Lord Dalhousie," read to the Agri-Horticultural Society of the Punjab on 20 February 1851, reproduced in G. Henderson, ed., *Select Papers of the Agro-Horticultural Society of the Punjab* (Lahore: Lahore United Press, 1868).

As early as 1847 Dalhousie had come firmly under the influence of Joseph D. Hooker, who both accompanied him in his entourage on arriving in India in 1847 and constantly plied the Governor-General with letters on the subject of forest conservation.[121] Hooker's main concern in India was to conduct a series of botanical surveys in the Himalayas. Forest conservation was an important secondary interest. He had, however, been firmly convinced since 1842 of a direct relationship between forest cover and rainfall incidence, mainly through his field observations on St. Helena and through his studies of the beneficial effect of Governor Alexander Beatson's plantations on the island hydrology. Hooker firmly believed, therefore, that afforestation and increased rainfall were closely linked. By 1850 Dalhousie had also become demonstrably convinced of the same phenomena, both through Hooker's influence and by reading reports submitted by Edward Balfour and his colleagues in the Madras government. Both in this respect, as well as in his voluminous correspondence with members of the medical service in India, Hooker had performed a decisive and persuasive part.

Despite the significance of the interaction between Dalhousie and Joseph Hooker, a distinctive model of colonial forest conservancy had, in fact, already been developed in Sind. This model was based firnly on an indigenous system, from which the installed colonial forestry system differed little in detail. Indeed the annexation of Sind by Sir Charles Napier had first provided an opportunity to set up an embryonic forest administration based directly upon the infrastructure of *shikargarhs* (forest and game reserves), which had been steadily built up along the Indus Valley by the Amirs of Sind since the 1750s. The shikargarhs were very extensive and served a multiplicity of state needs, including the widespread provision of a supply of firewood, for which the peasant population were required to pay heavily. The indigenous development of the shikargarhs in Sind and their influence on later colonial policy indicate that conventional historical interpretations of the primacy of European colonial rule in environmental control and conservation need to be carefully re-examined, particularly in the context of contemporary thinking about the erosion of forests as pre-colonial common property. Thus the pre-colonial Sind forest system, like many others in India by the mid-eighteenth century, was by no means amenable to traditions of community or village use. Forestry was a subject which much interested Napier, and he had recommended the immediate assignment of an officer "with some knowledge of forestry" to gain as much information as possible about the system developed by the Amirs.[122] The Amirs had, in fact, frequently carried out widespread

121 For detailed references to the correspondence on forest conservation carried on between Dalhousie and Hooker, see L. Huxley, *The Life and Letters of Sir Joseph Dalton Hooker*, vol. 1 (London: John Murray, 1918).

122 Sind File, (Political 201), Napier to Lord Ellenborough, 29 May 1843, Ellenborough Papers, Public Record Office, Kew. For details of the pre-colonial forest system of Sind and its later adaptation, see W. Scott, *Report on the Management of Canals and Forests in Scinde* (Bombay: Government Printer, 1853).

village removals to make space for their forest reserves and to curtail access for grazing and fuel collection. After the annexation of Sind, these measures were deliberately relaxed by Napier to elicit a degree of popular support for the new government.[123]

Alongside these examples of both indigenous and scientifically reasoned strategies for forest protection, Dalhousie was soon provided with the expert advice of another member of the medical service, Surgeon John McClelland, an ardent enthusiast of natural history and an advocate of rare plant preservation, in a survey of forests in Burma. This survey served as a further stimulus for action by the Governor-General, since McClelland specifically advised a rigorous system of state control to save the rapidly disintegrating forests of Pegu from private timber merchants.[124]

Although the formal structure of an all-India Forest Department was organised in 1865 specifically to exclude the activities of private capital from the forests, the colonial environmental debate, in which state forestry originated had already been going on in Southern India for three decades, was dominated by the conservationist thinking of the surgeons appointed to superintend the first two Presidency Conservancies in Bombay and Madras. The Madras Forest Department, established in 1856, was headed by Surgeon Hugh Cleghorn, who had earlier agitated with Alexander Gibson for conservation measures to be taken during the 1840s.[125]

Cleghorn, later made first Inspector-General of Forests in India,[126] was, like Gibson, primarily a botanist and even an early environmental historian.[127] Openly hostile to the kinds of development which promoted forest destruction, he campaigned against many of the contemporary plans for railway development in Southern India. In 1866 he also spoke out directly against the uncontrolled entry of plantation capital into the Nilgiris and Coorg, consistently claiming that the activities of European planters were far more destructive than the long-established shifting *koomri* cultivators.[128] These kinds of opinions, not uncommon with some of the Scottish surgeons of the time,

123 Accounts of pre-colonial evictions and fuelwood fees are in E. B. Eastwick, *Dry Leaves from Young Egypt*, James Madden (London, 1849), 24, and in J. Outram, *The Conquest of Scinde—a Commentary* (Edinburgh: Blackwood, 1846).

124 I. Burkill, *Chapters in the History of Botany in India*, 155.

125 Both of the new Presidency's Forestry Departments were run according to management methods based on the Sindhi, French, and Scottish models with which their medical founders were familiar. Additionally, Hugh Cleghorn was particularly interested in the history of Venetian attempts to control deforestation and soil erosion. In contrast the introduction of German forestry concepts constituted a much later development, and it was not until after 1878 that they were widely applied in India.

126 The appointment was made jointly with Dietrich Brandis, a German botanist.

127 An account of Cleghorn's activities in the early years of the Madras Forest Department is in H. Cleghorn's *The Forests and Gardens of South India* (Edinburgh, 1861).

128 Cleghorn, in "Discussion," following on the presentation of a paper by George Bidie, "On the Effects of Forest Destruction in Coorg," *Proceedings of the Royal Geographical Society* (1869), 74–75.

marked by their anti-English flavour. Indeed, there are strong reasons for understanding the Scottish medical enthusiasm for protecting forests not only as a consequence of their enthusiasm for scientific natural history but as a surrogate for a far more direct kind of critique of English attitudes to India and even of English attitudes towards the Scots at home. The public basis of Cleghorn's environmentalism, however, was based primarily on a fear of the climatic and physical effects of tropical deforestation as part of a wider global problem. These fears are articulated both in his report to the British Association for the Advancement of Science in 1851, which was probably the first report in a scientific forum to warn of environmental degradation on a global scale, and in his later books and articles on the development of tropical forest protection.[129]

CONCLUSION

An examination of the motivations underlying the emergence of environmentalism and conservation policies in the colonial context points to a heterodox legacy of Utopian, Physiocratic, and Hippocratic philosophical elements represented in the discourses of an influential minority of scientists. These men were distinctively empowered by virtue of the high status assigned to science by the colonial state, and their environmental preoccupations were closely connected to a variety of agendas of social reform. In surveying the whole colonial context of conservation one needs to ask exactly why the colonial state was prepared to accept the highly unorthodox prescriptions of the early conservation lobbyists and, in particular, to accept their view of the need to subjugate short-term economic benefits to nebulous long-term interests. To date, most accounts of the early history of colonial forestry have explained colonial conservationism as a rationalisation for intensifying the direct exploitation of erstwhile "common property resources" to serve the needs of the colonial state and the metropole. As we have seen, such explanations can now be perceived as insufficient. Much of the early colonial conservationism was, in fact, directly related to a growing awareness that western processes for economic development were inherently destructive to the environment. It also becomes appropriate to re-emphasise the political significance of the basic insecurity of the early colonial state faced with the risks of a little-understood tropical environment. The ecological changes brought about by the forces of western capitalist development apparently threatened the very existence of the state itself. Faced by the threat of famine and social unrest, the state became

129 H. Cleghorn, F. Royle, R. Baird-Smith and R. Strachey, "Report on the Committee Appointed by the British Association to Investigate the Probable Effects in an Economic and Physical Point of View of the Destruction of Tropical Forests," *Report of the Proceedings of the British Association for the Advancement of Science, Transactions* (1852), 82–100. See also H. Cleghorn, "On the Distribution of the Principal Forest Trees in India, and the Progress of Forest Conservancy," *Report of the British Association for the Advancement of Science* (1868), 91–94.

quite willing to accept those radical environmental prescriptions for survival proffered by the pioneering generation of scientific experts whom it had already learned to respect in specifically medical circumstances. This may help us to explain how conservation as a state response to ecological crisis emerged first at an economic periphery rather than at the metropolitan centre. It was, in fact, precisely the insecurity of the state at the colonial periphery which allowed a sensitivity to the dangers of ecological change to develop so precociously. In writing environmental history and in reconstructing the history of environmentalism we will need, perhaps, to pay more attention in the future to the fundamentally innovative role of the colonial periphery in terms of its relations with the metropolitan centre. Such a priority, largely dismissed by historians hitherto, would require a radical re-assessment both of the history of conservation and of the state's role in the development of western environmental concern.

Index

Please note page numbers which appear in italics are references to tables or illustrations.